Fluid Mec
Turbomachinery

Fluid Mechanics and Turbomachinery
Problems and Solutions

Bijay K. Sultanian

CRC Press
Taylor & Francis Group
Boca Raton London New York

CRC Press is an imprint of the
Taylor & Francis Group, an **Informa** business

First edition published 2021
by CRC Press
6000 Broken Sound Parkway NW, Suite 300, Boca Raton, FL 33487-2742

and by CRC Press
2 Park Square, Milton Park, Abingdon, Oxon, OX14 4RN

© 2022 Bijay Sultanian

CRC Press is an imprint of Taylor & Francis Group, LLC

Reasonable efforts have been made to publish reliable data and information, but the author and publisher cannot assume responsibility for the validity of all materials or the consequences of their use. The authors and publishers have attempted to trace the copyright holders of all material reproduced in this publication and apologize to copyright holders if permission to publish in this form has not been obtained. If any copyright material has not been acknowledged please write and let us know so we may rectify in any future reprint.

Except as permitted under U.S. Copyright Law, no part of this book may be reprinted, reproduced, transmitted, or utilized in any form by any electronic, mechanical, or other means, now known or hereafter invented, including photocopying, microfilming, and recording, or in any information storage or retrieval system, without written permission from the publishers.

For permission to photocopy or use material electronically from this work, access www.copyright. com or contact the Copyright Clearance Center, Inc. (CCC), 222 Rosewood Drive, Danvers, MA 01923, 978-750-8400. For works that are not available on CCC please contact mpkbookspermissions@ tandf.co.uk

Trademark notice: Product or corporate names may be trademarks or registered trademarks and are used only for identification and explanation without intent to infringe.

ISBN: 978-0-367-51475-4 (hbk)
ISBN: 978-0-367-51474-7 (pbk)
ISBN: 978-1-003-05399-6 (ebk)

Typeset in Times
by codeMantra

To my dearest friend Kailash Tibrewal whose mantra of "joy in giving" continues to inspire me; my wife, Bimla Sultanian; our daughter, Rachna Sultanian, MD; our son-in-law, Shahin Gharib, MD; our son, Dheeraj (Raj) Sultanian, JD, MBA; our daughter-in-law, Heather Benzmiller Sultanian, JD; and our grandchildren: Aarti Sultanian, Soraya Zara Gharib, Shayan Ali Gharib, Loki, and Millie for the privilege of their unconditional love, immensely enriching my life.

Contents

Preface

Fluid mechanics is at the core of aerospace propulsion and power generation, encompassing rocket engines, gas turbines, steam turbines, hydraulic turbines, wind turbines, land and marine power plants, compressors, fans, pumps, and others. In a typical graduate curriculum in the mechanical and aerospace engineering at most universities around the world, intermediate fluid mechanics is generally a required core course for those students who are pursuing their degree programs in thermal-fluids stream.

During a decade of teaching graduate courses on intermediate fluid mechanics and turbomachinery at the University of Central Florida (UCF), I realized that each class size was increasing each year. For example, in 2007, my EML 5713: Intermediate Fluid Mechanics class had only 15 students. The same class in 2016 had 65 students. The immediate impact of class-size increase was the reduction in the number of homework problems I asked the students to do during the entire semester. I did this primarily to reduce my load of grading homework from all students. Students certainly benefitted from various examples worked in the class and those solved in the textbook. During this period, I also realized that, while understanding fundamental concepts and various conservation laws in fluid mechanics was necessary, this alone did not ensure students' success in solving practical problems, unless the problems required straight-forward applications of the well-known equations in fluid mechanics. Whenever I had a closed-book exam in my class, students invariably asked me if I would provide them all the needed equations. I often reminded them of the words of wisdom from Albert Einstein, "Education is not the learning of facts, but the training of the mind to think." My most indelible experience happened when I once assigned the following homework problem in the first week of the Intermediate Fluid Mechanics class:

> A young boy helping his dad with the yard work and playing with a garden hose asked, "Hey dad, when I half cover the hose opening, the water jets out faster and goes further." The dad replied, "Obviously, son. For the same flow, if the area becomes half, the water jet velocity doubles." After a short while, the boy asked again, "Daddy, but now it takes longer to fill the bucket." To which the dad replied, "Hmm ..., that is something I need to think about." With your understanding of fluid flow, how would you resolve the dad's dilemma?

I was a bit disappointed that no student in the class had the correct solution to this relatively simple incompressible flow problem, which did not require the use of complicated equations—certainly not the dreaded Navier-Stokes equations! By way of teaching, testing, and grading over 300 graduate students at UCF, I learned that most of them, in addition to understanding various concepts of fluid mechanics and the governing conservation laws, needed a lot of practice in solving real-world problems. For the latter, they needed a good book of problems with solutions to supplement their assigned textbook.

Based on my over 45 years of industry experience focused primarily on thermal and fluids engineering of liquid rocket propulsion and design and development of world-class gas turbines for aircraft propulsion and power generation, I see a disappointing trend that the art of "the back of the envelope calculations" is becoming extinct in the new generation of otherwise talented thermal and fluids engineers. Key factors contributing to this trend are the overdependence on various design tools and commercial software, including their integration and automation to keep pace with the shrinking new product design cycle time. Certainly, a good book on practical problems with detailed solutions will go a long way in helping these practicing engineers maintain and reinforce their problem-solving skills outside the black-box applications of automated design tools in their analysis and design activities. Once, I asked a colleague of mine at a gas turbine company how he validated the results from a design tool that solved one-dimensional compressible air flow in a variable area duct with friction, heat transfer, and rotation for internal cooling design of turbine blades. This colleague answered that the tool correctly calculated limiting cases of an isentropic convergent-divergent nozzle flow, a Fanno flow, and a Rayleigh flow. My response was that, due to nonlinear coupling among various effects, the validation of the tool using one effect at a time was not sufficient, adding that we needed to numerically solve (hand-calculate using, for example, MS Excel) the duct flow for the combined effects of duct area change, friction, heat transfer, and rotation. This put my colleague at a loss, not knowing how to get started on my proposed primary validation of the design tool, which was essentially formulated for low Mach number flows—incompressible air flows—but was being used in design to compute turbine blade internal cooling flow with choking and normal shocks.

This book reflects years of my industry and teaching experience, featuring many innovative problems and their systematically worked detailed solutions using a physics-first approach. Although each chapter in this book briefly reviews some of the key concepts in support of the problems and their solutions included in the chapter, the book is not a substitute for a textbook on fluid mechanics and turbomachinery. Readers will need textbooks, such as Sultanian (2015, 2018, 2019), for a complete understanding of various topics covered in the book. As an ideal source of many practice problems with detailed solutions, senior-undergraduate and graduate students, teaching faculty, practicing engineers, and researchers engaged in various branches of fluid mechanics will find the book as an indispensable companion to review and reinforce their problem-solving skills in the subject matter. Being one of its kind with a broad global appeal, the book will hopefully serve its interested readers for a long time.

REFERENCES

Sultanian, B.K. 2015. *Fluid Mechanics: An Intermediate Approach.* Boca Raton, FL: Taylor & Francis.

Sultanian, B.K. 2018: *Gas Turbines: Internal Flow Systems Modeling* (Cambridge Aerospace Series). Cambridge: Cambridge University Press.

Sultanian, B.K. 2019. *Logan's Turbomachinery: Flowpath Design and Performance Fundamentals*, 3rd edition. Boca Raton, FL: Taylor & Francis.

Dr. Bijay K. Sultanian

Founder & Managing Member, Takaniki Communications, LLC
Former Adjunct Professor, University of Central Florida

Acknowledgments

This is my childhood-dream book! Growing up with no electricity in a small town in the State of Bihar (India) and learning the English alphabet at 11 in my sixth grade, I had two childhood dreams in my eighth grade—first, to become an engineer, and second, to publish a book to benefit other students. To realize my second dream, I prepared a manuscript on solutions for all problems in the advanced mathematics textbook taught in my class. However, the manuscript was rejected for publication on grounds that I had no credentials to write and publish a book at that age—no matter that the manuscript, when published, would have helped many students who were struggling in math, and all the solutions were correct as checked by my father, who was a genius in math. I had worked hard to handwrite the entire manuscript in the indispensable company of my best friend at night—a kerosene lantern! Although I was disappointed, the rejection of the manuscript became the nuclear fuel that propelled me through the rest of my life's journey with unyielding willpower, determination, and passion to realize both my childhood dreams, overcoming all headwinds in the long and winding journey ahead. Upon graduating with a B. Tech in mechanical engineering in the first division with distinction from the Indian Institute of Technology, Kanpur (IITK), I realized my first dream in 1971.

After advanced degrees in mechanical engineering (M.S. from IIT, Madras, 1978, and PhD from Arizona State University, Tempe, 1984); years of experience in liquid rocket propulsion and gas turbine engineering at world-leading companies, teaching graduate-level courses at the University of Central Florida, and becoming a registered Professional Engineer in the State of Ohio and an ASME Fellow, I finally earned enough credentials to publish my first dream book in 2015, followed by two more books, one in 2018 and the other in 2019. However, this book on problems and solutions is like my first manuscript, which I wanted to publish in my eighth grade as my childhood dream. I am so grateful to so many people—all my teachers, close friends, loving family members, and more than 300 students who participated in EML5713-Intermediate Fluid Mechanics and EML5402-Turbomachinery graduate-level courses I taught at UCF—who made this long journey toward realizing my childhood dreams possible. The bottom line is "Never Give Up on Your Dreams!" In the words of Dr. APJ Kalam, FAIL means First Attempt In Learning, and NO means Next Opportunity.

Last, but not least, I will ever remain grateful to all readers who would let me know of any inadvertently missed errors in the book and to those who will be inspired to write someday a better book dealing with many other problems and solutions in the most fascinating field of fluid mechanics and turbomachinery.

Author

Dr. Bijay (BJ) K. Sultanian, Ph.D., PE, MBA, ASME Life Fellow, is a recognized international authority in gas turbine heat transfer, aerodynamics, secondary air systems, and Computational Fluid Dynamics. Dr. Sultanian is founder and managing member of Takaniki Communications, LLC, and a provider of high-impact technical training programs for corporate engineering teams. Dr. Sultanian is also a former adjunct professor at the University of Central Florida, where he taught graduate-level courses in Turbomachinery and Intermediate Fluid Mechanics for 10 years. He has instructed several workshops at ASME Turbo Expos since 2009. He has been an active member of ASME IGTI's Heat Transfer Committee since 1994 and received the ASME IGTI Outstanding Service Award at ASME Turbo Expo 2018, Lillestrom, Norway. He is the author of graduate-level textbooks, *Fluid Mechanics: An Intermediate Approach*, which was published in 2015; *Gas Turbines: Internal Flow Systems Modeling* (Cambridge Aerospace Series #44), published in 2018, and *Logan's Turbomachinery: Flowpath Design and Performance Fundamentals*, Third Edition, published in 2019.

During his three decades in the gas turbine industry, Dr. Sultanian has worked in and led technical teams at several organizations, including Allison Gas Turbines (now Rolls-Royce), GE Aircraft Engines (now GE Aviation), GE Power Generation, and Siemens Energy. He has developed several physics-based improvements to legacy heat transfer and fluid systems design methods, including new tools to analyze critical high-temperature gas turbine components with and without rotation. He particularly enjoys training large engineering teams at prominent firms around the globe on cutting-edge technical concepts in engineering design and project management best practices.

During 1971–1981, Dr. Sultanian made several landmark contributions toward the design and development of India's first liquid rocket engine for a surface-to-air missile (Prithvi). He also developed the first numerical heat transfer model of steel ingots for optimal operations of soaking pits in India's steel plants.

Dr. Sultanian is a Life Fellow of the American Society of Mechanical Engineers, a registered Professional Engineer (PE) in the State of Ohio, a GE-certified Six-Sigma Green Belt, an Emeritus Member of Sigma Xi, The Scientific Research Honor Society, and a member of the American Society of Thermal and Fluids Engineers.

Dr. Sultanian received his B. Tech and M.S. in Mechanical Engineering from the Indian Institute of Technology, Kanpur, and the Indian Institute of Technology, Madras, respectively. He received his Ph.D. in Mechanical Engineering from the Arizona State University, Tempe, and MBA from the Lally School of Management and Technology at Rensselaer Polytechnic Institute, Troy.

1 Fluid Flow Kinematics and Key Concepts

REVIEW OF KEY CONCEPTS

We concisely present here some key concepts of fluid flow kinematics. More details on each topic are given, for example, in Sultanian (2015).

VELOCITY FIELD

All vector fields do not represent a velocity field; they must also satisfy the local mass conservation equation (continuity equation)—the first law of fluid mechanics

$$\frac{\partial \rho}{\partial t} + \frac{\partial (\rho V_x)}{\partial x} + \frac{\partial (\rho V_y)}{\partial y} + \frac{\partial (\rho V_z)}{\partial z} = 0 \tag{1.1}$$

which is valid for any unsteady laminar or turbulent flow. For a steady compressible flow, this equation reduces to

$$\frac{\partial (\rho V_x)}{\partial x} + \frac{\partial (\rho V_y)}{\partial y} + \frac{\partial (\rho V_z)}{\partial z} = 0 \tag{1.2}$$

For an incompressible flow with constant density, be it steady or unsteady, Equation 1.1 reduces to

$$\frac{\partial V_x}{\partial x} + \frac{\partial V_y}{\partial y} + \frac{\partial V_z}{\partial z} = 0 \tag{1.3}$$

STREAMLINE AND STREAM TUBE

In a fluid flow, a streamline is an instantaneous line to which the local velocity vectors are tangent everywhere—no flow crosses a streamline. While velocity magnitude may vary along a streamline, its direction always coincides with the local tangent to the streamline. A bundle of streamlines forms a stream tube—no flow crosses the surface of a stream tube. We write the following equation for a three-dimensional streamline

$$\frac{dx}{V_x} = \frac{dy}{V_y} = \frac{dz}{V_z} \tag{1.4}$$

Total Acceleration: Local and Convective Accelerations

We write the total acceleration in an unsteady fluid flow as

$$a = \frac{DV}{Dt} = \frac{\partial V}{\partial t} + V_x \frac{\partial V}{\partial x} + V_y \frac{\partial V}{\partial y} + V_z \frac{\partial V}{\partial z} \tag{1.5}$$

where the first term on the right-hand side is the local acceleration and the remaining three terms together make the convective acceleration. From Equation 1.5, we write the acceleration for each velocity component as

$$a_x = \frac{DV_x}{Dt} = \frac{\partial V_x}{\partial t} + V_x \frac{\partial V_x}{\partial x} + V_y \frac{\partial V_x}{\partial y} + V_z \frac{\partial V_x}{\partial z} \tag{1.6}$$

$$a_y = \frac{DV_y}{Dt} = \frac{\partial V_y}{\partial t} + V_x \frac{\partial V_y}{\partial x} + V_y \frac{\partial V_y}{\partial y} + V_z \frac{\partial V_y}{\partial z} \tag{1.7}$$

$$a_z = \frac{DV_z}{Dt} = \frac{\partial V_z}{\partial t} + V_x \frac{\partial V_z}{\partial x} + V_y \frac{\partial V_z}{\partial y} + V_z \frac{\partial V_z}{\partial z} \tag{1.8}$$

Vorticity and Rotation Vectors

Like the velocity, the vorticity is a kinematic vector property of a flow. We obtain vorticity from the curl of the local velocity. Each component of the vorticity vector along a coordinate direction represents twice the rate of local counterclockwise rotation of fluid particles about the coordinate direction. Thus, we write

$$\zeta = \text{curl } V = \nabla \times V = \left(\frac{\partial V_z}{\partial y} - \frac{\partial V_y}{\partial z} \right) \hat{i} + \left(\frac{\partial V_x}{\partial z} - \frac{\partial V_z}{\partial x} \right) \hat{j} + \left(\frac{\partial V_y}{\partial x} - \frac{\partial V_x}{\partial y} \right) \hat{k} \tag{1.9}$$

and

$$\zeta = 2\omega \tag{1.10}$$

which yields

$$\omega_x = \frac{1}{2} \left(\frac{\partial V_z}{\partial y} - \frac{\partial V_y}{\partial z} \right) \tag{1.11}$$

$$\omega_y = \frac{1}{2} \left(\frac{\partial V_x}{\partial z} - \frac{\partial V_z}{\partial x} \right) \tag{1.12}$$

$$\omega_z = \frac{1}{2} \left(\frac{\partial V_y}{\partial x} - \frac{\partial V_x}{\partial y} \right) \tag{1.13}$$

FREE VORTEX

While the rotation vector (ω) is a measure of the local rigid-body rotation of fluid particles, the vortex pertains to the bulk circular motion of the entire flow field with their streamlines as concentric circles. The angular momentum in a free vortex remains constant, giving

$$V_\theta = \frac{C}{r} \tag{1.14}$$

with its singularity at $r = 0$. As shown in Figure 1.1, the outer streamline in this vortex has lower tangential velocity than the inner one. Although counterintuitive, the vorticity in a free vortex away from the origin is zero.

FORCED VORTEX

A forced vortex behaves as solid-body rotation with constant angular velocity (radians/second) with all its streamlines as concentric circles. We write the equation for a forced vortex as

$$V_\theta = r\Omega \tag{1.15}$$

As shown in Figure 1.2, the outer streamline in a forced vortex features higher tangential or swirl velocity than the inner one.

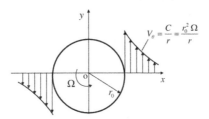

FIGURE 1.1   Free vortex.

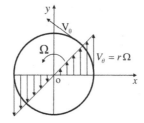

FIGURE 1.2   Forced vortex.

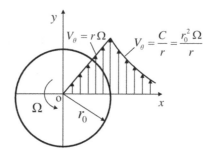

FIGURE 1.3 Rankine vortex.

RANKINE VORTEX

With its tangential velocity becoming infinity, an ideal free vortex flow features singularity at the origin (zero radius). This condition is physically not realizable for a real fluid with nonzero viscosity. A real free vortex flow, such as a tornado, features a core of forced vortex with zero tangential velocity at the origin—a Rankine vortex shown in Figure 1.3. Thus, for the Rankine vortex, we write $V_\theta = r\Omega$ for $r = r_0$ and $V_\theta = C/r$ for $r > r_0$.

CIRCULATION

Circulation is another property of a velocity field. For a closed contour, the circulation measures the total vorticity contained within the contour. To compute the circulation around a positively oriented closed contour, shown in Figure 1.4, we perform the line integral of the velocity as

$$\Gamma = \oint_C V \cdot dl \tag{1.16}$$

Applying the Stokes theorem, which relates the line and surface integrals, to this equation yields

$$\Gamma = \oint_C V \cdot dl = \iint_S (\nabla \times V) \cdot \hat{n} dS = \iint_S \zeta \cdot \hat{n} dS \tag{1.17}$$

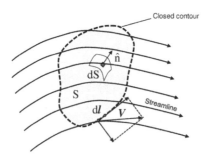

FIGURE 1.4 Circulation around a closed contour in a flow field.

TOTAL (STAGNATION) TEMPERATURE

For both incompressible and compressible flows, we obtain the total (stagnation) temperature as the sum of the static temperature T and the dynamic temperature $V^2/2c_p$ as

$$T_0 = T + \frac{V^2}{2c_p} \tag{1.18}$$

In terms of Mach number of a compressible flow, we write this equation as

$$\frac{T_0}{T} = 1 + \frac{\gamma - 1}{2} M^2 \tag{1.19}$$

where the Mach number $M = \sqrt{\gamma RT}$.

TOTAL (STAGNATION) PRESSURE

For an incompressible flow or a compressible flow at $M \leq 0.3$, we obtain the total (stagnation) pressure using the equation

$$p_0 = p + \frac{1}{2}\rho V^2 \tag{1.20}$$

where $\rho V^2/2$ is the incompressible dynamic pressure.

For a compressible flow, we assume an isentropic stagnation process to obtain the maximum total pressure by the equation

$$\frac{p_0}{p} = \left(\frac{T_0}{T}\right)^{\frac{\gamma}{\gamma-1}} \tag{1.21}$$

In terms of Mach number, we write this equation as

$$\frac{p_0}{p} = \left(1 + \frac{\gamma - 1}{2} M^2\right)^{\frac{\gamma}{\gamma-1}} \tag{1.22}$$

STREAM THRUST AND IMPULSE PRESSURE

At any section of a duct flow, for both incompressible and compressible flows, we obtain the stream thrust as the sum of the pressure force and the momentum flow (inertia force), that is,

$$S_T = pA \cdot \left(\frac{V}{V}\right) + \dot{m}V \tag{1.23}$$

which shows that S_T is a vector along the momentum direction, and it may be thought of as representing the total force (the sum of the pressure force and the inertia force) associated with a fluid flow.

When we divide the stream thrust at a section by its area normal to the momentum direction, we obtain a new pressure called the impulse pressure given by

$$p_i = p + \rho V^2 \tag{1.24}$$

Note the missing ½ in Equation 1.24 as compared to Equation 1.20 for computing total pressure in an incompressible flow. In a constant-area duct with no wall friction, the stream thrust and impulse pressure remain constant even if its static pressure, total pressure, static temperature, and total pressure may vary from inlet to outlet.

PROBLEM 1.1:   STREAMLINE IN A THREE-DIMENSIONAL STEADY INCOMPRESSIBLE FLOW

The velocity distribution in a three-dimensional steady incompressible flow is given by $V_x = 5xe^{-2z}$, $V_y = -3ye^{-2z}$, and $V_z = e^{-2z}$. Find the equation of the streamline that passes through the point $A(1,1,1)$.

SOLUTION FOR PROBLEM 1.1

Let us first verify that the given velocity components satisfy the continuity equation

$$\frac{\partial V_x}{\partial x} + \frac{\partial V_y}{\partial y} + \frac{\partial V_z}{\partial z} = 0$$

$$\frac{\partial\left(5xe^{-2z}\right)}{\partial x} + \frac{\partial\left(-3ye^{-2z}\right)}{\partial y} + \frac{\partial\left(e^{-2z}\right)}{\partial z}$$

$$5e^{-2z} - 3e^{-2z} - 2e^{-2z} = 0$$

Streamline Equation

$$\frac{dx}{V_x} = \frac{dy}{V_y} = \frac{dz}{V_z}$$

$$\frac{dx}{5xe^{-2z}} = \frac{dy}{-3ye^{-2z}} = \frac{dz}{e^{-2z}}$$

$$\frac{dx}{5x} = -\frac{1}{3}\frac{dy}{y} = dz$$

Integration of the first two terms of this equation yields

$$\frac{1}{5}\ln x = -\frac{1}{3}\ln y + \ln \tilde{C}_1$$

$$y = C_1 x^{-3/5}$$

For the streamline passing through A(1,1,1), we obtain $C_1 = 1$, giving $y = x^{-3/5}$.

Similarly, the integration of $\dfrac{dx}{5x} = dz$ yields

$$\frac{1}{5}\ln x = z + C_2$$

$$z = \frac{1}{5}\ln x - C_2$$

For the streamline passing through $A(1,1,1)$, this equation yields $C_2 = -1$, giving

$$z = 1 + \frac{1}{5}\ln x$$

Therefore, the coordinates of points on the required streamline passing through the point $A(1,1,1)$ satisfy the relations $y = x^{-3/5}$ and $z = 1 + \dfrac{1}{5}\ln x$.

PROBLEM 1.2: STREAMLINE IN A THREE-DIMENSIONAL STEADY FLOW WITH VARIABLE DENSITY

For the flow field given in Problem 1.1, the density varies as $\rho = e^{-z}$, which alters the velocity component in the z-direction only. Find the new distribution of V_z and the equation of the new streamline passing through the point $A(1,1,1)$.

SOLUTION FOR PROBLEM 1.2

The variable-density flow field must satisfy the continuity equation. Let us assume $V_z = ae^{-bz}$ where a and b are constants to be determined. We write the continuity equation with variable density as

$$\frac{\partial(\rho V_x)}{\partial x} + \frac{\partial(\rho V_y)}{\partial y} + \frac{\partial(\rho V_z)}{\partial z} = 0$$

$$\frac{\partial(5xe^{-2z}e^{-z})}{\partial x} + \frac{\partial(-3ye^{-2z}e^{-z})}{\partial y} + \frac{\partial(ae^{-bz}e^{-z})}{\partial z} = 0$$

$$5e^{-3z} - 3e^{-3z} - a(1+b)e^{-(1+b)z} = 0$$

from which we obtain $a = 2/3$ and $b = 2$, giving $V_z = 2/3e^{-2z}$.

STREAMLINE EQUATION

The streamline equation does not explicitly contain density. The density variation influences the streamlines only through the altered velocity field, which must always satisfy the continuity equation—the first law of fluid mechanics!

$$\frac{dx}{V_x} = \frac{dy}{V_y} = \frac{dz}{V_z}$$

$$\frac{dx}{5xe^{-2z}} = \frac{dy}{-3ye^{-2z}} = \frac{3}{2}\frac{dz}{e^{-2z}}$$

$$\frac{dx}{5x} = \frac{1}{3}\frac{dy}{y} = \frac{3}{2}dz$$

Integration of the first two terms of this equation yields

$$\frac{1}{5}\ln x = -\frac{1}{3}\ln y + \ln \tilde{C}_1$$

$$y = C_1 x^{-3/5}$$

For the streamline passing through $A\,(1,1,1)$, we obtain $C_1 = 1$, giving $y = x^{-3/5}$.

Integration of $\dfrac{dx}{5x} = \dfrac{3}{2}dz$ yields

$$\frac{1}{5}\ln x = \frac{3}{2}z + \tilde{C}_2$$

$$z = \frac{2}{15}\ln x - C_2$$

As the streamline passes through $A\big(1,1,1\big)$, we obtain $C_2 = -1$, giving

$$z = 1 + \frac{2}{15}\ln x$$

Therefore, the coordinates of points on the required streamline passing through $A\big(1,1,1\big)$ in the flow field with the given density variation satisfy the relations $y = x^{-3/5}$ and $z = 1 + \dfrac{2}{15}\ln x$.

PROBLEM 1.3: VELOCITY DISTRIBUTION FROM MEASUREMENTS

A team of graduate students carried out accurate measurements of V_x and temperature in a two-dimensional steady flow on a flat plate as shown in Figure 1.5. They correlated these measurements to $V_x = V_0 x^2$, where V_0 is a constant. Noting that V_y

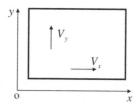

FIGURE 1.5 Two-dimensional flow on a flat plate (Problem 1.3).

is zero at $y = 0$, they calculated the distribution of V_y from their understanding of an incompressible fluid flow. Looking at the data and the calculations, the research adviser pointed out that the calculated V_y is 33% higher everywhere. She asked the students to recalculate it taking into account the temperature variation, which caused the density variation given by $\rho = \rho_0 xy$, where ρ_0 is a constant. Both x and y are dimensionless coordinates. Verify the error estimated by the research advisor in V_y computed by the graduate students.

SOLUTION FOR PROBLEM 1.3

Initially, the graduate students calculated the y-component of velocity $\left(\tilde{V}_y\right)$ using the incompressible form of the continuity equation for the two-dimensional flow as

$$\frac{\partial V_x}{\partial x} + \frac{\partial \tilde{V}_y}{\partial y} = 0$$

$$\frac{\partial \tilde{V}_y}{\partial y} = -\frac{\partial V_x}{\partial x} = -\frac{\partial (V_0 x^2)}{\partial x} = -2V_0 x$$

As $\tilde{V}_y = 0$ for $y = 0$, we obtain

$$\tilde{V}_y = -2V_0 xy$$

Considering the density variation due to nonuniform temperature distribution, the continuity equation becomes

$$\frac{\partial (\rho V_x)}{\partial x} + \frac{\partial (\rho V_y)}{\partial y} = 0$$

$$\frac{\partial (\rho V_y)}{\partial y} = -\frac{\partial (\rho V_x)}{\partial x} = -\frac{\partial (\rho_0 xy V_0 x^2)}{\partial x} = -3\rho_0 V_0 x^2 y$$

where V_y represents the y-component of velocity for the actual flow with the density variation. As $V_y = 0$ for $y = 0$, we obtain

$$\rho V_y = -\frac{3}{2}\rho_0 V_0 x^2 y^2$$

$$V_y = -\frac{3}{2}V_0 xy$$

We calculate the percentage error in the initial calculation of the y component of velocity as

$$\text{Error} = \frac{\left(\tilde{V}_y - V_y\right)}{V_y} \times 100 = 33.3\%$$

Therefore, the research adviser is right.

PROBLEM 1.4:   CONVECTIVE ACCELERATION IN A CONICAL DIFFUSER

For a steady one-dimensional incompressible flow with the volumetric flow rate Q in a conical diffuser shown in Figure 1.6, find the convective acceleration at a distance x from the inlet.

SOLUTION FOR PROBLEM 1.4

DIFFUSER DIAMETER

$$D_x = D_1 + x\left(\frac{D_2 - D_1}{L}\right) = D_1 + \alpha x$$

where

$$\alpha = \left(\frac{D_2 - D_1}{L}\right)$$

DIFFUSER FLOW AREA

$$A_x = \frac{\pi D_x^2}{4} = \frac{\pi}{4}(D_1 + \alpha x)^2$$

VELOCITY AT x

$$V_x = \frac{Q}{A_x} = \frac{4Q}{\pi}\frac{1}{(D_1 + \alpha x)^2}$$

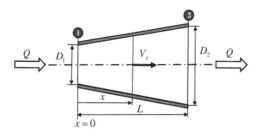

FIGURE 1.6   Steady one-dimensional incompressible flow in a conical diffuser (Problem 1.4).

CONVECTIVE ACCELERATION

$$a_x = V_x \frac{dV_x}{dx}$$

$$\frac{dV_x}{dx} = -\frac{8\alpha Q}{\pi} \frac{1}{(D_1 + \alpha x)^3}$$

$$a_x = -\left[\frac{4Q}{\pi} \frac{1}{(D_1 + \alpha x)^2}\right]\left[\frac{8\alpha Q}{\pi} \frac{1}{(D_1 + \alpha x)^3}\right]$$

$$a_x = -\left(\frac{32\alpha Q^2}{\pi^2}\right)\frac{1}{(D_1 + \alpha x)^5}$$

which is the required convective acceleration.

PROBLEM 1.5: CONVECTIVE ACCELERATION AND ROTATION VECTORS IN AN UNSTEADY THREE-DIMENSIONAL FLOW

The velocity vector in a three-dimensional unsteady incompressible flow is given by

$$\vec{V} = \left(x^2 y + 5xt + yz^2\right)\hat{i} + \left(xy^2 + 4t\right)\hat{j} - \left(4xyz - xy + 5zt\right)\hat{k}$$

a. Verify that the given velocity field satisfies the continuity equation.
b. Determine the acceleration vector at the point $A(1,1,1)$ at $t = 1.0$.
c. Determine the rotation vector at the point $A(1,1,1)$ at $t = 1.0$.

SOLUTION FOR PROBLEM 1.5

From each velocity component, we obtain
For $V_x = x^2 y + 5xt + yz^2$:

$$\frac{\partial V_x}{\partial t} = 5x; \quad \frac{\partial V_x}{\partial x} = 2xy + 5t; \quad \frac{\partial V_x}{\partial y} = x^2 + z^2; \text{and} \quad \frac{\partial V_x}{\partial z} = 2yz$$

For $V_y = xy^2 + 4t$:

$$\frac{\partial V_y}{\partial t} = 4; \quad \frac{\partial V_y}{\partial x} = y^2; \quad \frac{\partial V_y}{\partial y} = 2xy; \text{and} \quad \frac{\partial V_y}{\partial z} = 0$$

For $V_z = -(4xyz - xy + 5zt)$:

$$\frac{\partial V_z}{\partial t} = -5z; \quad \frac{\partial V_z}{\partial x} = -4yz + y; \quad \frac{\partial V_z}{\partial y} = -4xz + x; \text{and} \quad \frac{\partial V_z}{\partial z} = -4xy - 5t$$

(a) CONTINUITY EQUATION

$$\frac{\partial V_x}{\partial x} + \frac{\partial V_y}{\partial y} + \frac{\partial V_z}{\partial z} = 0$$

$$2xy + 5t + 2xy - 4xy - 5t = 4xy - 4xy + 5t - 5t = 0$$

(b) ACCELERATION VECTOR

$$\mathbf{a} = a_x\hat{i} + a_y\hat{j} + a_z\hat{k}$$

$$a_x = \frac{\partial V_x}{\partial t} + V_x\frac{\partial V_x}{\partial x} + V_y\frac{\partial V_x}{\partial y} + V_z\frac{\partial V_x}{\partial z}$$

Various terms in this equation evaluated at $A(1,1,1)$ become

$$\frac{\partial V_x}{\partial t} = 5x = 5$$

$$V_x\frac{\partial V_x}{\partial x} = \left(x^2y + 5xt + yz^2\right)\left(2xy + 5t\right) = 49$$

$$V_y\frac{\partial V_x}{\partial y} = \left(xy^2 + 4t\right)\left(x^2 + z^2\right) = 10$$

$$V_z\frac{\partial V_x}{\partial z} = \left(4xyz - xy + 5zt\right)\left(4xy + 5t\right) = 72$$

giving a_x as

$$a_x = 5 + 49 + 10 + 72 = 136$$

Similarly, we obtain for a_y

$$\frac{\partial V_y}{\partial t} = 4$$

$$V_x\frac{\partial V_y}{\partial x} = (x^2y + 5xt + yz^2)(y^2) = 7$$

$$V_y\frac{\partial V_y}{\partial y} = (xy^2 + 4t)(2xy) = 10$$

$$V_z\frac{\partial V_y}{\partial z} = (4xyz - xy + 5zt)(0) = 0$$

giving

$$a_y = 4 + 7 + 10 + 0 = 21$$

Similarly, we obtain for a_z

$$\frac{\partial V_z}{\partial t} = -5z = -5$$

$$V_x \frac{\partial V_z}{\partial x} = (x^2 y + 5xt + yz^2)(-4yz + y) = -21$$

$$V_y \frac{\partial V_z}{\partial y} = (xy^2 + 4t)(-4xz + x) = -15$$

$$V_z \frac{\partial V_z}{\partial z} = (4xyz - xy + 5zt)(-4xy - 5t) = -72$$

giving

$$a_z = -5 - 21 - 15 - 72 = -113$$

We write the total acceleration as

$$a = 136\hat{i} + 21\hat{j} - 113\hat{k}$$

(c) ROTATION VECTOR

For the rotation vector

$$\omega = \omega_x \hat{i} + \omega_y \hat{j} + \omega_z \hat{k}$$

we obtain at $A(1,1,1)$,

$$\omega_x = \frac{\partial V_z}{\partial y} - \frac{\partial V_y}{\partial z} = -4xz + x - 0 = x(1 - 4z) = -3$$

$$\omega_y = \frac{\partial V_x}{\partial z} - \frac{\partial V_z}{\partial x} = 2yz + 4yz = 6$$

$$\omega_z = \frac{\partial V_y}{\partial x} - \frac{\partial V_x}{\partial y} = y^2 - x^2 - z^2 = -1$$

finally giving

$$\omega = -3\hat{i} + 6\hat{j} - \hat{k}$$

which shows that the rotation vector or the vorticity vector, which is twice the rotation vector, is steady everywhere in the given unsteady three-dimensional velocity field.

PROBLEM 1.6:   CIRCULATION IN A FREE VORTEX, A FORCED VORTEX, AND A RANKINE VORTEX

The streamlines in a two-dimensional vortex are circles. Calculate the circulation around a circle of radius R from the origin for each vortex type: (1) forced vortex, (2) free vortex, and (3) Rankine vortex, which features a forced vortex at its inner core adjoined with a free vortex.

SOLUTION FOR PROBLEM 1.6

(a) CIRCULATION FOR A FORCED VORTEX

As the tangential flow velocity is tangent to the circle at each point, we calculate the circulation for the circle of radius R as

$$\Gamma_{\text{forced vortex}} = \oint V \cdot dl = \int_0^{2\pi} R\Omega(Rd\theta) = \pi R^2(2\Omega)$$

which shows that $\Gamma_{\text{forced vortex}}$ is equal to the product of the area of the circle and 2Ω, which is the constant vorticity of a forced vortex.

(b) CIRCULATION FOR A FREE VORTEX

For a free vortex, we write the circulation around a circle of radius R as

$$\Gamma_{\text{free vortex}} = \oint V \cdot dl = \int_0^{2\pi} \frac{C}{R}(Rd\theta) = 2\pi C$$

which shows that the circulation in a free vortex is constant and independent of the radius of the circle. Note that the annular region between two circles around the origin will have zero circulation (vorticity), the constant circulation being confined within the inner circle. Thus, in an ideal free vortex, the singularity at the origin contains the constant circulation and the related vorticity.

(c) CIRCULATION FOR A RANKINE VORTEX

As the Rankine vortex is a combination of a forced vortex and a free vortex, from the results in (a) and (b), we conclude that for a circle inside the forced vortex core $(R \leq r_0)$, the circulation equals $2\pi R^2\Omega$. For all circles with $R \geq r_0$, the circulation equals $2\pi r_0^2\Omega$.

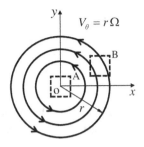

FIGURE 1.7 Circulation in a forced vortex (Problem 1.7).

PROBLEM 1.7: CIRCULATION AROUND A SQUARE ARBITRARILY LOCATED IN A FORCED VORTEX

Figure 1.7 shows a few streamlines (circles) of a forced vortex. Calculate the circulation around squares A and B, each of side a. The square A is centered at the origin and the square B at point $P(2a, a)$.

SOLUTION FOR PROBLEM 1.7

The vorticity a forced vortex with constant angular velocity Ω equals 2Ω everywhere. The circulation for any closed curve in a forced vortex equals the product of its enclosed area and the uniform value of its vorticity (2Ω). Accordingly, the circulation around both squares A and B equals $2\Omega a^2$.

PROBLEM 1.8: CIRCULATION AROUND A SQUARE ARBITRARILY LOCATED IN A FREE VORTEX

Figure 1.8 shows a few streamlines (circles) of a free vortex. Calculate the circulation around squares A and B, each of side a. The square A is centered at the origin and the square B at point $P(2a, a)$.

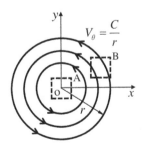

FIGURE 1.8 Circulation in a free vortex (Problem 1.8).

SOLUTION FOR PROBLEM 1.8

In a free vortex, the circulation for any closed curve that includes the origin has a constant value of $2\pi C$, and it is zero outside. Accordingly, the circulation around the square A is $2\pi C$ and that around square B is zero.

PROBLEM 1.9:   INCOMPRESSIBLE FLOW THROUGH A MANIFOLD WITH TWO INLETS AND ONE OUTLET

Figure 1.9 shows an incompressible flow of a fluid of density ρ through a manifold with two inlets and one outlet. The flow velocity at each inlet is uniform. The velocity exiting the outlet pipe has a parabolic profile with its average value denoted by V_3. Determine the force needed to hold the manifold in position under the given flow field.

SOLUTION FOR PROBLEM 1.9

The force needed to hold the manifold in position equals the force on the combined manifold and flow control volume, and it is given by the difference between the sum of steam thrusts at outlets minus the sum of stream thrusts at inlets.

We compute the stream thrust at the inlet at section 1 as

$$S_{T_1} = p_1 A_1 + \dot{m}_1 V_1 = p_1 A_1 + \rho A_1 V_1^2$$

Similarly, we compute the stream thrust at the inlet at section 2 as

$$S_{T_2} = p_2 A_2 + \dot{m}_2 V_2 = p_2 A_2 + \rho A_2 V_2^2$$

Similarly, we compute the stream thrust at the outlet at section 3 as

$$S_{T_3} = p_3 A_3 + \beta \dot{m}_3 V_2 = p_3 A_3 + \frac{4}{3} \rho A_3 V_3^2$$

where we have used the momentum correction factor $\left(\beta = 4/3\right)$ for the parabolic velocity profile in the outlet pipe.

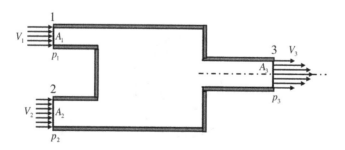

FIGURE 1.9 Incompressible flow through a manifold with two inlets and one outlet (Problem 1.9).

Thus, we calculate the required force as

$$F = S_{T_3} - S_{T_2} - S_{T_1} = (p_3 A_3 - p_2 A_2 - p_1 A_1) + \rho\left(\frac{4}{3} A_3 V_3^2 - A_2 V_2^2 - A_1 V_1^2\right)$$

NOMENCLATURE

a	Acceleration magnitude
$\boldsymbol{a}$	Acceleration vector
A	Area
$\boldsymbol{A}$	Area vector
c_P	Specific heat at constant pressure
c_v	Specific heat at constant volume
D	Diameter
F	Force
$\hat{i}$	Unit vector in x-direction
$\hat{j}$	Unit vector in y-direction
$\hat{k}$	Unit vector in z-direction
$\boldsymbol{l}$	Length vector locally tangent to a closed contour
L	Length
$\dot{m}$	Mass flow rate
M	Mach number
$\hat{n}$	Local outward-pointing unit normal to the surface within a closed contour
p	Static pressure
Q	Volumetric flow rate
r	Radial distance
R	Gas constant, circle radius
s	Entropy
S	Surface area within a closed contour
S_T	Stream thrust
t	Time
T	Temperature
u	Velocity in the x-direction
U	Freestream velocity
V	Velocity magnitude
$\boldsymbol{V}$	Velocity vector
W	Relative velocity
x	Cartesian coordinate x
y	Cartesian coordinate y
z	Cartesian coordinate z

SUBSCRIPTS AND SUPERSCRIPTS

i	Impulse
0	Total (stagnation)
x	Component in x-coordinate direction

y	Component in y-coordinate direction
z	Component in z-coordinate direction
θ	Tangential direction

GREEK SYMBOLS

α	Coefficient
Γ	Circulation
ζ	Vorticity vector
γ	Ratio of specific heats $\left(\gamma = c_p / c_v\right)$
ρ	Density
ω	Magnitude of ω
ω	Local angular velocity of a fluid particle
Ω	Flow angular velocity

REFERENCES

Sultanian, B.K. 2015. *Fluid Mechanics: An Intermediate Approach*. Boca Raton: Taylor & Francis.

Sultanian, B.K. 2019. *Logan's Turbomachinery: Flowpath Design and Performance Fundamentals*, 3rd edition. Boca Raton: Taylor & Francis.

BIBLIOGRAPHY

Lugt, H.J. 1995. *Vortex Flow in Nature and Technology*. Malabar: Krieger Publishing Company.

Samimy, M., K.S. Breuer, L.G. Leal et al. 2003. *A Gallery of Fluid Motion*. Cambridge: Cambridge University Press.

Van Dyke, M. 1982. *An Album of Fluid Motion*. Stanford: The Parabolic Press.

2 Control Volume Analysis

REVIEW OF KEY CONCEPTS

We summarize here integral mass conservation, energy conservation, and linear and angular momentum equations for control volume (CV) analyses. Details derivations of these equations may, for example, be found in Sultanian (2015).

REYNOLDS TRANSPORT THEOREM

Reynolds transport theorem provides a convenient means to transform the conservation equations of mass, momentum, and energy, which hold good for a system (Lagrangian viewpoint) into those valid for a CV analysis (Eulerian viewpoint). We call the resulting equations the integral CV equations. With uniform properties at inlets and outlets of a CV, these equations become algebraic equations. If the CV size is comparable to the size of the equipment of engineering interest, we call the analysis the large CV or the macroanalysis. If we subdivide a large CV into many small CVs, as is usually done in computational fluid dynamics, the corresponding analysis is referred to as small CV analysis or microanalysis.

INTEGRAL MASS CONSERVATION EQUATION: THE FIRST LAW OF FLUID MECHANICS

For a general CV with arbitrary inflows and outflows through its control surface (CS), we write the mass conservation equation as

$$\frac{\partial}{\partial t} \iiint_{CV} \rho \, d\mathcal{V} + \iint_{CS} \rho V \cdot dA = 0 \qquad (2.1)$$

For a CV with instantaneous fluid mass m_{cv} and multiple inflows and outflows, we can write this equation as

$$\frac{dm_{cv}}{dt} + \sum_{N_{outlets}} \dot{m} - \sum_{N_{inlets}} \dot{m} = 0 \qquad (2.2)$$

where N_{inlets} and $N_{outlets}$ are the total number of inlets and outlets piercing the CS, respectively, with no regard to the coordinate direction.

INTEGRAL LINEAR MOMENTUM EQUATION IN AN INERTIAL REFERENCE FRAME

For a nonaccelerating general CV with arbitrary inflows and outflows through its CS, we write the linear momentum equation as

$$F_s + F_b = \frac{\partial}{\partial t} \iiint_{cv} V\rho \, d\Psi + \iint_{cs} V(\rho V \cdot dA) \tag{2.3}$$

where the surface forces (F_s) typically consist of pressure forces at inlets and outlets and on walls, and shear forces on walls. Note that the pressure forces are always normal and compressive to the section they act upon. The shear forces are always parallel to the walls on which they are acting. We consider here body forces (F_b) due to gravity and rotation only.

If the flow properties are uniform at each inlet and outlet, we can replace the surface integral on the right-hand side of Equation 2.3 by algebraic summations. For example, we write the linear momentum equation in the x-coordinate direction as

$$F_{sx} + F_{bx} = \frac{dM_{cvx}}{dt} + \sum_{N_{\text{outlets}}} \dot{m}V_x - \sum_{N_{\text{inlets}}} \dot{m}V_x \tag{2.4}$$

where M_x is the total instantaneous x momentum within the CV. In this equation, identifying the mass velocity—the velocity that produces the mass flow rate—which is always positive, greatly simplifies the evaluation of the momentum flow rate with the correct sign at each inlet and outlet of a CV. The linear momentum flow rate at each inlet or outlet is positive if V_x is in the positive x-direction, otherwise it is negative. To deal with nonuniform velocity profiles at inlet and outlets, we use the momentum correction factor β presented in the following section.

MOMENTUM AND KINETIC ENERGY CORRECTION FACTORS

We define the momentum correction factor β as

$$\beta = \frac{\iint_A V(\rho V \cdot dA)}{\dot{m}\bar{V}} \tag{2.5}$$

which is the ratio of the actual momentum flux computed from integration over the inlet or outlet area to the momentum flux computed from the average velocity. Similarly, we define the kinetic energy correction factor α as

$$\alpha = \frac{\iint_A V^2(\rho V \cdot dA)}{\dot{m}V^2} \tag{2.6}$$

For an axisymmetric parabolic velocity profile, we obtain $\beta = 4/3$ and $\alpha = 2$.

INTEGRAL LINEAR MOMENTUM EQUATION IN A NONINERTIAL REFERENCE FRAME

For a general CV with linear and rotational velocities and accelerations, and with arbitrary inflows and outflows through its CS, we write the linear momentum equation as

$$F_s + F_b - \iiint\limits_{cv} \left(\ddot{R} + 2\Omega \times V_{xyz} + \Omega \times (\Omega \times r) + \dot{\Omega} \times r \right) \rho d\mathcal{V}$$

$$= \frac{\partial}{\partial t_{xyz}} \iiint\limits_{CV} \mathbf{V}_{xyz}(\rho d\mathcal{V}) + \iint\limits_{CS} \mathbf{V}_{xyz}(\rho \mathbf{V}_{xyz} \cdot d\vec{A}) \tag{2.7}$$

With $\Omega = \dot{\Omega} = 0$, this equation reduces to

$$F_s + F_b - \iiint\limits_{cv} \ddot{R} \rho \, d\mathcal{V} = F_s + F_b - m_{cv} \, \ddot{R}$$

$$= \frac{\partial}{\partial t_{xyz}} \iiint\limits_{cv} V_{xyz}(\rho d\mathcal{V}) + \iint\limits_{cs} V_{xyz}(\rho V_{xyz} \cdot d\vec{A}) \tag{2.8}$$

where m_{cv} is the instantaneous total mass of the CV.

INTEGRAL ANGULAR MOMENTUM EQUATION IN AN INERTIAL REFERENCE FRAME

We write the integral angular momentum equation for an inertial CV as

$$\Gamma_s + \Gamma_b = \frac{\partial}{\partial t_{XYZ}} \iiint\limits_{cv} (r_{XYZ} \times V_{XYZ})(\rho d\mathcal{V}) + \iint\limits_{cs} (r_{XYZ} \times V_{XYZ})(\rho V_{XYZ} \cdot dA) \tag{2.9}$$

INTEGRAL ANGULAR MOMENTUM EQUATION IN A NONINERTIAL REFERENCE FRAME

We write the integral angular momentum equation for a noninertial CV as

$$\Gamma_s + \Gamma_b - \iiint\limits_{cv} r \times \left\{ \ddot{R} + 2\Omega \times V_{xyz} + \Omega \times (\Omega \times r) + \dot{\Omega} \times r \right\} \rho d\mathcal{V}$$

$$= \iint\limits_{cv} r \times V_{xyz}(\rho V_{xyz} \cdot dA) + \frac{\partial}{\partial t_{xyz}} \iiint\limits_{cv} r \times V_{xyz}(\rho d\mathcal{V}) \tag{2.10}$$

With $\dot{\Omega} = 0$, this equation reduces to

$$\Gamma_s + \Gamma_b - \iiint\limits_{cv} r \times \left\{ \ddot{R} + 2\Omega \times V_{xyz} + \Omega \times (\Omega \times r) \right\} \rho d\mathcal{V}$$

$$= \iint\limits_{cs} r \times V_{xyz} \left(\rho V_{xyz} \cdot dA \right) + \frac{\partial}{\partial t_{xyz}} \iiint\limits_{cv} r \times V_{xyz} \left(\rho d\mathcal{V} \right) \tag{2.11}$$

As shown in Sultanian (2015), with $\ddot{R} = 0$ and uniform flow properties at CV inlets and outlets, we can further simplify this equation to

$$\Gamma_s + \Gamma_b = \underbrace{\sum \dot{m}(r \times V_{XYZ})}_{N_{\text{outlets}}} - \underbrace{\sum \dot{m}(r \times V_{XYZ})}_{N_{\text{inlets}}} = \underbrace{\sum \dot{H}_{XYZ}}_{N_{\text{outlets}}} - \underbrace{\sum \dot{H}_{XYZ}}_{N_{\text{inlets}}} \quad (2.12)$$

in which the velocities and angular momentum fluxes correspond to the inertial (absolute) reference frame.

INTEGRAL ENERGY CONSERVATION EQUATION

We write the energy equation for a CV as

$$\dot{Q} + \dot{W}_{\text{pressure}} + \dot{W}_{\text{shear}} + \dot{W}_{\text{rotation}} + \dot{W}_{\text{shaft}} + \dot{W}_{\text{other}} = \frac{\partial}{\partial t} \iiint_{cv} e\rho \, d\mathcal{V} + \iint_{cs} e\rho V \cdot dA \quad (2.13)$$

In a nondeformable CV, the pressure force can do work only at its inflows and outflows. Within the CV itself, the pressure force at any point cancels out by an equal and opposite pressure force, and does not contribute to the work transfer term $\dot{W}_{\text{pressure}}$. Because the pressure force at the CS is in the direction opposite to the surface normal, we can express $\dot{W}_{\text{pressure}}$ as

$$\dot{W}_{\text{pressure}} = -\iint_{cs} \frac{p}{\rho} (\rho V \cdot dA)$$

where p/ρ is the flow work per unit mass. Substituting this in Equation 2.13, we obtain

$$\dot{Q} + \dot{W}_{\text{shear}} + \dot{W}_{\text{rotation}} + \dot{W}_{\text{shaft}} + \dot{W}_{\text{other}}$$

$$= \frac{\partial}{\partial t} \iiint_{cv} e\rho \, d\mathcal{V} + \iint_{cs} \left(u + \frac{p}{\rho} + \frac{V^2}{2} + gz \right) \rho V \cdot dA \quad (2.14)$$

which yields the steady flow energy equation in terms of pressure as

$$\dot{Q} + \dot{W}_{\text{shear}} + \dot{W}_{\text{rotation}} + \dot{W}_{\text{shaft}} + \dot{W}_{\text{other}} = \iint_{cs} \left(u + \frac{p}{\rho} + \frac{V^2}{2} + gz \right) \rho V \cdot dA \quad (2.15)$$

and in terms of enthalpy as

$$\dot{Q} + \dot{W}_{\text{shear}} + \dot{W}_{\text{rotation}} + \dot{W}_{\text{shaft}} + \dot{W}_{\text{other}} = \iint_{cs} \left(h + \frac{V^2}{2} + gz \right) \rho V \cdot dA \quad (2.16)$$

PROBLEM 2.1: MAXIMUM WATER LEVEL IN A TANK WITH INFLOW AND OUTFLOW

As shown in Figure 2.1, water flows at a constant volumetric flow rate of Q_{in} into a large tank of cross-section area A_{tank} and exits the drainpipe of area A_{pipe} near the bottom of the tank at volumetric flow rate Q_{in} after the water level in the tank exceeds h_o. Derive the differential equation that governs the water level h in the tank as a function of time, and hence find the maximum water level in the tank under steady state.

SOLUTION FOR PROBLEM 2.1

The unsteady integral continuity equation applied to this flow system yields

$$A_{tank}\frac{dh}{dt} = Q_{in} - Q_{out} = Q_{in} - A_{pipe}\sqrt{2g(h-h_0)}$$

$$A_{tank}\frac{dh}{dt} + A_{pipe}\sqrt{2g(h-h_0)} = Q_{in}$$

where we have obtained $Q_{out} = A_{pipe}\sqrt{2g(h-h_0)}$ using the Bernoulli equation (see Chapter 3). Under steady state, the time-dependent term vanishes, giving $Q_{out} = Q_{in}$, and the water level in the tank attains its maximum value h_{max} given by

$$A_{pipe}\sqrt{2g(h_{max}-h_0)} = Q_{in}$$

$$h_{max} = h_0 + \frac{Q_{in}^2}{2gA_{pipe}^2}$$

PROBLEM 2.2: FORCE OF A JET IMPINGING ON A FLAT PLAT

In a qualifying exam, the professor asked a student a simple fluids question. If a vertical jet of fluid density ρ hits a horizontal plate with velocity V over the stagnation area A where the jet hits the plate, what force is exerted on the plate? Assume zero ambient pressure (vacuum). The student calculated the stagnation pressure on the plate as $\rho V^2/2$, and knowing that the force equals pressure times the area, quickly

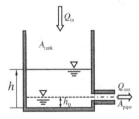

FIGURE 2.1 Water level in a tank with inflow and outflow (Problem 2.1).

responded with the answer $A\rho V^2/2$. The professor commented that the student's answer should be twice as much. Do you agree with the professor or the student and why?

SOLUTION FOR PROBLEM 2.2

The force exerted by the fluid stream on the plate plat needs to be calculated using the force-momentum balance over the stagnation area (A) while satisfying the continuity equation over the CV. An easy way to calculate this force is to consider vertical stream thrust at inlet and outlet, which in this case is zero. As the ambient pressure is zero, the inlet vertical stream thrust becomes

$$F = S_{T\_\text{inlet}} = \dot{m}V = (A\rho V)V = A\rho V^2$$

which is twice the value calculated by the student based on the flow stagnation pressure.

PROBLEM 2.3:   THRUST PRODUCED BY A FIREHOSE NOZZLE

Using the operating and geometric quantities for a firehose nozzle shown in Figure 2.2, find an expression to compute the force F that the firefighter will have to resist in handling the firehose. Will the firefighter experience a pull or a push force?

SOLUTION FOR PROBLEM 2.3

Continuity equation between sections 1 and 2 yields

$$\rho A_1 V_1 = \rho A_2 V_2$$

$$V_1 = V_2 \frac{A_2}{A_1}$$

Momentum equation over the CV with inlet at section 1 and outlet at section 2 yields

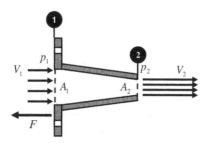

FIGURE 2.2   Thrust produced by a firehose nozzle (Problem 2.3).

$$-F + p_1 A_1 - p_2 A_2 - p_{amb}(A_1 - A_2) = \dot{m}V_2 - \dot{m}V_1$$

As $p_2 = p_{amb}$, the above equation reduces to

$$F = -(p_1 - p_{amb})A_1 + \dot{m}(V_2 - V_1)$$

$$F = -(p_1 - p_{amb})A_1 + \rho Q(V_2 - V_1)$$

where Q is the volumetric flow rate of water through the nozzle.
    Mechanical energy equation between sections 1 and 2 yields

$$\frac{p_1}{\rho} + \frac{V_1^2}{2} = \frac{p_2}{\rho} + \frac{V_2^2}{2}$$

$$p_1 - p_2 = p_1 - p_{amb} = \frac{1}{2}\rho\left(V_2^2 - V_1^2\right)$$

Substituting for $p_1 - p_{amb}$ from this equation into the momentum equation yields

$$F = -\frac{1}{2}\rho\left(V_2^2 - V_1^2\right)A_1 + \rho Q(V_2 - V_1)$$

$$F = \rho Q^2\left(\frac{1}{A_2} - \frac{1}{A_1}\right) - \frac{1}{2}\rho Q^2\left(\frac{1}{A_2^2} - \frac{1}{A_1^2}\right)A_1$$

which upon simplification yields

$$F = \frac{\rho Q^2}{2A_2}\frac{(\beta - 1)^2}{\beta}$$

where the nozzle area ratio $\beta = A_1/A_2$. We can rewrite this equation as

$$F = C_n\left(\frac{1}{2}\rho V_2^2\right)A_2$$

where $C_n = (\beta - 1)^2/\beta$. The positive value of F shows that the firefighter will experience a pull force.

PROBLEM 2.4:   FORCE ON A WOODEN LOG OF SQUARE CROSS-SECTION IN A TWO-DIMENSIONAL FLOW

Figure 2.3 shows a wooden log of square cross-section in a two-dimensional cross-flow with a uniform velocity in the x-direction. The figure also shows the x-direction linear velocity profile leaving the CV over the log. There is no flow across CD. The fluid has constant density ρ. Calculate the net force per unit depth acting on the CV in the x direction.

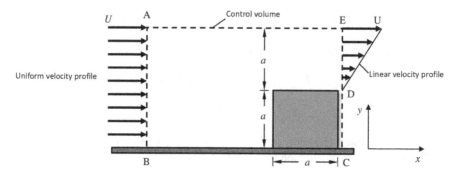

FIGURE 2.3 Force on a wooden log of square cross-section in a two-dimensional flow (Problem 2.4).

SOLUTION FOR PROBLEM 2.4

MASS CONSERVATION FOR THE CONTROL VOLUME (ASSUME UNIT DEPTH)

We obtain the mass rate through inlet AB as

$$\dot{m}_{AB} = 2a\rho U$$

and through outlet DE as

$$\dot{m}_{DE} = \int_0^a \rho \left(\frac{U}{a}\right) y\, dy = \frac{1}{2} a\rho U$$

giving the mass outflow rate through AE as

$$\dot{m}_{AE} = \dot{m}_{AB} - \dot{m}_{DE} = 2a\rho U - \frac{1}{2}a\rho\, U = \frac{3}{2}a\rho\, U$$

FORCE-MOMENTUM BALANCE OVER THE CONTROL VOLUME (ASSUME UNIT DEPTH)

We obtain the x-momentum flow rate entering AB as

$$\dot{M}_{AB} = 2a\rho\, U^2$$

leaving DE as

$$\dot{M}_{DE} = \int_0^a \rho \left(\frac{U}{a}\right)^2 y^2 dy = \frac{1}{3}a\rho U^2$$

and leaving AE as

$$\dot{M}_{AE} = \frac{3}{2}a\rho\, U^2$$

Thus, we write the x-direction force-momentum balance equation for the CV as

$$F_x = \dot{M}_{DE} + \dot{M}_{AE} - \dot{M}_{AB}$$

$$F_x = \frac{1}{3}a\rho\, U^2 + \frac{3}{2}a\rho\, U^2 - 2a\rho\, U^2$$

$$F_x = -\frac{1}{6}a\rho\, U^2$$

The negative sign for the force on the CV is consistent with the fact that the wooden log offers a drag to the incoming flow.

PROBLEM 2.5: MIXING OF TWO INCOMPRESSIBLE TURBULENT FLOWS IN AN EJECTOR

Figure 2.4 shows an ejector in which two incompressible turbulent flows of the same fluid mix between sections 1 and 2. At section 1, the central jet enters with uniform velocity V_j and the induced flow through the annulus enters with uniform velocity $(1/4)V_j$. Both flows at section 1 occupy equal cross-sectional areas. At the downstream section 2, the two flows have fully mixed with the resulting uniform velocity V_2. Neglecting wall shear force, calculate the increase in static pressure and the decrease in average total pressure between sections 1 and 2. The duct cross-section area (A) remains constant between these sections.

SOLUTION FOR PROBLEM 2.5

CONTINUITY EQUATION

We obtain the mass flow rate of central jet at section 1 as

$$\dot{m}_{1j} = \frac{1}{2}A\rho V_j$$

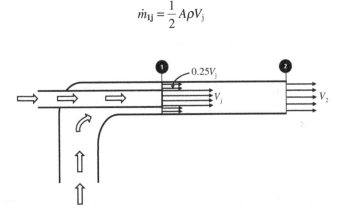

FIGURE 2.4   Mixing of two incompressible turbulent flows in an ejector (Problem 2.5).

through the annulus at section 1 as

$$\dot{m}_{1\_annulus} = \left(\frac{1}{2}A\right)\rho\left(\frac{1}{4}V_j\right) = \frac{1}{8}A\rho V_j$$

and through section 2 as

$$\dot{m}_2 = A\rho V_2$$

Equating total mass flow rate entering section 1 to that leaving section 2 yields

$$\dot{m}_2 = \dot{m}_{1j} + \dot{m}_{1\_annulus}$$

$$A\rho V_2 = \frac{1}{2}A\rho V_j + \frac{1}{8}A\rho V_j$$

$$V_2 = \frac{5}{8}V_j$$

MOMENTUM EQUATION

We obtain the momentum flow rate of the central jet at section 1 as

$$\dot{m}_{1j} = \dot{m}_{1ij}\,V_j = \frac{1}{2}A\rho V_j^2$$

through the annulus section 1 as

$$\dot{m}_{1i\_annulus} = \dot{m}_{1ij}\left(\frac{1}{4}V_j\right) = \frac{1}{32}A\rho V_j^2$$

and through section 2 as

$$\dot{M}_{1j} = A\rho\left(\frac{5}{8}V_j\right)^2 = \frac{25}{64}A\rho V_j^2$$

Assuming uniform static pressure at section 1, we write the momentum equation for the CV between sections 1 and 2 as

$$(p_1 - p_2)A = \dot{M}_2 - \dot{M}_{1j} - \dot{M}_{1\_annulus} = \frac{25}{64}A\rho V_j^2 - \frac{1}{2}A\rho V_j^2 - \frac{1}{32}A\rho V_j^2$$

which yields the increase in static pressure from section 1 to section 2 as

$$(p_2 - p_1) = \frac{9}{64}\rho V_j^2$$

In an incompressible flow, we obtain the total pressure at any point as sum of the static pressure and the dynamic pressure

$$p_0 = p + p_{dyn} = p + \frac{1}{2}\rho V^2$$

At section 1, the dynamic pressure associated with the two streams is different. We first calculate the average dynamic pressure at this section. Because the dynamic pressure in an incompressible flow equals the kinetic energy per unit volume, it is physically consistent to calculate mass-weighted average, not area-weighted average, dynamic pressure of the two streams at section 1. The fractions of total inlet mass flow rate associated with the central jet and annulus flows are 4/5 and 1/5, respectively.

We obtain the dynamic pressure in the central jet at section 1 as

$$p_{1j\_dyn} = \frac{1}{2}\rho V_j^2$$

and in the flow through the annulus as

$$p_{1\_annulus\_dyn} = \frac{1}{2}\rho\left(\frac{1}{4}V_j\right)^2 = \frac{1}{32}\rho V_j^2$$

giving the average dynamic pressure and average total pressure at this section as

$$\overline{p}_{1\_dyn} = \left(\frac{4}{5}\right)\frac{1}{2}\rho V_j^2 + \left(\frac{1}{5}\right)\frac{1}{32}\rho V_j^2 = \frac{13}{32}\rho V_j^2$$

and

$$\overline{p}_{01} = p_1 + \frac{13}{32}\rho V_j^2$$

We obtain the dynamic pressure and the average total pressure at section 2 as

$$p_{2\_dyn} = \frac{1}{2}\rho V_2^2 = \frac{1}{2}\rho\left(\frac{5}{8}V_j\right)^2 = \frac{25}{128}\rho V_j^2$$

and

$$\overline{p}_{02} = p_2 + \frac{25}{128}\rho V_j^2$$

We now obtain the decrease in the average total pressure between sections 1 and 2 as

$$\overline{p}_{01} - \overline{p}_{02} = \left(p_1 + \frac{13}{32}\rho V_j^2\right) - \left(p_2 + \frac{25}{128}\rho V_j^2\right)$$

$$\overline{p}_{01} - \overline{p}_{02} = (p_1 - p_2) + \frac{13}{32}\rho V_j^2 - \frac{25}{128}\rho V_j^2$$

Substituting in this equation for $(p_1 - p_2)$ from the foregoing yields

$$\bar{p}_{01} - \bar{p}_{02} = -\frac{9}{64}\rho V_j^2 + \frac{13}{32}\rho V_j^2 - \frac{25}{128}\rho V_j^2$$

$$\bar{p}_{01} - \bar{p}_{02} = \frac{9}{128}\rho V_j^2$$

PROBLEM 2.6: DAD'S GARDEN HOSE DILEMMA

A young boy helping his dad with the yard work and playing with the garden hose asked, "Hey dad, when I half cover the hose opening, the water jets out faster and goes farther. The dad replied, "Obviously, son. For the same flow, if the area becomes half, the jet velocity doubles." After a short while, the boy asked again, "Dad, but now it takes longer to fill the bucket." To which the dad replied, "Hmm …, that's something I need to think about." With your understanding of fluid flow, how would you resolve dad's dilemma?

SOLUTION FOR PROBLEM 2.6

The jet velocity at the garden hose exit depends on the difference between the total pressure and the ambient pressure, which remains constant. With the fixed water supply pressure at the garden hose inlet, the total pressure at the hose exit equals this supply pressure minus the hose system pressure loss, which is proportional to the square of the water flow rate (for a turbulent flow). The fact that it takes longer to fill the bucket indicates that blocking the hose exit reduces the water flow rate. This results in a lower pressure loss in the hose system and higher total pressure at the hose exit, in turn increasing the jet velocity. Completely closing the hose exit results in no flow through the hose, its pressure becoming the supply pressure everywhere.

PROBLEM 2.7: RESTRAINING FORCE ON CARTS WITH ONE INLET AND ONE OUTLET

Figure 2.5 shows four flow devices resting on frictionless wheels. These devices are restricted to move in the x-direction only and are initially held stationary. The pressure at the inlet and the outlet of each device is atmospheric and the flows entering and leaving them are incompressible. When released, which device will move to the right and which to the left?

SOLUTION FOR PROBLEM 2.7

The force acting on each device to keep it in stationary position is the force acting on the CV. When released, each device will move in the direction opposite to this the restraining force. Let us assume that this force acts on each device in the positive x-direction (to the right). Further note that each device has one inlet and one outlet with equal static pressure (atmospheric pressure) at both sections.

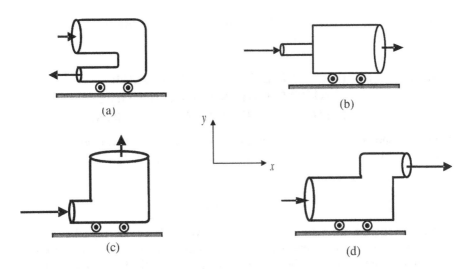

FIGURE 2.5 Four flow devices constrained to move along the x-direction only when released (Problem 2.7).

As the static pressure is uniform all over the device surface, its effect in the momentum CV analysis may be neglected and assumed to be zero. To satisfy mass conservation (continuity equation), the product of velocity and area at inlet and outlet must be equal in each device. This means higher flow area results in lower velocity. In the following, the momentum equation is used on each device to determine the restraining force acting on it.

For the device shown in Figure 2.5a, we write

$$F = -\dot{m}V_{\text{outlet}} - \dot{m}V_{\text{inlet}} = -\dot{m}\left(V_{\text{outlet}} + V_{\text{inlet}}\right)$$

As the restraining force F is in the negative x-direction, the device will move to the right when released.

For the device shown in Figure 2.5b, we write

$$F = \dot{m}V_{\text{outlet}} - \dot{m}V_{\text{inlet}} = \dot{m}\left(V_{\text{outlet}} - V_{\text{inlet}}\right)$$

As we have $V_{\text{outlet}} < V_{\text{inlet}}$, the restraining force F is in the negative x-direction, and the device will move to the right when released.

For the device shown in Figure 2.5c, we write

$$F = 0 - \dot{m}V_{\text{inlet}} = -\dot{m}V_{\text{inlet}}$$

As the restraining force F is in the negative x-direction, the device will move to the right when released.

For the device shown in Figure 2.5d, we write

$$F = \dot{m}V_{\text{outlet}} - \dot{m}V_{\text{inlet}} = \dot{m}(V_{\text{outlet}} - V_{\text{inlet}})$$

As we have $V_{outlet} > V_{inlet}$, the restraining force F is in the positive x-direction, and the device will move to the left when released.

PROBLEM 2.8:   FORCE NEEDED TO HOLD A WATER TANK IN POSITION UNDER THE GIVEN INFLOW AND OUTFLOW CONDITIONS

Figure 2.6 shows a cylindrical water tank of diameter 1.0 m having an exit pipe of diameter 0.25 m. The velocity profile at the tank inlet is given by

$$V_x = 1 - \left(\frac{r}{0.5}\right)^2$$

and that at the pipe exit corresponds to a fully developed turbulent pipe flow with the 1/7th power law profile. What is the net force needed to hold the tank in position? Use a momentum correction factor of 1.333 at the inlet to the large tank and 1.02 in the exit pipe. The density of water is 1000 kg/m³, the gauge pressure at tank inlet is 1.4 bar, and the ambient pressure is 1.0 bar.

SOLUTION FOR PROBLEM 2.8

We compute the following quantities at section 1 (inlet):

$$A_1 = \frac{\pi(1)^2}{4} = 0.785 \text{ m}^2$$

$$V_{1\_max} = 1.0 \text{ m/s}$$

For a parabolic velocity profile, the average velocity is half of the centerline (maximum) velocity, giving

$$\overline{V}_1 = 0.5 \text{ m/s}$$

$$\dot{m}_1 = A_1 \rho \overline{V}_1 = 0.785 \times 1000 \times 0.5 = 392.699 \text{ kg/s}$$

$$\dot{M}_1 = \beta_1 \dot{m}_1 \overline{V}_1 = 1.333 \times 392.699 \times 0.5 = 264.734 \text{ N}$$

At section 2 (outlet), we obtain

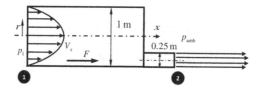

FIGURE 2.6   Force needed to hold a water tank in position under the given inflow and outflow conditions (Problem 2.8).

$$A_2 = \frac{\pi(0.25)^2}{4} = 0.0491 \text{ m}^2$$

From mass conservation (continuity equation), the average velocity in the exit pipe is

$$\bar{V}_2 = \frac{m_1}{\rho A_2} = \frac{392.699}{1000 \times 0.0491} = 8 \text{ m/s}$$

giving the momentum flow rate as

$$\dot{M}_1 = \beta_1 \dot{m}_1 \bar{V}_1 = 1.02 \times 392.699 \times 8 = 3204.425 \text{ N}$$

The force momentum balance on the tank-pipe CV yields

$$F + p_1 A_1 - p_{\text{amb}} A_1 = \dot{M}_2 - \dot{M}_1$$

$$F = \dot{M}_2 - \dot{M}_1 - (p_1 - p_{\text{amb}}) A_1$$

$$F = 3204.425 - 264.734 - 1.4 \times 10^5 \times 0.785$$

$$F = -28,473 \text{ N}$$

The negative sign indicates that the force is acting in the direction opposite to the direction shown in Figure 2.6.

PROBLEM 2.9: LAMINAR FLOW ENTERING A SUDDEN PIPE EXPANSION

Figure 2.7 shows an incompressible flow in a sudden pipe expansion with the pipe diameters D_1 and D_2. The laminar flow in the smaller pipe is fully developed with a parabolic velocity profile. The flow exiting the larger pipe is turbulent with a uniform

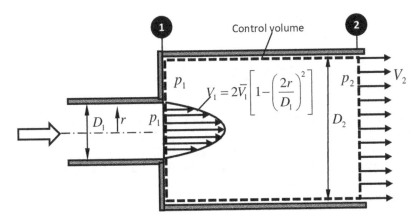

FIGURE 2.7 Sudden expansion pipe flow with a parabolic velocity profile at inlet and uniform velocity at outlet (Problem 2.9).

velocity. The static pressures at sections 1 and 2 are uniform. Find the change in the static pressure and the total pressure between these sections. Neglect any shear force from the downstream pipe wall.

SOLUTION FOR PROBLEM 2.9

In this problem, the laminar velocity profile entering the larger pipe is parabolic, given by

$$V_1 = 2\bar{V}_1 \left[1 - \left(\frac{2r}{D_1} \right)^2 \right]$$

We obtain the mass flow rate entering section 1 as $\dot{m}_1 = A_1 \rho \bar{V}_1$ and that exiting section 2 as $\dot{m}_2 = A_2 \rho V_2$ where $A_1 = \pi D_1^2/4$ and $A_2 = \pi D_2^2/4$. Equating $\dot{m}_1$ and $\dot{m}_2$ for a steady flow, we write

$$\frac{A_2}{A_1} = \frac{\bar{V}_1}{V_2}$$

For applying the momentum equation to the CV shown in Figure 2.7, we assume that the static pressure p_1 at the exit of the smaller pipe is uniform over the entire section 1, which includes the annulus area between the two pipes. With the axial momentum flow rates $\dot{M}_1 = \beta_1 \dot{m}_1 \bar{V}_1 = \frac{4}{3} \dot{m}_1 \bar{V}_1$ entering the CV through section 1 and $\dot{M}_2 = \dot{m}_2 V_2 = \dot{m}_1 V_2$ exiting the CV through section 2, we write the force-momentum balance as

$$\left(p_1 - p_2 \right) A_2 = \dot{M}_2 - \dot{M}_1 = \dot{m}_1 \left(V_2 - \frac{4}{3} \bar{V}_1 \right) = \rho A_1 \bar{V}_1 \left(V_2 - \frac{4}{3} \bar{V}_1 \right)$$

$$\left(p_2 - p_2 \right) = \rho \bar{V}_1^2 \left(\frac{A_1}{A_2} \right) \left(\frac{4}{3} - \frac{A_1}{A_2} \right)$$

$$C_p = \frac{\left(p_2 - p_1 \right)}{\frac{1}{2} \rho \bar{V}_1^2} = 2 \left(\frac{A_1}{A_2} \right) \left(\frac{4}{3} - \frac{A_1}{A_2} \right)$$

where we have defined the pressure-rise coefficient C_p in terms of the dynamic pressure based on the average inlet velocity $\bar{V}_1$. Note that for $A_1/A_2 = 2/3$, C_p reaches the maximum value of 8/9. A part of the recovery in static pressure in this problem happens when the nonuniform (parabolic) velocity profile with its higher momentum content at the inlet becomes uniform with lower momentum at the exit.

With the total pressure at section 1 as

$$\bar{p}_{01} = p_1 + \frac{1}{2} \alpha \rho \bar{V}_1^2 = p_1 + \frac{1}{2} 2\rho \bar{V}_1^2 = p_1 + \rho \bar{V}_1^2$$

where α is the kinetic energy correction factor, which equals 2 for the parabolic velocity profile in a circular pipe. We obtain the total pressure at section 2 as

$$p_{02} = p_2 + \frac{1}{2}\rho V_2^2$$

and the total pressure loss between these sections as

$$\bar{p}_{01} - p_{02} = p_1 + \rho \bar{V}_1^2 - p_2 - \frac{1}{2}\rho V_2^2 = \rho \bar{V}_1^2 - \frac{1}{2}\rho V_2^2 - (p_2 - p_1)$$

Substituting for $(p_2 - p_1)$ yields

$$\bar{p}_{01} - p_{02} = \rho \bar{V}_1^2 - \frac{1}{2}\rho V_2^2 - \rho \bar{V}_1^2 \left(\frac{A_1}{A_2}\right)\left(\frac{4}{3} - \frac{A_1}{A_2}\right)$$

With $A_1/A_2 = V_2/\bar{V}_1$, this equation becomes

$$\bar{p}_{01} - p_{02} = \frac{1}{2}\rho(\bar{V}_1 - V_2)^2 + \frac{1}{2}\rho \bar{V}_1^2 \left(1 - \frac{2}{3}\frac{V_2}{\bar{V}_1}\right)$$

which we can express in terms of the loss coefficient K as

$$K = \frac{(\bar{p}_{01} - p_{02})}{\frac{1}{2}\rho \bar{V}_1^2} = 2 - \frac{8}{3}\left(\frac{A_1}{A_2}\right) + \left(\frac{A_1}{A_2}\right)^2$$

which shows that for $A_1/A_2 \ll 1$, we obtain $K \approx 2.0$, that is, the dynamic pressure of the incoming flow with its parabolic velocity profile is fully lost in the downstream pipe, acting as a plenum.

PROBLEM 2.10: AN ACCELERATING CART WITH A CONSTANT OUTFLOW

Figure 2.8 shows an accelerating cart containing fluid of constant density ρ. The slow motion of a heavy plate generates a constant mass flow rate $\dot{m}$ through the nozzle with an exit velocity W_0 relative to the cart. The total initial mass of the cart system is m_0. Assuming the air drag and contact friction to be negligible, calculate the time-dependent velocity U_c of the cart if it starts from rest.

SOLUTION FOR PROBLEM 2.10

This is a variable mass problem like the rocket problem with constant exhaust velocity relative to the rocket. The mass conservation for the CV encompassing the cart system yields

$$m_c(t) = m_0 - \dot{m}t$$

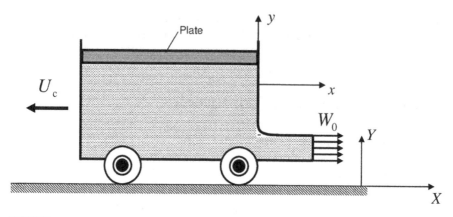

FIGURE 2.8   An accelerating cart with constant outflow (Problem 2.10).

In the absence of surface and body forces on the CV, the x-momentum equation in the noninertial (accelerating) x-y coordinate system attached to the cart yields

$$m_c \frac{dU_c}{dt} = \dot{m}W_0$$

Substituting for m_c from the continuity equation, we obtain

$$\frac{dU_c}{dt} = \frac{\dot{m}W_0}{m_c} = \frac{\dot{m}W_0}{m_0 - \dot{m}t}$$

With $U_c = 0$ at $t=0$, the solution to this ordinary differential equation yields

$$U_c = W_0 \ln\left(\frac{m_0}{m_0 - \dot{m}t}\right)$$

PROBLEM 2.11:   AN ACCELERATING CART UNDER A DEFLECTED WATER JET

As shown in Figure 2.9, a horizontal water jet of constant velocity V_j and area A_j enters a vane-cart system of mass m_c and leaves it at an angle θ, accelerating the cart-vane system. The mass of water with velocity V_j within the cart CV remains constant at m_w. Assume that A_j and V_j remain constant from vane inlet to exit. Neglecting any frictional force resisting the cart motion, find the angle at which the jet leaves the vane when the velocity of the cart is U_{c1} at time t_1.

SOLUTION FOR PROBLEM 2.11

In this problem, the mass of the cart-vane system and that of water within the CV remains constant. As the cart is accelerating, the x-y coordinate system attached to the cart becomes noninertial. In this coordinate system, although the water jet

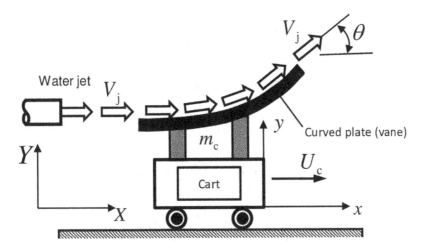

FIGURE 2.9   An accelerating cart under a deflected water jet (Problem 2.11).

velocity relative to cart is time-dependent, the x-momentum of water within the CV does not change with time.

We express the water jet velocity relative to the cart as

$$W_j = V_j - U_c$$

which yields

$$\frac{dW_j}{dt} = -\frac{dU_c}{dt}$$

From the continuity equation, we deduce that the water mass flow rate at vane inlet and exit remain constant at

$$\dot{m} = \rho A_j W_j$$

In the absence of surface and body forces on the CV, write the x-momentum equation in the noninertial (accelerating) x-y coordinate system attached to the cart as

$$-(m_c + m_w)\frac{dU_c}{dt} = \dot{m} W_j \cos\theta - \dot{m} W_j$$

Substituting for $\dot{m}$ from the continuity equation yields

$$(m_c + m_w)\frac{dU_c}{dt} = \rho A_j W_j^2 (1 - \cos\theta)$$

$$\frac{dU_c}{dt} = \frac{\rho A_j (1 - \cos\theta)}{(m_c + m_w)} W_j^2$$

$$-\frac{dW_j}{dt} = \frac{\rho A_j (1-\cos\theta)}{(m_c + m_w)} W_j^2$$

$$\frac{dW_j}{W_j^2} = \frac{\rho A_j (1-\cos\theta)}{(m_c + m_w)} dt$$

whose integration from $t = 0$ to $t = t_1$ yields

$$-\int_{V_j}^{(V_j - U_{c1})} \frac{dW_j}{W_j^2} = \frac{\rho A_j (1-\cos\theta)}{(m_c + m_w)} \int_0^{t_1} dt$$

$$\frac{1}{V_j - U_{c1}} - \frac{1}{V_j} = \frac{\rho A_j (1-\cos\theta)}{(m_c + m_w)} t_1$$

$$\frac{U_{c1}}{V_j (V_j - U_{c1})} = \frac{\rho A_j (1-\cos\theta)}{(m_c + m_w)} t_1$$

which yields

$$\cos\theta = 1 - \frac{U_{c1}(m_c + m_w)}{\rho A_j V_j (V_j - U_{c1}) t_1}$$

$$\theta = \cos^{-1}\left[1 - \frac{U_{c1}(m_c + m_w)}{\rho A_j V_j (V_j - U_{c1}) t_1}\right]$$

PROBLEM 2.12:   AN ACCELERATING CART WITH VARIABLE OUTFLOW

Find the time-dependent acceleration of the cart shown in Figure 2.10. The initial total mass of the cart containing the fluid of density ρ is m_0. Other quantities needed for the solution are shown in the figure. Neglect any frictional force opposing the cart movement.

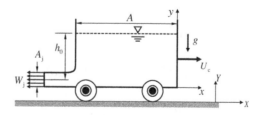

FIGURE 2.10   An accelerating cart with variable outflow (Problem 2.12).

SOLUTION FOR PROBLEM 2.12

In this variable-mass problem, the time-dependent jet velocity (assumed uniform over the jet cross-section) relative to the cart depends on the height (geodetic head) of fluid level in the cart from the jet centerline (datum). Neglecting the velocity of fluid free surface in the cart, we obtain the jet velocity as

$$W_j = \sqrt{2gh}$$

From mass conservation, we obtain

$$\dot{m}_j = \rho A_j W_j = \frac{d}{dt}\{m_0 - \rho(h_0 - h)A\}$$

$$A\frac{dh}{dt} = A_j\sqrt{2gh}$$

$$\frac{dh}{dt} = \frac{A_j\sqrt{2g}}{A}\sqrt{h}$$

$$\int_{h_0}^{h}\frac{dh}{\sqrt{h}} = \frac{A_j\sqrt{2g}}{A}\int_0^t dt$$

$$2\left(\sqrt{h} - \sqrt{h_0}\right) = \frac{A_j\sqrt{2g}}{A}t$$

$$\sqrt{h} = \sqrt{h_0} + \frac{A_j\sqrt{2g}}{2A}t$$

$$h = \left(\sqrt{h_0} + \frac{A_j\sqrt{g/2}}{A}t\right)^2$$

The momentum equation in the x-direction in the noninertial reference frame yields (in the absence of surface and body forces)

$$-\{m_0 - \rho(h_0 - h)A\}\dot{U}_c = -\dot{m}_j W_j = -\rho A_j W_j^2$$

$$\dot{U}_c = \frac{\rho A_j W_j^2}{\{m_0 - \rho(h_0 - h)A\}}$$

$$\dot{U}_c = \frac{2\rho A_j gh}{\{m_0 - \rho(h_0 - h)A\}}$$

where we have h from the foregoing as

$$h = \left(\sqrt{h_0} + \frac{A_j\sqrt{g/2}}{A}t\right)^2$$

PROBLEM 2.13:   A WATER TANKER ACCELERATING DUE TO AN INCOMING WATER JET

As shown in Figure 2.11, a water jet of constant velocity V_j and cross-section area A_j is entering a water tanker, which is free to move with negligible friction. Assuming that the tanker is initially at rest with mass m_o, find its resulting acceleration $U_t(t)$, velocity $U_t(t)$, and mass $m(t)$.

SOLUTION FOR PROBLEM 2.13

In this problem, the water tanker is gaining mass at a decreasing rate until it asymptotically reaches the jet velocity at which point the water jet does not enter the tanker. Applying the unsteady form of the continuity equation (Equation 2.1) to the tanker CV shown in Figure 2.11, we obtain

$$\frac{\partial}{\partial t} \iiint_{cv} \rho \, d\Psi = \rho \, W_j \, A_j$$

$$\frac{dm}{dt} = \rho \, W_j \, A_j$$

where $W_j = V_j - U_t$ is the jet velocity relative to the noninertial coordinate axes attached to the tanker. With a constant V_j, we obtain

$$\dot{W}_j = -\dot{U}_t$$

LINEAR MOMENTUM EQUATION IN THE NONINERTIAL REFERENCE FRAME

With no body and surface forces acting on the tanker CV, the left-hand side of Equation 2.8 becomes $\{-(m+m_j)\dot{U}_t\}$. The CV in this case has one inflow and no

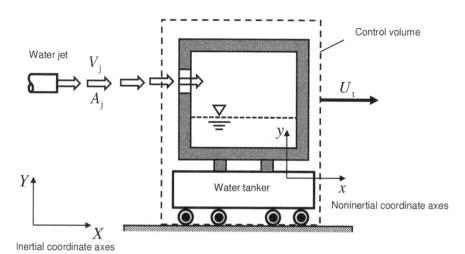

FIGURE 2.11   A water tanker accelerating due to an incoming water jet (Problem 2.13).

outflow. The surface integral on the right-hand side of the equation, therefore, yields $\{-\dot{m}W_j\}$. The unsteady term in this equation needs a careful evaluation. We assume that the mass of the water jet, after it enters the tanker, instantly assumes the tanker velocity U_t, resulting in zero velocity relative to the tanker. The total instantaneous mass within the tanker CV consists of mass m of the tanker with water and a part of the water jet having a constant mass m_j and velocity V_j before it enters the tanker. Thus, we can write the instantaneous momentum of the entire mass of the tanker CV in the noninertial reference frame as $m_j(V_j - U_t)$, giving

$$\frac{\partial}{\partial t_{xyz}} \iiint_{cv} V_{xyz}(\rho\, d\mathcal{V}) = \frac{d}{dt}(m_j V_j - m_j U_t) = -\frac{d}{dt}(m_j U_t) = -m_j \dot{U}_t$$

where we have used the fact that both m_j and V_j remain constant. Thus, Equation 2.8 becomes

$$-(m + m_j)\dot{U}_t = -\dot{m}W_j - m_j\dot{U}_t$$

$$-m\dot{U}_t = -\dot{m}W_j$$

Substituting $\dot{W}_j = -\dot{U}_t$ from the continuity equation yields

$$m\dot{W}_j = -\dot{m}W_j$$

$$\frac{\dot{m}}{m} = -\frac{\dot{W}_j}{W_j}$$

We now derive this equation using the linear momentum equation in the inertial reference frame.

LINEAR MOMENTUM EQUATION IN THE INERTIAL REFERENCE FRAME

With no body and surface forces acting on the tanker CV, the left-hand side of Equation 2.3 becomes zero. The water jet velocity $V_j = W_j + U_t$ is the momentum-velocity at the inlet to the water tanker. With only one inflow into the tanker CV and no outflow from it, we obtain the net momentum outflow rate as $\{-\dot{m}V_j\}$.

The unsteady term on the right-hand side of Equation 2.3 needs a careful evaluation. The total instantaneous mass within the tanker CV consists of the mass m of the tanker with water and a part of the water jet with constant m_j and velocity V_j before it enters the tanker. We assume that the mass of the water jet, upon entering the tanker, instantly assumes its velocity U_t. Thus, we write the instantaneous momentum of the entire mass of the tanker CV in the inertial reference frame as $mU_t + m_jV_j$. The unsteady term in Equation 2.3 becomes

$$\frac{\partial}{\partial t_{XYZ}} \iiint_{cv} V_{XYZ}(\rho\, d\mathcal{V}) = \frac{d}{dt}(mU_t + m_jV_j) = \frac{d}{dt}(mU_t) = U_t\frac{dm}{dt} + m\frac{dU_t}{dt}$$

where we have used the fact that $m_j V_j$ remains constant. Thus, Equation 2.3 reduces to

$$0 = -\dot{m}V_j + U_t \dot{m} + m\dot{U}_j$$

$$\dot{m}\left(V_j - U_t\right) = m\dot{U}_t$$

$$\dot{m}W_j = -m\dot{W}_j$$

$$\frac{\dot{m}}{m} = -\frac{\dot{W}_j}{W_j}$$

which is identical to the equation we obtained in the foregoing in a noninertial reference frame.

When the water tanker is at rest, we have $m = m_0$ and $W_j = V_j$. Integrating this equation yields

$$\int_{m_0}^{m} \frac{\mathrm{d}m}{m} = -\int_{V_j}^{W_j} \frac{\mathrm{d}W_j}{W_j}$$

$$\ln\left(\frac{m}{m_0}\right) = -\ln\left(\frac{W_j}{V_j}\right) = \ln\left(\frac{V_j}{W_j}\right)$$

Equating expressions under the natural log on both sides and rearranging terms, we obtain

$$W_j = \frac{m_0 V_j}{m}$$

Substituting this in the continuity equation $\left(\mathrm{d}m/\mathrm{d}t = \rho\, W_j\, A_j\right)$ yields

$$\frac{\mathrm{d}m}{\mathrm{d}t} = \rho A_j \frac{m_0 V_j}{m}$$

$$m\,\mathrm{d}m = \left(\rho A_j\, V_j\right)m_0\mathrm{d}t = \dot{m}_j\, m_0\, \mathrm{d}t$$

where $m_j = \rho A_j\, V_j$ is the constant water jet mass flow rate exiting the nozzle. By integrating this equation, we obtain

$$\int_{m_0}^{m} m\,\mathrm{d}m = \int_0^t \dot{m}_j m_0\, \mathrm{d}t$$

$$\frac{m^2 - m_0^2}{2} = \dot{m}_j m_0 t$$

$$m^2 = m_0^2 + 2\dot{m}_j m_0 t$$

$$m = \left(m_0^2 + 2\dot{m}_j m_0 t\right)^{1/2}$$

$$m = m_0\left(1 + 2\zeta t\right)^{1/2}$$

where

$$\zeta = \frac{\dot{m}_j}{m_0} = \frac{\rho A_j V_j}{m_0}$$

Substituting $m = m_0 \left(1 + 2\zeta t\right)^{1/2}$ in $W_j = m_0 V_j / m$ yields

$$W_j = \frac{m_0 V_j}{m} = V_j \left(1 + 2\zeta t\right)^{-1/2}$$

Further, we can write the time-varying tanker velocity as

$$U_t = V_j - W_j = V_j - V_j \left(1 + 2\zeta t\right)^{-1/2}$$

$$U_t = V_j \left\{ 1 - \left(1 + 2\zeta t\right)^{-1/2} \right\}$$

Differentiating this equation with time yields the tanker acceleration

$$U_t = V_j \zeta \left(1 + 2\zeta t\right)^{-3/2}$$

PROBLEM 2.14:   A LAWN SPRINKLER WITH TWO UNEQUAL ARMS

Figure 2.12 shows a lawn sprinkler with two unequal arms. For the given geometric quantities and the water jet velocity at the exit of the nozzle attached to each sprinkler arm, calculate the sprinkler rotational speed under negligible frictional torque. Assume that the jet area at outlet 1 is A_1 and that at outlet 2 is A_2.

SOLUTION FOR PROBLEM 2.14

In this sprinkler problem there is one inlet at the axis of rotation and two outlets at different radii. The flow at the inlet will have zero angular momentum. The mass velocity each outlet corresponds to the jet velocity. Let us use the convention that the angular momentum is positive in the counterclockwise direction and negative in the clockwise direction. We can write various quantities at each outlet as follows.

At outlet 1, we compute the mass flow rate as

$$\dot{m}_1 = \rho A_1 W_1$$

the specific angular momentum as

$$R_1 V_{\theta 1} = R_1 \left(R_1 \Omega + W_1 \right)$$

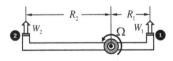

FIGURE 2.12   A lawn sprinkler with two unequal arms (Problem 2.14).

and the outflow rate of angular momentum as

$$\dot{H}_1 = \dot{m}_1 R_1 \left(R_1 \Omega + W_1 \right)$$

Similarly, at outlet 2 we calculate the mass flow rate as

$$\dot{m}_2 = \rho A_2 W_2$$

the specific angular momentum as

$$R_2 V_{\theta 2} = R_2 \left(R_2 \Omega - W_2 \right)$$

and the outflow rate of angular momentum as

$$\dot{H}_2 = \dot{m}_2 R_2 \left(R_2 \Omega - W_2 \right)$$

In the absence of any torque acting on the sprinkler CV, the net outflow rate of angular momentum must be zero, giving

$$\dot{H}_1 + \dot{H}_2 = 0$$

$$\dot{m}_1 R_1 \left(R_1 \Omega + W_1 \right) + \dot{m}_2 R_2 \left(R_2 \Omega - W_2 \right) = 0$$

$$\Omega = \frac{\dot{m}_2 R_2 W_2 - \dot{m}_1 R_1 W_1}{\dot{m}_1 R_1^2 + \dot{m}_2 R_2^2}$$

Substituting in this equation $\dot{m}_1 = \rho A_1 W_1$ and $\dot{m}_2 = \rho A_2 W_2$ from the foregoing, we finally obtain

$$\Omega = \frac{A_2 R_2 W_2^2 - A_1 R_1 W_1^2}{A_1 W_1 R_1^2 + A_2 W_2 R_2^2}$$

which shows that Ω does not depend on the sprinkler fluid density.

PROBLEM 2.15: EFFECTIVE AVERAGE MOODY FRICTION FACTOR FOR AN ANNULUS FLOW WITH DIFFERENT RELATIVE ROUGHNESS ON INNER AND OUTER WALLS

Consider a fully developed incompressible turbulent flow of a fluid of density ρ having the average velocity V through an annulus of inner diameter D_1, outer diameter D_2, and length L. The surface roughness of the annulus inner wall is higher than that of its outer wall. For the prevailing flow Reynolds number and the relative roughness parameters, the test data indicate that the Moody friction factor for the inner wall is

f_1 and that for the outer wall is f_2. If the static pressure drop across the annulus is given by

$$p_1 - p_2 = \frac{f^* L}{D_h}\left(\frac{1}{2}\rho V^2\right)$$

Find an expression to compute f^* in this equation in terms of the given quantities.

SOLUTION FOR PROBLEM 2.15

For a fully developed pipe flow, the net outflow of axial momentum is zero for the CV of length L. Accordingly, the net pressure force between inlet and outlet must balance the net shear force generated at the annulus walls, giving

$$\left(p_1 - p_2\right)\frac{\pi}{4}\left(D_2^2 - D_1^2\right) = \pi L\left(D_1\tau_{w1} + D_2\tau_{w2}\right)$$

As

$$\tau_{w1} = C_{f1}\left(\frac{1}{2}\rho V^2\right)$$

and

$$\tau_{w2} = C_{f2}\left(\frac{1}{2}\rho V^2\right)$$

we obtain

$$\left(p_1 - p_2\right)\frac{\pi}{4}\left(D_2^2 - D_1^2\right) = \pi L\left(\frac{1}{2}\rho V^2\right)\left(D_1 C_{f1} + D_2 C_{f2}\right)$$

$$\left(p_1 - p_2\right) = \frac{L}{\left(D_2^2 - D_1^2\right)}\left(\frac{1}{2}\rho V^2\right)\left(D_1 4C_{f1} + D_2 4C_{f2}\right)$$

With the mean hydraulic diameter $D_h = \left(D_2 - D_1\right)$, $4C_{f1} = f_1$, and $4C_{f2} = f_2$, rewriting this equation as

$$p_1 - p_2 = \frac{f^* L}{D_h}\left(\frac{1}{2}\rho V^2\right)$$

yields

$$f^* = \left(\frac{D_1 f_1 + D_2 f_2}{D_2 + D_1}\right)$$

NOMENCLATURE

a	Acceleration magnitude
$\boldsymbol{a}$	Acceleration vector
A	Area
$\boldsymbol{A}$	Area vector
c_p	Specific heat at constant pressure
c_v	Specific heat at constant volume
C_p	Static pressure-rise (recovery) coefficient
D	Pipe diameter
e	Specific total energy of a system
$\hat{e}$	Unit vector
E	Total energy of a control system
F	Force magnitude
$\boldsymbol{F}$	Force vector
g	Acceleration due to gravity
h	Height measured from a datum, specific enthalpy, heat transfer coefficient
$\dot{H}$	Flow rate of angular momentum
$\boldsymbol{H}$	Angular momentum vector
K	Loss coefficient
L	Length
m	Mass
$\dot{m}$	Mass flow rate
M	Linear momentum; Mach number
$\boldsymbol{M}$	Linear momentum vector
$\dot{M}$	Flow rate of linear momentum
N	Number of inlets or outlets of a CV
p	Static pressure
Q	Volumetric flow rate
$\dot{Q}$	Heat transfer rate
r	Radial distance
$\boldsymbol{r}$	Displacement vector in noninertial coordinate system
R	Pipe radius; gas constant
$\boldsymbol{R}$	Displacement vector in inertial coordinate system
s	Specific entropy
S_T	Stream thrust
t	Time
T	Temperature
u	Specific internal energy
U	Uniform free stream velocity; cart velocity
V	Uniform velocity at pipe inlet; absolute velocity
V	Velocity vector
$\cancel{V}$	Volume
W	Relative velocity
x	Cartesian coordinate x
xyz	Noninertial Cartesian coordinate axes

XYZ	Inertial Cartesian coordinate axes
y	Cartesian coordinate y
z	Cartesian coordinate z

SUBSCRIPTS AND SUPERSCRIPTS

1	Location 1; section 1
2	Location 2; section 2
3	Location 3; section 3
amb	Ambient
b	Body
c	Cart
cv	Control volume
cs	Surface of the CV
f	Friction
h	Solution of the homogeneous part
in	Inlet
j	Jet
max.	Maximum
out	Outlet
0	Total (stagnation)
s	Surface
sh	Shear
t	Water tanker
x	Component in x-coordinate direction
xyz	In noninertial Cartesian coordinate system
XYZ	In inertial Cartesian coordinate system
y	Component in y-coordinate direction
z	Component in z-coordinate direction
θ	Tangential direction
$(\bar{\ })$	Section-average value

GREEK SYMBOLS

α	Kinetic energy correction factor
β	Momentum correction factor
ε	Dimensionless electrical heat generation parameter
Γ	Torque
γ	Ratio of specific heats ($\gamma = c_p/c_v$)
η	Number of transfer units (NTU)
θ	Angle made with the x-direction, dimensionless temperature
ξ	Dimensionless axial distance
ρ	Density
ω	Magnitude of $\boldsymbol{\omega}$
ω	Local angular velocity of a fluid particle

Ω Angular velocity
Ω Rotation vector for the noninertial coordinate axes xyz
$\dot{\Omega}$ Angular acceleration

REFERENCE

Sultanian, B.K. 2015. *Fluid Mechanics: An Intermediate Approach*. Boca Raton, FL: Taylor & Francis.

BIBLIOGRAPHY

Sultanian, B.K. 2018: *Gas Turbines: Internal Flow Systems Modeling* (Cambridge Aerospace Series #44). Cambridge: Cambridge University Press.
Sultanian, B.K. 2019. *Logan's Turbomachinery: Flowpath Design and Performance Fundamentals*, 3rd edition. Boca Raton, FL: Taylor & Francis.
White, F. 2015. *Fluid Mechanics*, 8th edition. New York: McGraw-Hill Education.

3 Bernoulli Equation
Mechanical Energy Equation

REVIEW OF KEY CONCEPTS

In this chapter we consider steady incompressible flow problems requiring applications of the Bernoulli equation and the mechanical energy equation (MEE), which is also known as the extended Bernoulli equation. These equations and related concepts are briefly presented here. Readers may find their detailed derivations in Sultanian (2015).

BERNOULLI EQUATION

In terms of specific energy, we write the Bernoulli equation, which is valid for an inviscid flow along a streamline or between any two points in a potential (irrotational) flow, as

$$\frac{p}{\rho} + \frac{V^2}{2} + gz = E_{\mathrm{B}} \tag{3.1}$$

where p/ρ, $V^2/2$, gz, and E_{B} are, respectively, the specific flow work, the specific kinetic energy, the specific potential energy in the conservative gravitational force field, and the Bernoulli total specific energy (mechanical).

Multiplying Equation 3.1 by ρ, we obtain the Bernoulli equation in terms of pressure as

$$p + \frac{\rho V^2}{2} + \rho gz = p_{\mathrm{B}} \tag{3.2}$$

where p, $\rho V^2/2$, ρgz, p_{B} are, respectively, the static pressure, the dynamic pressure, the hydrostatic pressure, and the Bernoulli total pressure, which is different from the total (stagnation) pressure $p_0 = p + \rho V^2/2$.

Dividing Equation 3.1 by g, we obtain the Bernoulli equation in terms of head (length unit)

$$\frac{p}{\rho g} + \frac{V^2}{2g} + z = Z_{\mathrm{B}} \tag{3.3}$$

where $p/(\rho g)$, $V^2/(2g)$, z, and Z_{B} are, respectively, the pressure, the velocity head, the geodetic head (potential head or hydrostatic head), and Bernoulli constant in terms of head.

49

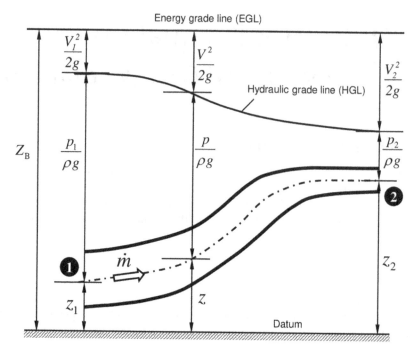

FIGURE 3.1   Delineation of terms in the Bernoulli equation.

Figure 3.1 shows how three heads vary at each section along an arbitrary frictionless duct (stream tube). The energy grade line (EGL) is the top line parallel to the datum. This line represents the total Bernoulli head Z_B, which in this case remains constant due to zero loss of flow energy along the duct. The hydraulic grade line (HGL) is the line representing the sum of the geodetic (potential or hydrostatic) head and the pressure head. As shown in the figure, the difference between EGL and HGL at any section of the duct flow equals the velocity head at that section.

MECHANICAL ENERGY EQUATION

The MEE is a generalized Bernoulli equation—also called the extended Bernoulli equation (see, e.g., Sultanian, 2015). MEE is grounded in the first law of thermodynamics in the form of the steady flow energy equation, where the flow energy changes are due to work transfer (e.g., shaft work) only. All other forms of energy transfer are neglected. This equation is widely used in the analysis of incompressible internal flows with nonuniform velocity distributions at flow cross-sections, featuring losses of mechanical energy to friction, and entropy generated by other means. Between any two sections of an internal flow, the extended Bernoulli equation in its most general form is given by Equation 3.4. Note that this equation has no direct relation to the linear momentum equation and, unlike the original Bernoulli equation, is not restricted to an inviscid incompressible flow along a streamline or between any points in a potential (irrotational) flow.

$$\frac{p_1}{\rho g} + \alpha_1 \frac{\bar{V}_1^2}{2g} + z_1 + \sum \Delta h_{\text{gain}} - \sum \Delta h_{\text{loss}} = \frac{p_2}{\rho g} + \alpha_2 \frac{\bar{V}_2^2}{2g} + z_2 \qquad (3.4)$$

In this equation, the static pressure and the flow velocity are the section-average values. To account for a nonuniform velocity profile, we use the kinetic energy correction factor defined at each section as

$$\alpha_1 = \frac{\displaystyle\int_{A_1} V_1^3 \, dA}{A_1 \bar{V}_1^3}$$

and

$$\alpha_2 = \frac{\displaystyle\int_{A_2} V_2^3 \, dA}{A_2 \bar{V}_2^3}$$

For a fully developed laminar flow with its parabolic velocity profile in a circular pipe, we obtain $\alpha = 2.0$. For a turbulent flow with nearly uniform velocity profile, we assume $\alpha = 1.0$. Equation 3.4 has a simple interpretation. On the left-hand side of the equation, the first three terms representing the pressure head, the velocity head, and the geodetic head form the total head at section 1. Similarly, the corresponding three terms on the right-hand side of the equation constitute the total head at section 2. The difference between the total heads at these sections arises from the total gain in head $\sum \Delta h_{\text{gain}}$ due to work transfer (e.g., by a pump) and the loss in head $\sum \Delta h_{\text{loss}}$ (e.g., due to friction) over the flow path connecting these sections. For a flow through an arbitrary duct with wall friction, Figure 3.2 shows various terms in Equation 3.4, excluding $\sum \Delta h_{\text{gain}}$. Compared to an ideal EGL, the actual EGL drops monotonically downstream along the duct. The static pressure loss in the duct due to friction makes both the HGL and EGL for the flow lower than the values shown in Figure 3.1, which corresponds to the duct flow with no friction.

Let us write p_0 as

$$p_0 = p + \frac{\alpha \rho \bar{V}^2}{2} = \rho \left(\frac{p}{\rho} + \frac{\alpha \bar{V}^2}{2} \right) \qquad (3.5)$$

In this equation, the first term within parentheses is the specific flow work, and the second term is the specific kinetic energy. Total pressure p_0 may therefore be interpreted as the total mechanical energy per unit volume.

In terms of total pressure, we write Equation 3.4 as

$$p_{02} = p_{01} + \sum \Delta p_{0\_\text{gain}} - \sum \Delta p_{0\_\text{loss}} - \rho g (z_2 - z_1) \qquad (3.6)$$

where z_1 and z_2 are measured from a fixed datum plane, just as in the original Bernoulli equation. Note that in Equation 3.36, $\sum \Delta p_{0\_\text{gain}}$ includes all increases in

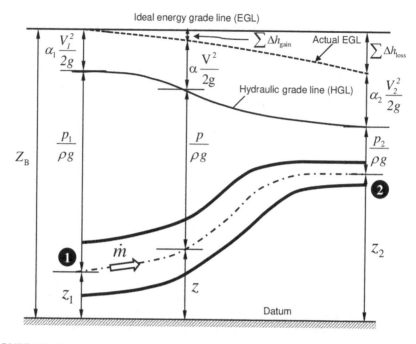

FIGURE 3.2 Delineation of terms in the extended Bernoulli equation.

total pressure except gravity, for example, by pump work; and $\sum \Delta p_{0\_loss}$ includes all decreases in total pressure except gravity, for example, due to turbine work, due to friction, and entropy generation by other means—e.g., the mixing loss.

Knowing total pressures p_{01} and p_{02}, we can compute the hydraulic power P_{hyd} needed between sections 1 and 2 by the equation

$$P_{hyd} = Q\left(p_{02} - p_{01}\right) \tag{3.7}$$

where Q is the volumetric flow rate between these sections.

Major Loss

The major loss arises from viscous shear stress at the duct wall. To calculate the head loss Δh_{loss} over the length L of a fully developed flow in a circular pipe of diameter D, we use the Darcy-Weisbach equation

$$\Delta h_{loss\_major} = f \frac{L}{D} \frac{\bar{V}^2}{2g} \tag{3.8}$$

where the Darcy friction factor for a fully developed laminar flow with Reynolds number $Re = \rho \bar{V} D / \mu < 2300$ is given by

$$f = \frac{64}{Re} \tag{3.9}$$

For a fully developed turbulent flow with Reynolds number $Re \geq 2300$, we obtain f from the curves presented by Moody (1944), found in many undergraduate texts in fluid mechanics such as Fox and McDonald (2010). Three widely used equations to compute f are as follows:

Colebrook (1938–1939) proposed the equation

$$\frac{1}{\sqrt{f}} = -2.0 \log_{10} \left(\frac{e/D}{3.7} + \frac{2.51}{Re\sqrt{f}} \right) \tag{3.10}$$

where e is the absolute roughness (in length units) of the pipe wall and e/D is its relative roughness.

Swamee and Jain (1976) proposed the approximate equation

$$\frac{1}{\sqrt{f}} = -2.0 \log_{10} \left(\frac{e/D}{3.7} + \frac{5.74}{Re^{0.9}} \right) \tag{3.11}$$

Haaland (1983) proposed the approximate equation

$$\frac{1}{\sqrt{f}} = -1.8 \log_{10} \left[\left(\frac{e/D}{3.7} \right)^{1.11} + \frac{6.9}{Re} \right] \tag{3.12}$$

For a noncircular duct, we replace the pipe diameter D used in these equations by the duct hydraulic mean diameter $D_h = 4A/P_w$, where A is the flow area and P_w is the wetted perimeter.

MINOR LOSS

A minor loss, also known as local or form loss, arises from the local features in the flow through a duct. These features include the change in duct-shape and size such as a sudden-expansion or a sudden-contraction in duct flow area. We obtain the minor loss using the equation

$$\Delta h_{\text{loss\_minor}} = K \frac{\overline{V}^2}{2g} \tag{3.13}$$

where K is the loss coefficient.

PROBLEM 3.1: FREE FALL OF STEADY WATER FLOW UNDER GRAVITY FROM A VERTICAL PIPE

Figure 3.3 shows the free fall of a steady water flow under gravity from a vertical pipe of area A_1. The constant mass flow rate is $\dot{m}$. Using various quantities shown in the figure and neglecting any drag force on the water jet, find the area ratio A_2/A_1.

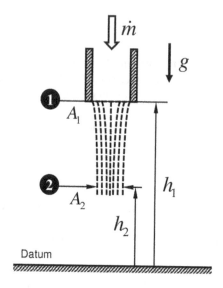

FIGURE 3.3   Free fall of steady water flow from a vertical pipe (Problem 3.1).

SOLUTION FOR PROBLEM 3.1

In this problem, the gravity is the only force acting on the falling water. The constant ambient pressure cancels on both sides of the Bernoulli equation applied between sections 1 and 2, giving

$$\frac{V_1^2}{2} + gh_1 = \frac{V_2^2}{2} + gh_2$$

which states that the sum of the potential energy and the kinetic energy of the water fall remains constant at each section.

From the continuity equation we write

$$\dot{m} = \rho A_1 V_1 = \rho A_2 V_2$$

giving

$$V_1 = \frac{\dot{m}}{\rho A_1}$$

and

$$V_2 = \frac{\dot{m}}{\rho A_2}$$

Substituting for V_1 and V_2 into the reduced Bernoulli equation, we obtain

$$\frac{\dot{m}^2}{2\rho^2 A_1^2} + gh_1 = \frac{\dot{m}^2}{2\rho^2 A_2^2} + gh_2$$

$$\frac{1}{A_2^2} = \frac{1}{A_1^2} + \frac{2\rho^2 g(h_1 - h_2)}{\dot{m}^2}$$

$$\frac{A_1^2}{A_2^2} = 1 + \frac{2\rho^2 A_1^2 g(h_1 - h_2)}{\dot{m}^2} = 1 + \frac{2g(h_1 - h_2)}{V_1^2}$$

$$\frac{A_2}{A_1} = \frac{1}{\sqrt{1 + \dfrac{2g(h_1 - h_2)}{V_1^2}}}$$

This result shows that $A_2 < A_1$. The reduction in water jet flow area depends upon the initial flow velocity V_1 and the fall $(h_1 - h_2)$.

PROBLEM 3.2: OPEN CHANNEL WATER FLOW AS A FLOW MEASURING DEVICE

Figure 3.4 shows an open channel flow of water. The cross-section of the channel parallel to the flow, shown in Figure 3.4a, remains uniform from top to bottom. Over the length L in the middle, the width of this cross-section reduces from b_1 to b_2. The side walls of the open channel have constant height h. For a volumetric flow rate Q, Figure 3.4b shows that the water depth in the open channel varies from z_1 to z_2.

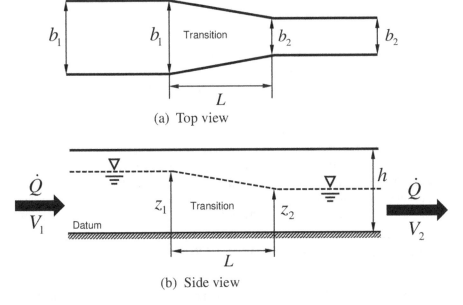

FIGURE 3.4 An open channel water flow as a flow measuring device: (a) top view and (b) side view (Problem 3.2).

Assuming no energy loss in this flow system, find an equation to express Q in terms of other quantities shown in the figure. Does the water depth vary linearly in the transition section?

SOLUTION FOR PROBLEM 3.2

We assume that the flow velocity is uniform over each cross-section of the open channel. The volumetric flow rate Q remains constant through the channel. At the beginning of the transition, the flow area is $b_1 z_1$ and the velocity is V_1; at the end of the transition, they are $b_2 z_2$ and V_2. For a steady flow, the continuity equation over the transition yields

$$V_1 = \frac{Q}{b_1 z_1}$$

$$V_2 = \frac{Q}{b_2 z_2}$$

For a streamline along the channel open surface with constant ambient pressure, the Bernoulli equation yields

$$\frac{V_1^2}{2} + g z_1 = \frac{V_2^2}{2} + g z_2$$

Substituting for V_1 and V_2, we obtain

$$\frac{Q^2}{2 b_1^2 z_1^2} + g z_1 = \frac{Q^2}{2 b_2^2 z_2^2} + g z_2$$

$$Q^2 \left(\frac{1}{2 b_2^2 z_2^2} - \frac{1}{2 b_1^2 z_1^2} \right) = g(z_1 - z_2)$$

$$Q = \sqrt{\frac{g(z_1 - z_2)}{\left(\dfrac{1}{2 b_2^2 z_2^2} - \dfrac{1}{2 b_1^2 z_1^2} \right)}}$$

which we can use to calculate Q by measuring z_1 and z_2 in the steady flow in an open channel for which b_1 and b_2 are known. This result shows that the decrease in water depth in the transition is not linear. The open channel device analyzed in this problem is often used to measure flow rate in the cold-flow calibration of various equipment.

PROBLEM 3.3:   WATER FLOW IN A VARIABLE-AREA FRICTIONLESS DUCT

In 1738, Daniel Bernoulli asked a young fluids engineer to perform some calculations on water flow in a variable area duct shown in Figure 3.5. The engineer was asked to neglect any frictional force in the duct. Reviewing the calculation results shown in

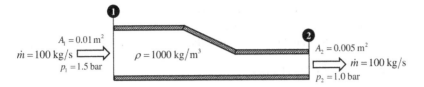

FIGURE 3.5   Water flow in a frictionless duct (Problem 3.3).

the figure, Bernoulli commented that young engineer's calculated results are physically not possible. Why do you agree or disagree with Bernoulli's conclusion?

SOLUTION FOR PROBLEM 3.3

The solution provided by the young engineer clearly satisfies the continuity equation as the mass flow rate entering and leaving the duct is 100 kg/s. Let us examine if the solution also satisfies the momentum equation.

SECTION 1 (INLET)

Velocity

$$V_1 = \frac{\dot{m}}{A_1 \rho} = \frac{100}{0.01 \times 1000} = 10 \text{ m/s}$$

Stream Thrust

$$S_{T1} = p_1 A_1 + \dot{m} V_1 = 1.5 \times 10^5 \times 0.01 + 100 \times 10 = 2500 \text{ N}$$

SECTION 2 (OUTLET)

Velocity

$$V_2 = \frac{\dot{m}}{A_2 \rho} = \frac{100}{0.005 \times 1000} = 20 \text{ m/s}$$

Stream Thrust

$$S_{T2} = p_2 A_2 + \dot{m} V_2 = 1.0 \times 10^5 \times 0.005 + 100 \times 20 = 2500 \text{ N}$$

These results show that there is no change in the stream thrust between the duct inlet and outlet. Although we neglect wall friction in this problem, the convergent upper wall in the mid-section of the duct will create an opposing pressure force on the fluid control volume. As a result, the stream thrust at the duct exit would be lower than that at its inlet.

From the energy consideration, in the absence of any energy loss, the sum of flow work and kinetic energy must be conserved along the duct. Accordingly, let us now examine how the calculated total pressure changes between duct inlet and outlet.

Total Pressure at the Duct Inlet

$$p_{01} = 1.5 \times 10^5 + \frac{1000 \times 10 \times 10}{2} = 2.0 \text{ bar}$$

Total Pressure at the Duct Outlet

$$p_{02} = 1.0 \times 10^5 + \frac{1000 \times 20 \times 20}{2} = 3.0 \text{ bar}$$

The total pressure calculated at the duct inlet and outlet shows that it increases from inlet to outlet. This is in violation of the flow physics. For a duct flow with friction, the total pressure must decrease downstream; for a frictionless flow, it must remain constant. Thus, our analysis here fully supports the conclusion reached by Bernoulli.

PROBLEM 3.4:   OPERATION OF A SIPHON

Figure 3.6 shows the operation of a siphon using water with density 1000 kg/m^3. The ambient pressure is 1.0 bar. Calculate the discharge velocity at D and the static pressures inside the constant-area tube at B and C, which is the highest point on the siphon centerline. Neglect any frictional loss in the siphon system.

SOLUTION FOR PROBLEM 3.4

The Bernoulli equation (Equation 3.1) governs the operation of an ideal siphon shown in Figure 3.4, yielding constant total head at each point A, B, C, and D.

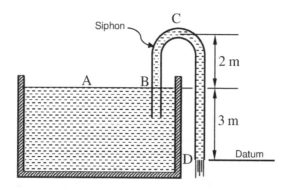

FIGURE 3.6   Operation of a siphon (Problem 3.4).

DISCHARGE VELOCITY AT D

Assigning the datum at D for measuring the geodetic head, we can write the Bernoulli equation between points A and D as

$$\frac{p_{amb}}{\rho g} + \frac{V_A^2}{2g} + 3.0 = \frac{p_{amb}}{\rho g} + \frac{V_D^2}{2g}$$

Neglecting V_A in comparison to V_D, this equation reduces to

$$\frac{V_D^2}{2g} = 3.0$$

which yields

$$V_D = \sqrt{2 \times 9.81 \times 3.0} = 7.672 \text{ m/s}$$

STATIC PRESSURE AT B INSIDE THE SIPHON TUBE

Writing the Bernoulli equation between points B and D yields

$$\frac{p_B}{\rho g} + \frac{V_B^2}{2g} + 3.0 = \frac{p_{amb}}{\rho g} + \frac{V_D^2}{2g}$$

As the continuity equation between points B and D inside the tube with constant flow area yields $V_B = V_D$, this equation reduces to

$$\frac{p_B}{\rho g} = \frac{p_{amb}}{\rho g} - 3$$

$$p_B = p_{amb} - 3\rho g = 1.0 \times 10^5 - 3 \times 1000 \times 9.81$$

$$p_B = 70,570 \text{ Pa}$$

which is subambient.

STATIC PRESSURE AT C INSIDE THE SIPHON TUBE

Writing the Bernoulli equation between points C and D yields

$$\frac{p_C}{\rho g} + \frac{V_C^2}{2g} + 5.0 = \frac{p_{amb}}{\rho g} + \frac{V_D^2}{2g}$$

As the continuity equation between points C and D inside the tube with constant flow area yields $V_C = V_D$, this equation reduces to

$$\frac{p_C}{\rho g} = \frac{p_{amb}}{\rho g} - 5$$

$$p_C = p_{amb} - 5\rho g = 1.0 \times 10^5 - 5 \times 1000 \times 9.81$$

$$p_C = 50,950 \, \text{Pa}$$

which is subambient and lower than that at B. Thus, the static pressure in the siphon tube reaches its lowest value at C, which is the highest point. The static pressure increases from C to D, where it equals the ambient pressure. As we increase the height of C, the static pressure in the tube becomes more and more subambient until it reaches the vapor pressure of the liquid used in the siphon. This sets the limit on how high C can go. Going higher results in the liquid vapor breaking the liquid flow, and the siphon ceases to operate.

PROBLEM 3.5: PUMPING POWER NEEDED IN A WATER FLOW SYSTEM

Figure 3.7 shows a flow system that uses a centrifugal pump to pump oil of density $850 \, \text{kg/m}^3$ at the volumetric flow rate $Q = 0.015 \, \text{m}^3/\text{s}$. The gauge pressure measured at point A is −30 kPa and that at point B is 300 kPa. The inner diameter of the pipe upstream of the pump is 7.62 cm and that of the downstream pipe is 5.08 cm. At the

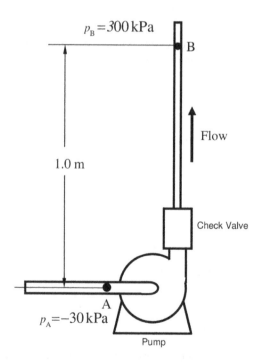

FIGURE 3.7   A pump flow system (Problem 3.5).

given flow rate, the minor head loss of the check valve is 0.75 m and the major head loss due to friction in pipes is 1.25 m. Calculate the power delivered by the pump in this flow system.

SOLUTION FOR PROBLEM 3.5

From the given inner diameters of pipes upstream and downstream of the pump, we obtain $A_A = 4.560 \times 10^{-3}$ m^2 and $A_A = 2.027 \times 10^{-3}$ m^2.

VELOCITY AT A

$$V_A = \frac{\dot{Q}}{A_A} = \frac{0.015}{4.560 \times 10^{-3}} = 3.289 \text{ m/s}$$

VELOCITY AT B

$$V_B = \frac{\dot{Q}}{A_B} = \frac{0.015}{2.027 \times 10^{-3}} = 7.401 \text{ m/s}$$

We write the extended Bernoulli equation (Equation 3.4) between points A and B as

$$\frac{p_A}{\rho g} + \frac{V_A^2}{2g} + z_A + H_{\text{pump}} - \sum \Delta h_{\text{loss}} = \frac{p_B}{\rho g} + \frac{V_B^2}{2g} + z_B$$

$$H_{\text{pump}} = \frac{p_B - p_A}{\rho g} + \frac{V_B^2 - V_A^2}{2g} + (z_B - z_A) + \sum \Delta h_{\text{loss}}$$

Each term on the right-hand side of this equation is calculated as follows:

$$\frac{p_B - p_A}{\rho g} = \frac{300 \times 10^3 - \left(-30 \times 10^3\right)}{850 \times 9.81} = 39.575 \text{ m}$$

$$\frac{V_B^2 - V_A^2}{2g} = \frac{(7.401)^2 - (3.289)^2}{2 \times 9.81} = 2.240 \text{ m}$$

$$(z_B - z_A) = 1.0 - 0.0 = 1.0 \text{ m}$$

$$\sum \Delta h_{\text{loss}} = 0.75 + 1.25 = 2.0 \text{ m}$$

Hence, the head imparted by the pump becomes

$$H_{\text{pump}} = 39.575 + 2.240 + 1.0 + 2.0 = 44.815 \text{ m}$$

PUMPING POWER

$$P_{\text{pump}} = H_{\text{pump}}\dot{m}g = (H_{\text{pump}}\rho g)\, Q = \Delta p_{0\_\text{pump}} Q$$

$$P_{\text{pump}} = 44.815 \times 850 \times 9.81 \times 0.015 = 5605\ \text{W} = 5.605\ \text{kW}$$

PROBLEM 3.6: TUNING OF THE CARBURETOR IN TOM'S OLD SCOOTER

Tom, an electrical engineer, loves his old scooter equipped with a carburetor. Without fully understanding how a carburetor works, he often fine-tunes the scooter-carburetor to obtain an optimal air-fuel mixture ratio. Figure 3.8 shows the schematic of a carburetor where the air flows through a Venturi tube in which a fuel jet is placed. As the air velocity increases in the Venturi throat, according to Bernoulli equation, its static pressure drops. This reduction in static pressure causes the fuel to jet into the air stream creating a homogeneous air-fuel mixed that undergoes combustion downstream. Assuming K_a and K_g as the minor loss coefficients, respectively, in the air and gasoline flow passages and neglecting all related major losses, use the extended Bernoulli equation to find an expression to compute the air-fuel mixture ratio for this carburetor. Use your result to explain to Tom how the air-fuel mixture ratio changes with the change in Venturi throat diameter D.

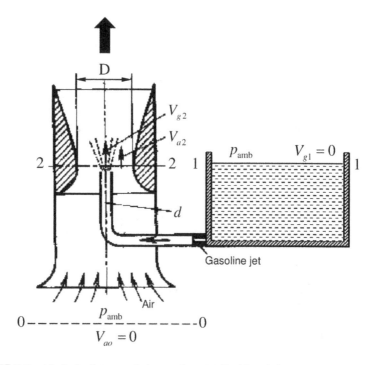

FIGURE 3.8 Air-fuel mixture ratio in a carburetor (Problem 3.6).

SOLUTION FOR PROBLEM 3.6

At various sections, we define the following quantities.

SECTION 0-0

$V_{a0} \equiv$ Uniform velocity of ambient air $\left(V_{a0} = 0\right)$
$p_{amb} \equiv$ Ambient static pressure

SECTION 1-1

$V_{g1} \equiv$ Uniform gasoline velocity $\left(V_{g1} = 0\right)$

SECTION 2-2

$D \equiv$ Throat diameter
$p_2 \equiv$ Static pressure at Venturi throat
$V_{a2} \equiv$ Uniform air velocity
$V_{g2} \equiv$ Uniform gasoline velocity

For the air flow between sections 0-0 and 2-2, we can write the extended Bernoulli equation as

$$\frac{p_{amb}}{\rho_a g} + \frac{V_{a1}^2}{2g} = \frac{p_2}{\rho_a g} + \frac{V_{a2}^2}{2g} + K_a \frac{V_{a2}^2}{2g}$$

With $V_{a1} = 0$, this equation reduces to

$$\frac{p_{amb}}{\rho_a g} = \frac{p_2}{\rho_a g} + \left(1 + K_a\right)\frac{V_{a2}^2}{2g}$$

$$p_{amb} - p_2 = \left(1 + K_a\right)\frac{\rho_a V_{a2}^2}{2g}$$

where ρ_a is the air density and K_a is the minor loss coefficient in the air flow line.

For the gasoline flow between sections 1-1 and 2-2, we write the extended Bernoulli equation as

$$\frac{p_{amb}}{\rho_g g} + \frac{V_{g1}^2}{2g} = \frac{p_2}{\rho_g g} + \frac{V_{g2}^2}{2g} + K_g \frac{V_{g2}^2}{2g}$$

With $V_{g1} = 0$, this equation reduces to

$$\frac{p_{amb}}{\rho_g g} = \frac{p_2}{\rho_g g} + \left(1 + K_g\right)\frac{V_{g2}^2}{2g}$$

$$p_{amb} - p_2 = \left(1 + K_g\right)\frac{\rho_g V_{g2}^2}{2g}$$

where ρ_g is the gasoline density and K_g is the minor loss coefficient in the gasoline flow line.

From the foregoing equations for $(p_{amb} - p_2)$ for the air and gasoline flow lines, we obtain

$$(1 + K_a)\frac{\rho_a V_{a2}^2}{2g} = (1 + K_g)\frac{\rho_g V_{g2}^2}{2g}$$

$$\frac{V_g}{V_a} = \sqrt{\frac{\rho_a(1 + K_a)}{\rho_g(1 + K_g)}}$$

$$\frac{\dot{m}_g}{\dot{m}_a} = \frac{\rho_g V_g d^2}{\rho_a V_a D^2}$$

$$\frac{\dot{m}_g}{\dot{m}_a} = \left(\frac{d}{D}\right)^2 \sqrt{\frac{\rho_g(1 + K_a)}{\rho_a(1 + K_g)}}$$

This result shows how we can change the gasoline-air mixture ratio by adjusting the Venturi throat diameter.

NOMENCLATURE

A	Pipe flow area
b	Open channel width
D	Pipe diameter
D_h	Hydraulic mean diameter
e	Absolute roughness
E_B	Bernoulli constant in terms of specific energy
EGL	Energy grade line
f	Darcy friction factor
g	Acceleration due to gravity
h	Height from a datum (geodetic head)
H	Pump head
HGL	Hydraulic grade line
K	Minor-loss coefficient
L	Pipe length
m	Mass
$\dot{m}$	Mass flow rate
M	Mach number
p	Static pressure
p_B	Bernoulli constant in terms of pressure
P	Power
P_w	Wetted perimeter
Q	Volumetric flow rate

R	Gas constant
Re	Reynolds number
V	Total velocity (magnitude)
x	Cartesian coordinate x
y	Cartesian coordinate y
z	Cartesian coordinate z; geodetic head in Bernoulli equation
Z_B	Bernoulli constant in terms of head

SUBSCRIPTS AND SUPERSCRIPTS

1	Location 1; Section 1
2	Location 2; Section 2
amb	Ambient
gain	Gain in total pressure
hyd	Hydraulic
loss	Loss in total pressure or head
loss-major	Major loss
loss-minor	Minor loss
pump	Pump
0	Total (stagnation)
x	Component in x-coordinate direction
y	Component in y-coordinate direction
z	Component in z-coordinate direction
$(\bar{\ })$	Average

GREEK SYMBOLS

α	Kinetic energy correction factor
μ	Dynamic viscosity
ρ	Density

REFERENCES

Colebrook, C. 1938–1939. Turbulent flow in pipes, with particular reference to the transition region between the smooth and rough pipe laws. *Journal of the Institution of Civil Engineers*. 11: 133–156.

Fox, W., P. Prichard, and A. McDonald. 2010. *Introduction to Fluid Mechanics*, 7th edition. New York: John Wiley & Sons.

Haaland, S. 1983. Simple and explicit formulas for friction factor in turbulent flow. Transactions of ASME. *Journal of Fluids Engineering*. 103:89–90.

Moody, L. 1944. Friction factors for pipe flow. *Transactions of the ASME*. 66(8):671–684.

Sultanian, B.K. 2015. *Fluid Mechanics: An Intermediate Approach*. Boca Raton, FL: Taylor & Francis.

Swamee, P. and A. Jain. 1976. Explicit equations for pipe-flow problems. *Journal of the Hydraulics Division*. 102(5):657–664.

BIBLIOGRAPHY

Blevin, R.D. 2003. *Applied Fluid Dynamics Handbook*. Malabar: Krieger Publishing Company.

Idelchik, I.E. 2005. *Handbook of Hydraulic Resistance*, 3rd edition. New Delhi: Jaico Publishing House.

Schlichting, H. 1979. *Boundary Layer Theory*, 7th edition. New York: McGraw-Hill.

Miller, R.W. 1996. *Flow Measurement Engineering Handbook*, 3rd edition. New York: McGraw-Hill.

Mott, R.L. 2006. *Applied Fluid Mechanics*, 6th edition. Upper Saddle River, NJ: Pearson Prentice Hall.

4 Compressible Flow

REVIEW OF KEY CONCEPTS

We summarize here the concepts and equations used in solving problems on one-dimensional compressible flows with area change, friction, heat transfer, and rotation, including normal and oblique shocks. Details derivations of these equations are presented in Sultanian (2015). Mach number characterizes a compressible flow—the higher the Mach number, the higher the flow compressibility with two-way exchange between internal and external flow energies. Low Mach number $(M \leq 0.3)$ flows of a compressible fluid may be modeled as incompressible.

ISENTROPIC FLOWS

We define the flow Mach number as the ratio of flow velocity and the speed of sound, that is,

$$M = \frac{V}{C} \tag{4.1}$$

where $C = \sqrt{\gamma R T}$. Using the Mach number, we relate the static and total temperatures by the equation

$$T_0 = T + \frac{V^2}{2c_p} = T\left(1 + \frac{\gamma+1}{2}M^2\right)$$

$$\frac{T_0}{T} = 1 + \frac{\gamma+1}{2}M^2 \tag{4.2}$$

For a perfect gas undergoing an isentropic stagnation process, we write

$$\Delta s = 0 = c_p \ln\left(\frac{T_0}{T}\right) - R\ln\left(\frac{p_0}{p}\right)$$

$$\frac{p_0}{p} = \left(\frac{T_0}{T}\right)^{\frac{c_p}{R}} = \left(\frac{T_0}{T}\right)^{\frac{\gamma}{\gamma-1}} \tag{4.3}$$

Using Equation 4.2, we express Equation 4.3 as

$$\frac{p_0}{p} = \left(1 + \frac{\gamma-1}{2}M^2\right)^{\frac{\gamma}{\gamma-1}} \tag{4.4}$$

For a perfect gas, using the equation of state given by $p = \rho R T$, $p_0 = \rho_0 R T_0$, and Equation 4.2, we obtain

$$\frac{\rho_0}{\rho} = \left(\frac{T_0}{T}\right)^{\frac{1}{\gamma-1}} = \left(1 + \frac{\gamma-1}{2}M^2\right)^{\frac{1}{\gamma-1}} \tag{4.5}$$

In an isentropic compressible flow, stagnation properties remain constant. When its velocity equals the speed of sound, the flow becomes sonic with $M = 1$ for which Equations 4.2, 4.4, and 4.5 yield

$$\frac{T_0}{T^*} = \frac{\gamma+1}{2} \tag{4.6}$$

$$\frac{p_0}{p^*} = \left(\frac{\gamma+1}{2}\right)^{\frac{\gamma}{\gamma-1}} \tag{4.7}$$

$$\frac{\rho_0}{\rho^*} = \left(\frac{\gamma+1}{2}\right)^{\frac{1}{\gamma-1}} \tag{4.8}$$

in which the static properties (superscripted by star) are also known as the characteristic (critical) properties of an isentropic flow.

CHARACTERISTIC MACH NUMBER

We define characteristic Mach number as

$$M^* = \frac{V}{C^*} = \frac{V}{\sqrt{\gamma R T^*}} \tag{4.9}$$

It is easy to show, see Sultanian (2015), that M and M^* are related by the equation

$$M^* = \sqrt{\frac{(\gamma+1)M^2}{2+(\gamma-1)M^2}} \tag{4.10}$$

For a normal shock in a compressible flow, Prandtl gave the following equation relating the characteristic Mach numbers before and after a normal shock:

$$M_2^* = \frac{1}{M_1^*} \tag{4.11}$$

In a normal shock, the upstream Mach number must always be supersonic—$M_1^* > 1$. According to Equation 4.11, therefore, the Mach number downstream of a normal shock must always be subsonic—$M_2^* < 1$.

TOTAL-PRESSURE MASS FLOW FUNCTIONS

Using total-pressure mass flow functions, we compute mass flow rate any section of uniform flow properties as

$$\dot{m} = \frac{F_{f0} A p_0}{\sqrt{T_0}} = \frac{\hat{F}_{f0} A p_0}{\sqrt{R T_0}} \tag{4.12}$$

where the total-pressure mass flow function F_{f0} and its dimensionless counterpart $\hat{F}_{f0}$ are given as follows:

$$F_{f0} = M \sqrt{\frac{\gamma}{R\left(1 + \dfrac{\gamma - 1}{2} M^2\right)^{\frac{\gamma + 1}{\gamma - 1}}}} \tag{4.13}$$

$$\hat{F}_{f0} = M \sqrt{\frac{\gamma}{\left(1 + \dfrac{\gamma - 1}{2} M^2\right)^{\frac{\gamma + 1}{\gamma - 1}}}} \tag{4.14}$$

Equation 4.13 shows that the units of F_{f0} are those of $1/\sqrt{R}$. Together with Equation 4.14, we can write $F_{f0} = \hat{F}_{f0}/\sqrt{R}$. For $M = 1$ at a section, we obtain

$$F_{f0}^* = \sqrt{\frac{\gamma}{R\left(\dfrac{\gamma + 1}{2}\right)^{\frac{\gamma + 1}{\gamma - 1}}}} \tag{4.15}$$

and

$$\hat{F}_{f0}^* = \sqrt{\frac{\gamma}{\left(\dfrac{\gamma + 1}{2}\right)^{\frac{\gamma + 1}{\gamma - 1}}}} \tag{4.16}$$

From these equations, for air with $\gamma = 1.4$ and $R = 287\ \text{J}/(\text{kg K})$, we obtain $\hat{F}_{f0}^* = 0.6847$ and $F_{f0}^* = 0.0404$ in the units of $\sqrt{(\text{kg K})/\text{J}}$.

Static-Pressure Mass Flow Function

Using the static pressure, instead of the total pressure, we compute the mass flow rate at a section as

$$\dot{m} = \frac{F_f A p}{\sqrt{T_0}} = \frac{\hat{F}_f A p}{\sqrt{R T_0}} \tag{4.17}$$

where the static-pressure mass flow function F_f and its dimensionless counterpart $\hat{F}_f$ are expressed as follows:

$$F_f = F_{f0}\left(\frac{p_0}{p}\right) \tag{4.18}$$

and

$$\hat{F}_f = \hat{F}_{f0}\left(\frac{p_0}{p}\right) \tag{4.19}$$

Using Equations 4.4, 4.13, and 4.14, we rewrite Equations 4.18 and 4.19 as

$$F_f = M\sqrt{\frac{\gamma\left(1 + \dfrac{\gamma - 1}{2} M^2\right)}{R}} \tag{4.20}$$

and

$$\hat{F}_f = M\sqrt{\gamma\left(1 + \frac{\gamma - 1}{2} M^2\right)} \tag{4.21}$$

For $M = 1$, these equations yield

$$F_f^* = \sqrt{\frac{\gamma(\gamma + 1)}{2R}} \tag{4.22}$$

and

$$\hat{F}_f^* = \sqrt{\frac{\gamma(\gamma + 1)}{2}} \tag{4.23}$$

For air with $\gamma = 1.4$ and $R = 287 \, \text{J}/(\text{kg K})$, we obtain $\hat{F}_f^* = 1.2961$ and $F_f^* = 0.07651$ in the units of $\sqrt{(\text{kg K})/\text{J}}$.

For a given value of $\hat{F_f}$, we can directly compute the Mach number as follows: Squaring both sides of Equation 4.21 yields

$$\hat{F_f}^2 = \gamma M^2 + 0.5\gamma(\gamma - 1)M^4$$

$$0.5\gamma(\gamma - 1)M^4 + \gamma M^2 - \hat{F_f}^2 = 0$$

which is a quadratic equation in M^2, giving

$$M = \left(\frac{-\gamma + \sqrt{\gamma^2 + 2\gamma(\gamma - 1)\hat{F_f}^2}}{\gamma(\gamma - 1)} \right)^{\frac{1}{2}} \tag{4.24}$$

Similarly, we obtain

$$M = \left(\frac{-\gamma + \sqrt{\gamma^2 + 2\gamma(\gamma - 1)R F_f^2}}{\gamma(\gamma - 1)} \right)^{\frac{1}{2}} \tag{4.25}$$

In Equations 4.12 and 4.17, we have assumed that V is normal to the flow area. If, instead, only a component of V equal to V_n is normal to the flow area, we rewrite these equations as

$$\dot{m} = \frac{C_V F_{f0} A p_0}{\sqrt{T_0}} = \frac{C_V \hat{F}_{f0} A p_0}{\sqrt{R T_0}} \tag{4.26}$$

$$\dot{m} = \frac{C_V F_f A p}{\sqrt{T_0}} = \frac{C_V \hat{F}_f A p}{\sqrt{R T_0}} \tag{4.27}$$

where $C_V = V_n/V$. Note that in Equations 4.26 and 4.27, the Mach number is based on V.

STATIC-PRESSURE IMPULSE FUNCTION

In terms of Mach number, at a section with uniform properties, we express the static-pressure impulse pressure as follows:

$$p_i = p + \rho V^2 = p\left(1 + \frac{\rho V^2}{p} \right) = p\left(1 + \frac{\gamma V^2}{\gamma R T} \right) \tag{4.28}$$

$$p_i = p(1 + \gamma M^2) = p I_f$$

where I_f is the static-pressure impulse function given by

$$I_f = (1 + \gamma M^2) \tag{4.29}$$

For $M = 1$, this equation yields

$$I_f^* = (1 + \gamma) \tag{4.30}$$

TOTAL-PRESSURE IMPULSE FUNCTION

In terms of total pressure, we can write Equation 4.28 as

$$p_i = p_0 \frac{1 + \gamma M^2}{\left(1 + \dfrac{\gamma - 1}{2} M^2\right)^{\frac{\gamma}{\gamma-1}}} = p_0 I_{f0} \tag{4.31}$$

where I_{f0} is the total-pressure impulse function given by

$$I_{f0} = \frac{1 + \gamma M^2}{\left(1 + \dfrac{\gamma - 1}{2} M^2\right)^{\frac{\gamma}{\gamma-1}}} \tag{4.32}$$

For $M = 1$, Equation 4.32 yields

$$I_{f0}^* = \frac{1 + \gamma}{\left(\dfrac{\gamma + 1}{2}\right)^{\frac{\gamma}{\gamma-1}}} \tag{4.33}$$

Using impulse functions, we can express the stream thrust at a duct flow area as

$$S_T = pAI_f = p_0 AI_{f0} \tag{4.34}$$

which, using mass flow functions, yields

$$S_T = \frac{\dot{m}\sqrt{RT_0}}{\left(\hat{F}_f / I_f\right)} = \frac{\dot{m}\sqrt{RT_0}}{\left(\hat{F}_{f0} / I_{f0}\right)}$$

$$S_T = \frac{\dot{m}\sqrt{RT_0}}{N(M, \gamma)} \tag{4.35}$$

where $N(M, \gamma) = \hat{F}_f / I_f = \hat{F}_{f0} / I_{f0}$.

Normal Shock Function

The mass flow rate, the stream thrust, and the total temperature remain constant across a normal shock. Equation 4.35 reveals that $N(M,\gamma)$ must also remain constant across a normal shock. We call $N(M,\gamma)$ the normal shock function, which we can write in terms of M and γ as

$$N(M,\gamma) = \frac{\hat{F}_f}{I_f} = \frac{\hat{F}_{f0}}{I_{f0}} = \frac{M}{1+\gamma M^2}\sqrt{\gamma\left(1+\frac{\gamma-1}{2}M^2\right)} \qquad (4.36)$$

$$N^*(1,\gamma) = \sqrt{\frac{\gamma}{2(1+\gamma)}} \qquad (4.37)$$

which for $\gamma = 1.4$ equals 0.540. For $M \to \infty$, Equation 4.36 yields the asymptotic value of $N(M,\gamma)$ as

$$N^\infty(\infty,\gamma) = \sqrt{\frac{\gamma-1}{2\gamma}} \qquad (4.38)$$

which for $\gamma = 1.4$ equals 0.378.

Nonisentropic Flow with Friction: Fanno Flow

A Fanno flow is nonisentropic with its total temperature remaining constant throughout the duct. In this flow, for a subsonic inlet, the outlet Mach number can at most equal 1.0 (frictional choking) with the pipe length L_{max}. Increasing the pipe length beyond L_{max} will result in a lower flow rate, the outlet remaining choked. If, however, the inlet flow is supersonic, the downstream Mach number decreases until it reaches $M = 1$ at the pipe exit when the pipe length equals L_{max}. Increasing the pipe length beyond L_{max}, while keeping the mass flow rate unchanged and exit choked, will result in a normal shock in the pipe. Table 4.1 summarizes a set of equations to evaluate changes in various properties in this flow. Detailed derivations of these equations are given in Sultanian (2015).

TABLE 4.1
Fanno Flow Equations

$$\frac{T}{T^*} = \frac{\gamma+1}{2+(\gamma-1)M^2} \qquad \frac{\rho}{\rho^*} = \frac{1}{M}\sqrt{\frac{2+(\gamma-1)M^2}{\gamma+1}} \qquad \frac{p}{p^*} = \frac{1}{M}\sqrt{\frac{\gamma+1}{2+(\gamma-1)M^2}}$$

$$\frac{p_0}{p_0^*} = \frac{1}{M}\left(\frac{2+(\gamma-1)M^2}{\gamma+1}\right)^{\frac{\gamma+1}{2(\gamma-1)}} \qquad \frac{p_i}{p_i^*} = \frac{p(1+\gamma M^2)}{p^*(\gamma+1)}$$

$$\frac{fL_{max}}{D_h} = \frac{\gamma+1}{2\gamma}\ln\left\{\frac{(\gamma+1)M^2}{2+(\gamma-1)M^2}\right\} + \frac{1}{\gamma}\left(\frac{1}{M^2}-1\right)$$

TABLE 4.2
Rayleigh Flow Equations

$$\frac{T}{T^*} = M^2 \left(\frac{\gamma+1}{1+\gamma M^2} \right)^2 \qquad \frac{\rho}{\rho^*} = \frac{1}{M^2} \left(\frac{1+\gamma M^2}{\gamma+1} \right) \qquad \frac{p}{p^*} = \frac{\gamma+1}{1+\gamma M^2}$$

$$\frac{p_0}{p_0^*} = \left(\frac{\gamma+1}{1+\gamma M^2} \right) \left(\frac{2+(\gamma-1)M^2}{\gamma+1} \right)^{\frac{\gamma}{\gamma-1}} \qquad \frac{T_0}{T_0^*} = M^2 \left(\frac{\gamma+1}{1+\gamma M^2} \right)^2 \left(\frac{2+(\gamma-1)M^2}{\gamma+1} \right)$$

TABLE 4.3
Isothermal Constant-Area Flow with Friction

$$\frac{p}{p^*} = \frac{1}{M\sqrt{\gamma}} \qquad \frac{p_0}{p_0^*} = \frac{1}{M\sqrt{\gamma}} \left(\frac{\gamma\{2+(\gamma-1)M^2\}}{(3\gamma-1)} \right)^{\frac{\gamma}{\gamma-1}}$$

$$\frac{T_0}{T_0^*} = \frac{\gamma\{2+(\gamma-1)M^2\}}{(3\gamma-1)} \qquad \frac{fL_{max}}{D_h} = \ln(\gamma M^2) + \frac{1}{\gamma M^2} - 1$$

Nonisentropic Flow with Heat Transfer: Rayleigh Flow

The Rayleigh flow is nonisentropic. For the case of heating the flow with subsonic inlet, the entropy increases continuously from pipe inlet to outlet, where it reaches its maximum value with $M = 1$ (thermal choking). For a supersonic inlet flow with heating, the Mach number decreases downstream, again reaching the limit of thermal choking at the exit. Cooling the flow results in the reduction in entropy, showing trends opposite to those for heating the flow. In all cases, the flow stream thrust in a Rayleigh flow remains constant throughout. Table 4.2 summarizes a set of equations to evaluate changes in various properties in a Rayleigh flow. Detailed derivations of these equations are given in Sultanian (2015).

Isothermal Constant-Area Flow with Friction

Like the Fanno and Rayleigh flows, the isothermal flow in a constant-area duct with friction is nonisentropic with its static temperature remaining constant throughout the duct. Table 4.3 summarizes a set of equations to evaluate changes in various properties in this flow. Detailed derivations of these equations are given in Sultanian (2015).

Normal Shock

The normal shock, featuring abrupt changes in flow properties across it, represents a compression wave normal to the flow direction. In modeling a normal shock, we assume no area change, friction, and heat transfer across it. The flow upstream of a normal shock is always supersonic, and its downstream flow is

TABLE 4.4
Normal Shock Equations

$$M_{n2}^2 = \frac{2+(\gamma-1)M_{n1}^2}{2\gamma M_{n1}^2-(\gamma-1)} \qquad \frac{p_2}{p_1} = \frac{2\gamma M_{n1}^2-(\gamma-1)}{(\gamma+1)} \qquad \frac{\rho_2}{\rho_1} = \frac{(\gamma+1)M_{n1}^2}{2+(\gamma-1)M_{n1}^2}$$

$$\frac{T_2}{T_1} = \left\{\frac{2\gamma M_{n1}^2-(\gamma-1)}{(\gamma+1)}\right\}\left\{\frac{2+(\gamma-1)M_{n1}^2}{(\gamma+1)M_{n1}^2}\right\}$$

$$\frac{p_{02}}{p_{01}} = \left\{\frac{(\gamma+1)}{2\gamma M_{n1}^2-(\gamma-1)}\right\}^{\frac{1}{\gamma-1}}\left[\frac{(\gamma+1)M_{n1}^2}{\{2+(\gamma-1)M_{n1}^2\}}\right]^{\frac{\gamma}{\gamma-1}}$$

$$\frac{p_{02}}{p_1} = \left\{\frac{2\gamma M_{n1}^2-(\gamma-1)}{(\gamma+1)}\right\}^{\frac{-1}{\kappa-1}}\left\{\frac{(\gamma+1)}{2}M_{n1}^2\right\}^{\frac{\gamma}{\gamma-1}}$$

always subsonic. Note the normal shock function $N(M,\gamma)$ remains constant across a normal shock. Table 4.4 summarizes a set of equations relating flow properties before and after the normal shock. Detailed derivations of these equations are given in Sultanian (2015).

OBLIQUE SHOCK

When we have a geometric protrusion, like a bump on the duct wall, or an obstruction in an external flow, like a wedge sitting in an oncoming supersonic gas flow, an oblique shock wave occurs and provides the necessary mechanism for the flow to negotiate these geometric features. In an oblique shock, the velocity component normal to the shock wave undergoes a normal shock and the velocity component along the shock wave passes unchanged. As a result, the flow downstream of the oblique shock turns and flows along the wall of the obstruction. Like a normal shock wave, an oblique shock wave is also compressive.

Figure 4.1 illustrates the flow geometry of an oblique shock wave. Table 4.5 summarizes a set of equations relating flow properties before and after an oblique shock. Detailed derivations of these equations are given in Sultanian (2015).

Equations Relating Wave Angle β, Deflection Angle θ, and Upstream Mach Number M_1

For the given β and M_1, we determine θ from the equation

$$\tan\theta = 2\cot\beta\left[\frac{M_1^2\sin^2\beta-1}{M_1^2(\gamma+\cos 2\beta)+2}\right] \tag{4.39}$$

For the given θ and M_1, we determine β from the equation

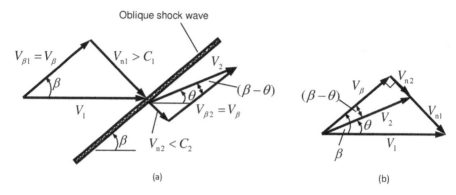

FIGURE 4.1 (a) Flow geometry of an oblique shock wave and (b) upstream and downstream velocity triangles with V_β as the common base.

TABLE 4.5
Oblique Shock Equations

$$M_2^2 = \frac{1}{\sin^2(\beta-\theta)} \left[\frac{2+(\gamma-1)M_1^2 \sin^2\beta}{2\gamma M_1^2 \sin^2\beta - (\gamma-1)} \right]$$

$$\frac{p_2}{p_1} = \frac{2\gamma M_1^2 \sin^2\beta - (\gamma-1)}{(\gamma+1)} \qquad\qquad \frac{p_2}{\rho_1} = \frac{(\gamma+1)M_1^2 \sin^2\beta}{2+(\gamma-1)M_1^2 \sin^2\beta}$$

$$\frac{T_2}{T_1} = \frac{\{2\gamma M_1^2 \sin^2\beta - (\gamma-1)\}\{2+(\gamma-1)M_1^2 \sin^2\beta\}}{(\gamma+1)^2 M_1^2 \sin^2\beta}$$

$$\frac{p_{02}}{p_{01}} = \left\{ \frac{(\gamma+1)}{2\gamma M_1^2 \sin^2\beta - (\gamma-1)} \right\}^{\frac{1}{\kappa-1}} \left[\frac{(\gamma+1)M_1^2 \sin^2\beta}{\{2+(\gamma-1)M_1^2 \sin^2\beta\}} \right]^{\frac{\gamma}{\gamma-1}}$$

$$\frac{(s_2-s_1)}{R} = -\ln\left[\left\{ \frac{(\gamma+1)}{2\gamma M_1^2 \sin^2\beta - (\gamma-1)} \right\}^{\frac{1}{\gamma-1}} \left[\frac{(\gamma+1)M_1^2 \sin^2\beta}{\{2+(\gamma-1)M_1^2 \sin^2\beta\}} \right]^{\frac{\gamma}{\gamma-1}} \right]$$

$$\tan\beta = \frac{M_1^2 - 1 + 2\lambda\cos\left(\dfrac{4\pi\delta + \cos^{-1}\chi}{3}\right)}{3\left(1+\dfrac{\gamma-1}{2}M_1^2\right)\tan\theta} \tag{4.40}$$

where $\delta = 0$ yields the strong shock solution and $\delta = 1$ corresponds to the weak shock solution. In both cases we compute λ using the equation

$$\lambda = \sqrt{\left(M_1^2-1\right)^2 - 3\left(1+\frac{\gamma-1}{2}M_1^2\right)\left(1+\frac{\gamma+1}{2}M_1^2\right)\tan^2\theta} \tag{4.41}$$

and χ using the equation

$$\chi = \frac{\left(M_1^2 - 1\right)^3 - 9\left(1 + \frac{\gamma - 1}{2}M_1^2\right)\left(1 + \frac{\gamma - 1}{2}M_1^2 + \frac{\gamma + 1}{4}M_1^4\right)\tan^2\theta}{\lambda^3} \tag{4.42}$$

PRANDTL-MEYER FLOW

Oblique shocks are compression waves, and except for an infinitesimal deflection angle, they are nonisentropic. When an oncoming supersonic flow needs to negotiate a convex corner, it happens through a series of expansion waves, which may be assumed isentropic. Assuming the Prandtl-Meyer expansion angle $\phi = 0$ for $M = 1$, we compute ϕ for $M > 1$ by the equation

$$\phi = \sqrt{\frac{\gamma + 1}{\gamma - 1}}\,\tan^{-1}\sqrt{\frac{\gamma - 1}{\gamma + 1}\left(M^2 - 1\right)} - \tan^{-1}\sqrt{M^2 - 1} \tag{4.43}$$

PROBLEM 4.1:   FORCE ON AN AIR COMPRESSOR OPERATING IN A CONSTANT-AREA DUCT

Figure 4.2 shows an air compressor operating in a duct of constant area $A = 0.0645$ m^2. For the one-dimensional flow in the duct, the total pressure and the Mach number at sections 1 and 2 are given as $p_{01} = 1.5$ bar, $M_1 = 0.3$, $p_{02} = 2.0$ bar, and $M_2 = 0.4$. Assuming $\gamma = 1.4$ for air, find (a) the force acting on the compressor, (b) the ratio of total temperatures T_{02}/T_{01}, and (c) the entropy change $(s_2 - s_1)/R$.

SOLUTION FOR PROBLEM 4.1

In the solution of this problem, we further assume that the frictional force acting on the duct wall is negligible compared to the drag force of the compressor body.

(a) FORCE ACTING ON THE COMPRESSOR IN THE FLOW DIRECTION

If F is the drag force acting on the compressor in the flow direction, the force acting on the flow will be $-F$, giving

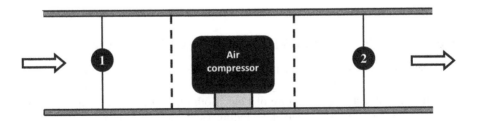

FIGURE 4.2   Air compressor operating in a constant-area duct (Problem 4.1).

$$-F = S_{T2} - S_{T1}$$

$$F = S_{T1} - S_{T2} = p_{01}AI_{f01} - P_{t_2}AI_{f02}$$

$$F = 2.0 \times 10^5 \times 0.0645 \times 1.0962 - 1.5 \times 10^5 \times 0.0645 \times 1.0578$$

$$F = 14{,}141.4 - 10{,}234.6 = 3906.8 \text{ N}$$

(b) Ratio of Total Temperatures T_{02}/T_{01}

As the mass flow rates at sections 1 and 2 are equal, we can write

$$\dot{m} = \frac{A\hat{F}_{f01}p_{01}}{\sqrt{RT_{01}}} = \frac{A\hat{F}_{f02}p_{02}}{\sqrt{RT_{02}}}$$

which yields

$$\frac{T_{02}}{T_{01}} = \left(\frac{\hat{F}_{f02}p_{02}}{\hat{F}_{f01}p_{01}} \right)^2$$

$$\frac{T_{02}}{T_{01}} = \left(\frac{0.4306 \times 3 \times 10^5}{0.3365 \times 1.5 \times 10^5} \right)^2 = 2.912$$

(c) Entropy Change $(s_2 - s_1)/R$

In terms of the total properties, we can compute change in entropy between sections 1 and 2 as

$$s_2 - s_1 = c_p \ln\left(\frac{T_{02}}{T_{01}} \right) - R \ln\left(\frac{p_{02}}{p_{01}} \right)$$

$$\frac{s_2 - s_1}{R} = \frac{c_p}{R} \ln\left(\frac{T_{02}}{T_{01}} \right) - \ln\left(\frac{p_{02}}{p_{01}} \right) = \frac{\gamma}{\gamma - 1} \ln\left(\frac{T_{02}}{T_{01}} \right) - \ln\left(\frac{p_{02}}{p_{01}} \right)$$

$$\frac{s_2 - s_1}{R} = \frac{1.4}{1.4 - 1} \ln(2.912) - \ln\left(\frac{2 \times 10^5}{1.5 \times 10^5} \right) = 3.453$$

PROBLEM 4.2:   CHOKING IN A COMPRESSED AIR FLOW SYSTEM

As shown in Figure 4.3, air at the total pressure of 8 bar and at the total temperature of 300 K enters the plenum through the inlet pipe at A and leaves the outlet pipe at D to the ambient at the pressure of 1 bar. Both pipes are identical in length and diameter. Going from inlet to outlet, including a sudden expansion at B and a sudden contraction at C (neglect any vena contracta effect in the flow entering at C), the flow

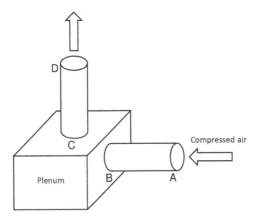

FIGURE 4.3 A compressed air flow system (Problem 4.2).

suffers an overall loss of 0.5 bar in total pressure. The entire flow system is adiabatic. Based on your understanding of a compressible flow, choose a section (A, B, C, or D) at which the air flow will choke ($M = 1$). Give the reason for your choice.

SOLUTION FOR PROBLEM 4.2

Under steady state, the mass flow rate through A, B, C, and D must be equal. At any section, which is choked flow ($M = 1$), the mass flow rate is directly proportional to the section area and the total pressure and inversely proportional to the square root of the total temperature. As all four sections have equal areas and equal total temperature (adiabatic flow), the choked mass flow rate is minimum where the total pressure is minimum, which is 7.5 bar at D. Therefore, the air flow in the system will choke at D.

PROBLEM 4.3: ADIABATIC AIR FLOW THROUGH TWO CONVERGENT NOZZLES AND ONE DIVERGENT NOZZLE CONNECTED TO A PLENUM

As shown in Figure 4.4, air flows through two identical convergent nozzles into a large plenum and exits from it sideways through a choked divergent nozzle. The throat area of the divergent nozzle equals twice the throat area of each convergent nozzle. For the supply total pressure and total temperature of 8 bar and 436.5K, respectively, for each convergent nozzle, find the mass flow rate through the choked divergent nozzle. All walls are adiabatic and frictionless. Assume air as a perfect gas with $\gamma = 1.4$ and $R = 287 \text{ J}/(\text{kg K})$.

SOLUTION FOR PROBLEM 4.3

The dynamic pressure of the flow from each convergent nozzle is lost in the plenum. As a result, the inlet total pressure of the choked divergent nozzle equals the static pressure at the exit of each convergent nozzle. As the flow system is adiabatic, the air

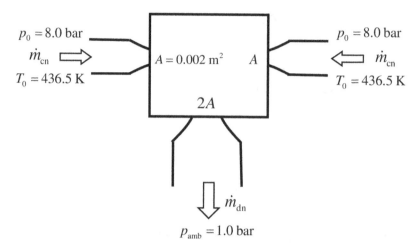

FIGURE 4.4 Adiabatic air flow through two convergent nozzles and one divergent nozzle connected to a plenum (Problem 4.3).

total temperature remains constant throughout the flow. For a steady flow, the mass flow rate exiting the divergent nozzle equals the sum of the mass flow rates through both convergent nozzles, that is, $\dot{m}_{dn} = 2\dot{m}_{cn}$.

We obtain the mass flow rate through each convergent nozzle as

$$\dot{m}_{cn} = \frac{A\hat{F}_{f}p}{\sqrt{RT_0}}$$

and through the choked divergent nozzle as

$$\dot{m}_{dn} = \frac{2A\hat{F}_{f0}^{*}p}{\sqrt{RT_0}}$$

giving

$$\dot{m}_{dn} = \frac{2A\hat{F}_{f0}^{*}p}{\sqrt{RT_0}} = 2\dot{m}_{cn} = 2\frac{A\hat{F}_{f}p}{\sqrt{RT_0}}$$

$$\hat{F}_{f0}^{*} = 0.6847 = \hat{F}_{f}$$

which yields the Mach number at each convergent nozzle throat as

$$M = \left\{ \frac{-\gamma + \sqrt{\gamma^2 + 2\gamma(\gamma-1)\hat{F}_f^2}}{\gamma(\gamma-1)} \right\}^{\frac{1}{2}} = \left\{ \frac{-1.4 + \sqrt{1.4^2 + 2\times1.4\times(1.4-1)\times(0.6847)^2}}{1.4\times(1.4-1)} \right\}^{\frac{1}{2}}$$

$$M = 0.561$$

from which we calculate the static pressure at each convergent nozzle throat as

$$\frac{p_0}{p} = \left(1 + \frac{\gamma-1}{2}M^2\right)^{\frac{\gamma}{\gamma-1}} = \left(1 + \frac{1.4-1}{2}(0.561)^2\right)^{\frac{1.4}{1.4-1}} = 1.238$$

$$p = p_0\frac{p}{p_0} = \frac{8}{1.238} = 6.46 \text{ bar} = 6.46 \times 10^5 \text{ Pa}$$

giving the choked mass flow rate through the divergent nozzle

$$\dot{m}_{dn} = \frac{2A\hat{F}_{f0}^*p}{\sqrt{RT_0}} = \frac{2\times0.002\times0.6847\times6.46\times10^5}{\sqrt{287\times436.5}} = 4.999 \text{ kg/s}$$

PROBLEM 4.4: AIR FLOW RATE THROUGH A PUNCHED AUTOMOBILE TIRE

The pressure and temperature of air in an automobile tire are 2.4 bar and 298 K, respectively. A hole of 1.0 mm diameter is accidentally punched in the tire. The hole assumes the shape of a convergent nozzle. Neglecting frictional effects, find the air mass flow rate through the hole. What would the air velocity and Mach number be if the hole was correctly shaped for an isentropic expansion to the ambient pressure of 1 bar? Assume air as a perfect gas with $\gamma = 1.4$ and $R = 287$ J/(kg K).

SOLUTION FOR PROBLEM 4.4

For the hole of diameter $d = 1.0$ mm $= 0.001$ m, we calculate the flow area as

$$A = \frac{\pi d^2}{4} = 7.854 \times 10^{-7} \text{ m}^2$$

For the given $p_0 = 2.4$ bar $= 2.4 \times 10^5$ Pa and $p_{amb} = 1$ bar $= 10^5$ Pa, we obtain

$$\frac{p_0}{p_{amb}} = \frac{2.4\times10^5}{10^5} = 2.4$$

As $p_0/p_{amb} = 2.4$ is greater than the value required for choking, the air flow at the punched hole is choked with $M = 1.0$. We compute the maximum mass flow rate as

$$\dot{m} = \frac{\hat{F}_{f0}^*Ap_0}{\sqrt{RT_0}} = \frac{0.6847\times7.854\times10^{-7}\times2.4\times10^5}{\sqrt{287\times298}} = 4.413\times10^{-4} \text{ kg/s}$$

For the exit pressure to equal the ambient pressure, the hole must be shaped as a convergent-divergent nozzle, giving

$$\frac{T_0}{T_{\text{exit}}} = \left(\frac{p_0}{p_{\text{exit}}}\right)^{\frac{\gamma-1}{\gamma}} = (2.4)^{\frac{1.4-1}{1.4}} = 1.284$$

$$T_{\text{exit}} = \frac{298}{1.284} = 232.1\,\text{K}$$

$$M_{\text{exit}} = \sqrt{\left(\frac{2}{\gamma-1}\right)\left(\frac{T_0}{T_{\text{exit}}}-1\right)} = \sqrt{\left(\frac{2}{1.4-1}\right)(1.284-1)} = 1.192$$

$$C_{\text{exit}} = \sqrt{\gamma R T_{\text{exit}}} = \sqrt{1.4 \times 287 \times 232.1} = 305.349\,\text{m/s}$$

and

$$V_{\text{exit}} = 305.349 \times 1.192 = 364\,\text{m/s}$$

PROBLEM 4.5:   ISENTROPIC AIR FLOW THROUGH A DEFORMABLE RUBBER PIPE

Figure 4.5 shows an isentropic flow of air in a rubber pipe of constant diameter 0.100 m with the inlet total pressure and total temperature of 1.2 bar and 300 K, respectively. The ambient pressure is 1.0 bar. The pipe is slowly deformed into a convergent-divergent nozzle, as shown by the dotted line, until the flow just chokes $(M = 1.0)$ at the throat. For the given boundary conditions, calculate the throat diameter. Assume air as a perfect gas with $\gamma = 1.4$ and $R = 287\,\text{J/(kg K)}$.

SOLUTION FOR PROBLEM 4.5

We calculate the pipe inlet and exit areas as

$$A_{\text{inlet}} = A_{\text{exit}} = \frac{\pi D_{\text{exit}}^2}{4} = \frac{\pi (0.1)^2}{4} = 7.854 \times 10^{-3}\,\text{m}^2$$

pressure ratio at the pipe exit as

$$\frac{p_{0\_\text{exit}}}{p_{\text{exit}}} = \frac{p_{0\_\text{exit}}}{p_{\text{amb}}} = \frac{1.2}{1.0} = 1.2$$

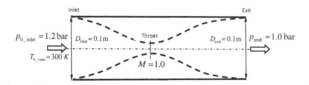

FIGURE 4.5   Isentropic air flow through a deformable rubber pipe (Problem 4.5).

and the exit Mach number M_{exit} by first calculating the temperature ratio as

$$\frac{T_{0\_\text{exit}}}{T_{\text{exit}}} = \left(\frac{p_{0\_\text{exit}}}{p_{\text{exit}}}\right)^{\frac{\gamma-1}{\gamma}} = (1.2)^{0.286} = 1.0535$$

$$\frac{T_{0\_\text{exit}}}{T_{\text{exit}}} = 1 + \frac{\gamma-1}{2} M_{\text{exit}}^2$$

which yields

$$M_{\text{exit}} = \sqrt{\frac{2}{\gamma-1}\left(\frac{T_{0\_\text{exit}}}{T_{\text{exit}}} - 1\right)} = 0.517$$

and

$$\hat{F}_{\text{f0\_exit}} = M_{\text{exit}}\sqrt{\frac{\gamma}{R\left(1+\frac{\gamma-1}{2}M_{\text{exit}}^2\right)^{\frac{\gamma+1}{\gamma-1}}}} = 0.517\sqrt{\frac{1.4}{287\times\left(1+\frac{1.4-1}{2}M_{\text{exit}}^2\right)^{\frac{1.4+1}{1.4-1}}}} = 0.5233$$

giving

$$\dot{m} = \frac{\hat{F}_{\text{f0\_exit}}A_{\text{exit}}p_{0\_\text{exit}}}{\sqrt{RT_{0\_\text{exit}}}} = \frac{0.5233\times7.854\times10^{-3}\times1.2\times10^5}{\sqrt{287\times300}} = 1.681\,\text{kg/s}$$

For the choked flow at the throat, we have $M_{\text{throat}} = 1.0$ and $\hat{F}_{\text{f0}}^* = 0.6847$, giving

$$A_{\text{throat}} = \frac{\dot{m}\sqrt{RT_{0\_\text{throat}}}}{\hat{F}_{\text{f0}}^* p_{0\_\text{throat}}} = \frac{1.681\times\sqrt{287\times300}}{0.6847\times1.2\times10^5} = 6.003\times10^{-3}\,\text{m}^2$$

and

$$D_{\text{throat}} = 8.742\times10^{-2}\,\text{m}$$

PROBLEM 4.6: LOCATING THE NORMAL SHOCK IN A CONVERGENT-DIVERGENT NOZZLE

For the air flow in a convergent-divergent nozzle shown in Figure 4.6, a normal shock stands in the divergent section. The exit-to-throat area ratio $\left(A_{\text{exit}}/A_{\text{throat}}\right)$ of the C-D nozzle is known. For the given exit Mach number M_{exit}, write a step-by-step non-graphical and noniterative procedure to determine the ratio of nozzle area A_{ns}, where the normal shock is located, to the throat area A^*.

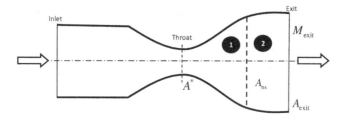

FIGURE 4.6 Normal shock in a convergent-divergent nozzle (Problem 4.6).

SOLUTION FOR PROBLEM 4.6

Step 1. For the given value of M_{exit}, use isentropic flow table (Sultanian, 2015) to determine A_{exit}/A_2^*.

Step 2. As $A_1^* = A^*$, calculate the ratio A_1^*/A_2^* as

$$\frac{A_1^*}{A_2^*} = \frac{\dfrac{A_{\text{exit}}}{A_2^*}}{\dfrac{A_{\text{exit}}}{A_1^*}}$$

Step 3. As the mass flow rate and the total temperature remain constant across a normal shock, we can write $A_1^*/A_2^* = p_{02}/p_{01}$.

Step 4. Knowing p_{02}/p_{01}, find M_1 and M_2 from the normal shock table (Sultanian, 2015).

Step 5. Use M_1 to find A_{ns}/A_1^* from the isentropic flow table (Sultanian, 2015).

PROBLEM 4.7: HIGH-PRESSURE INLET BLEED HEAT SYSTEM OF A LAND-BASED GAS TURBINE FOR POWER GENERATION

Figure 4.7 shows a high-pressure inlet bleed heat (IBH) system of a land-based gas turbine used for power generation. When the ambient air is cold, the IBH system is used to raise its temperature to prevent ice formation in the compressor IGVs (inlet guide vanes). In this system, the hot air is bled from an intermediate compressor stage and mixed uniformly with air flow at the engine inlet. Let us consider such a system. The ambient pressure and temperature are 1 bar and 20°C, respectively. The Mach number of the air flow entering the IGV is 0.6. To prevent ice formation, the static temperature at this section is required to be 2°C. The air mass flow rate entering the engine inlet system (before the high-pressure bleed heat section) at low Mach number ($M < 0.2$) is 275 kg/s. The total temperature and pressure of the air bled from the compressor are 269°C and 8 bar, respectively. Neglect any changes in the pressure and temperature in the bleed air supply system. Assuming air as a perfect gas with $\gamma = 1.4$ and $R = 287 \text{ J}/(\text{kg K})$.

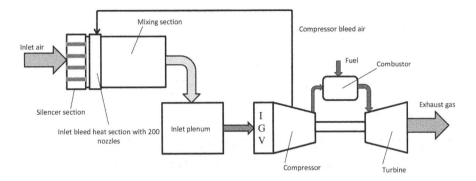

FIGURE 4.7 High-pressure IBH system of a land-based gas turbine for power generation (Problem 4.7).

a. Calculate the mass flow rate of the compressor bleed air to meet the design objective at compressor IGV inlet.
b. To promote uniform mixing of the hot compressor bleed air with the cold inlet ambient air, 200 nozzles for bleed air injection are uniformly placed in the inlet duct cross-section. Find the effective flow area of each nozzle.

SOLUTION FOR PROBLEM 4.7

(a) COMPRESSOR BLEED AIR MASS FLOW RATE

For $T_{\text{IGV}} = 2 + 273 = 275$ K and $M_{\text{IGV}} = 0.6$, we obtain

$$\frac{T_{0\_\text{IGV}}}{T_{\text{IGV}}} = 1 + \frac{\gamma - 1}{2} M_{\text{IGV}}^2 = 1 + \frac{1.4 - 1}{2}(0.6)^2 = 1.072$$

$$T_{0\_\text{IGV}} = 1.072 \times T_{\text{IGV}} = 1.072 \times 275 = 294.8 \text{ K}$$

For $T_{0\_\text{inlet}} = 20 + 273 = 293$ K, $T_{0\_\text{bleed}} = 269 + 273 = 542$ K, and $\dot{m}_{\text{inlet}} = 275 \text{ kg/s}$, from energy conservation, the mixing of the bleed flow and the inlet air yields

$$\dot{m}_{\text{bleed}} c_p \left(T_{0\_\text{bleed}} - T_{0\_\text{IGV}}\right) = \dot{m}_{\text{inlet}} c_p \left(T_{0\_\text{IGV}} - T_{0\_\text{inlet}}\right)$$

$$\dot{m}_{\text{bleed}} = \frac{\dot{m}_{\text{inlet}}\left(T_{0\_\text{IGV}} - T_{0\_\text{inlet}}\right)}{\left(T_{0\_\text{bleed}} - T_{0\_\text{IGV}}\right)} = \frac{275 \times (294.8 - 293)}{(542 - 294.8)} = 2.002 \text{ kg/s}$$

(b) EFFECTIVE FLOW AREA OF EACH NOZZLE

Number of nozzles = 200
  We obtain the mass flow rate per nozzle $(\dot{m}_{\text{nozzle}})$ as

$$\dot{m}_{\text{nozzle}} = \frac{\dot{m}_{\text{bleed}}}{200} = \frac{2.002}{200} = 1.002 \times 10^{-2} \text{ kg/s}$$

We express the maximum mass flow rate through each nozzle as

$$\dot{m}_{\text{nozzle}} = \frac{\hat{F}_{f0}^* A_{\text{nozzle}} p_{0\_\text{bleed}}}{\sqrt{RT_{0\_\text{bleed}}}}$$

$$A_{\text{nozzle}} = \frac{\dot{m}_{\text{nozzle}} \sqrt{RT_{0\_\text{bleed}}}}{\hat{F}_{f0}^* p_{0\_\text{bleed}}} = \frac{1.002 \times 10^{-2} \sqrt{287 \times 542}}{0.6847 \times 8 \times 10^5} = 7.209 \times 10^{-6} \text{ m}^2$$

$$A_{\text{nozzle}} = 7.209 \text{ mm}^2$$

PROBLEM 4.8: A CONVERGENT-DIVERGENT NOZZLE AIR FLOW WITH A NORMAL SHOCK

Air flows through a frictionless adiabatic convergent-divergent nozzle. The stagnation pressure and temperature at the nozzle inlet are 6.0×10^5 Pa and 600 K, respectively. The diverging section of the nozzle has an area ratio $A_{\text{exit}}/A_{\text{throat}} = 12$. A normal shock wave stands in the diverging section of the nozzle where the Mach number is 3.3. Determine the Mach number, static pressure, and static temperature at the nozzle exit plane. Assume air as a perfect gas with $\gamma = 1.4$ and $R = 287$ J/(kg K).

SOLUTION FOR PROBLEM 4.8

The normal shock in the C-D divides the flow in two isentropic flows, one upstream and another downstream of the normal shock. Both the mass flow rate and the total temperature remain constant across the normal shock.

For $M_1 = 3.3$, either using the normal shock equations or the normal shock table (Sultanian, 2015), we obtain $M_2 = 0.4596$ and $p_{02}/p_{01} = 0.2533$. As the ratio of critical areas is inversely proportional to the ratio of total pressures across a normal shock, we obtain $A_1^*/A_2^* = 0.2533$ where $A_1^* = A_{\text{throat}}$. Thus, we write

$$\frac{A_{\text{exit}}}{A_2^*} = \frac{A_{\text{exit}}}{A_1^*} \times \frac{A_1^*}{A_2^*} = 12 \times 0.2533 = 3.0393$$

which from the isentropic flow equations or isentropic flow table (Sultanian, 2015) yields $M_{\text{exit}} = 0.1948$, $p_{02}/p_{\text{exit}} = 1.0268$, and $T_{02}/T_{\text{exit}} = 1.0076$.

From the values in the foregoing, we compute

$$p_{\text{exit}} = p_{01} \frac{p_{02}}{p_{01}} \frac{p_{\text{exit}}}{p_{02}} = 6.0 \times 10^5 \times \frac{0.2533}{1.0268} = 1.48 \times 10^5 \text{ Pa}$$

$$T_{\text{exit}} = T_{02} \frac{T_{\text{exit}}}{T_{02}} = \frac{600}{1.0076} = 595.5 \text{ K}$$

PROBLEM 4.9: FINDING TOTAL-PRESSURE MASS FLOW FUNCTIONS FROM ISENTROPIC FLOW TABLE

For $\gamma = 1.4$, the maximum value of the dimensionless total-pressure mass flow function is 0.6847. Use the isentropic flow table (Sultanian, 2015) to find (1) the dimensionless total-pressure mass flow function for $M = 0.5$ and (2) supersonic Mach number for which the value of the dimensionless total-pressure mass flow function equals that obtained in (1). Verify your answers from the table on compressible flow functions (Sultanian, 2015).

SOLUTION FOR PROBLEM 4.9

(a) VALUE OF $\hat{F}_{f0}$ FOR $M = 0.5$

For an isentropic internal compressible flow, where the total pressure and total temperature remain constant, the continuity equation written in terms of total-pressure mass flow function yields

$$\hat{F}_{f0} = \frac{\hat{F}_{f0}^*}{\left(A/A^* \right)}$$

For $M = 0.5$, from isentropic flow table (Sultanian, 2015), we obtain $A/A^* = 1.3398$, which yields

$$\hat{F}_{f0} = \frac{0.6847}{1.3398} = 0.5110$$

which from the table on compressible flow functions (Sultanian, 2015) is 0.5111.

(b) SUPERSONIC MACH NUMBER FOR $\hat{F}_{f0} = 0.5110$

Knowing $\hat{F}_{f0} = 0.5110$, we can obtain the area ratio as

$$A/A^* = \frac{\hat{F}_{f0}^*}{\hat{F}_{f0}} = \frac{0.6847}{0.5110} = 1.3398$$

which, using linear interpolation, yields $M = 1.7023$ from the isentropic flow table (Sultanian, 2015). For $\hat{F}_{f0} = 0.5110$, using linear interpolation between tabular values, we also obtain $M = 1.7023$ from the table on compressible flow functions (Sultanian, 2015).

PROBLEM 4.10:   PRANDTL-MEYER EQUATION VERSUS NORMAL SHOCK FUNCTION

For $\gamma = 1.4$, use the table on compressible flow functions (Sultanian, 2015) to find the Mach number M_2 downstream of a normal shock when the upstream Mach number $M_1 = 3.5$, first by using the Prandtl-Meyer equation and second by using the tabulated normal shock function N. Verify your answer from the normal shock table (Sultanian, 2015).

SOLUTION FOR PROBLEM 4.10

According to Prandtl-Meyer equation the product of the characteristic Mach numbers across a normal shock is always unity. From the flow functions table (Sultanian, 2015), we obtain $M_1^* = 2.0642$ for $M_1 = 3.5$. We compute the characteristic Mach number downstream of the normal shock as

$$M_2^* = \frac{1}{M_1^*} = \frac{1}{2.0642} = 0.4844$$

which by linear interpolation between tabular values yields $M_2 = 0.4511$.

Across a normal shock, the function N remains constant. For $M_1 = 3.5$, we obtain $N = 0.4238$ from the table on flow functions (Sultanian, 2015), which further yields the subsonic Mach number $M_2 = 0.4511$ by linear interpolation. From the normal shock table (Sultanian, 2015), for $M_1 = 3.5$, we obtain $M_2 = 0.4512$.

PROBLEM 4.11:   NORMAL SHOCK ANALYSIS USING FANNO AND RAYLEIGH FLOW LINES

From the normal shock (Sultanian, 2015), for $M_1 = 2.0$, we obtain $M_2 = 0.5774$ and $p_{02}/p_{01} = 0.7209$. Use the Fanno flow and Rayleigh flow tables to show that the points of intersection of the Fanno and Rayleigh flow lines have the same total pressure ratio $\left(p_{02}/p_{01} = 0.7209\right)$. Also compute the entropy change $\Delta s/R$ between these two points.

SOLUTION FOR PROBLEM 4.11

From the Fanno flow table (Sultanian, 2015), for $M_1 = 2.0$, we obtain $p_{01}/p_{01} = 1.6875$ and, for $M_2 = 0.5774$, $p_{02}/p_0^* = 1.2165$ by linear interpolation between tabular values, together giving $p_{02}/p_{01} = 0.7209$. Similarly, from the Rayleigh flow table (Sultanian, 2015), for $M_1 = 2.0$, we obtain $p_{01}/p_0^* = 1.5031$ and, for $M_2 = 0.5774$, $p_{02}/p_0^* = 1.0836$ by linear interpolation, together giving $p_{02}/p_{01} = 0.7209$.

As the total temperature remains constant across a normal shock, we obtain

$$\frac{\Delta s}{R} = -\ln\left(\frac{p_{02}}{p_{01}}\right) = -\ln(0.7209) = 0.3273$$

PROBLEM 4.12:   MAXIMUM LENGTH OF A FANNO FLOW PIPE TO DELIVER THE SPECIFIED MASS FLOW RATE

A straight pipe of 0.05 m diameter is attached to a large air reservoir at pressure 14×10^5 Pa and 300 K. The pipe exit is open to atmosphere. Assuming adiabatic flow with an average Darcy friction factor of 0.02, calculate the maximum pipe length to deliver a mass flow rate of 2.5 kg/s. Assume air as a perfect gas with $\gamma = 1.4$ and $R = 287 \text{ J}/(\text{kg K})$.

SOLUTION FOR PROBLEM 4.12

We calculate the pipe area at inlet and exit as

$$A = \frac{\pi D^2}{4} = \frac{\pi (0.05)^2}{4} = 1.9635 \times 10^{-3} \text{ m}^2$$

The maximum pipe length corresponds to frictional choking at the pipe exit with $M_{exit} = 1.0$ and $\hat{F}_{f0}^* = 0.6847$.
   We compute the exit total pressure as

$$p_0^* = \frac{\dot{m}\sqrt{RT_0}}{A\hat{F}_{f0}^*} = \frac{2.5\sqrt{287 \times 300}}{1.9635 \times 10^{-3} \times 0.6847} = 5.4565 \times 10^5 \text{ Pa}$$

For

$$\frac{p_0}{p_0^*} = \frac{14 \times 10^5}{5.4565 \times 10^5} = 2.5658$$

we obtain $f L_{max}/D = 10.1003$ from the Fanno flow table (Sultanian, 2015) with linear interpolation between the tabular values, giving

$$L_{max} = 10.1003 \times \frac{D}{f} = 10.1003 \times \frac{0.05}{0.02} = 25.251 \text{ m}$$

PROBLEM 4.13:   FANNO FLOW THROUGH A LONG PIPE

Air at static pressure 3.5×10^5 Pa and at total temperature 300 K is to be transported at the rate of 0.11 kg/s over 500 m through a constant-diameter pipe. The static pressure at the pipe exit is 1.50×10^5 Pa. Assuming adiabatic flow and average Darcy friction factor of 0.015, determine the pipe diameter. Assume air as a perfect gas with $\gamma = 1.4$ and $R = 287 \text{ J}/(\text{kg K})$.

SOLUTION FOR PROBLEM 4.13

In this problem, the adiabatic air flow in the constant-diameter pipe with friction is a Fanno flow. As the overall solution method to solve this problem needs to be iterative, we will use the compressible flow equations rather than the tabular values given, for example, in Sultanian (2015). In the stepwise iterative solution method using "Goal Seek" in MS Excel presented here, the values within parentheses are the final converged values.

SECTION 1 (INLET)

Step 1. Assume an inlet Mach number M_1 (0.05676)

Step 2. Calculate $\hat{F}_{f1}$ from the equation

$$\hat{F}_{f1} = M_1 \sqrt{\gamma \left(1 + \frac{\gamma - 1}{2} M_1^2\right)} \quad (0.06719)$$

Step 3. Calculate the pipe flow area

$$A = \frac{\dot{m}\sqrt{RT_0}}{\hat{F}_{f1} p_1} = \frac{0.11 \times \sqrt{287 \times 300}}{0.06719 \times 3.50 \times 10^5} \quad \left(1.3726 \times 10^{-3}\ \text{m}^2\right)$$

and diameter

$$D = \sqrt{\frac{4A}{\pi}} = \sqrt{\frac{4 \times 1.3726 \times 10^{-3}}{\pi}} \quad \left(0.04181\text{m}\right)$$

Step 4. Calculate $f\,L/D$

$$\frac{fL}{D} = \frac{0.015 \times 500}{4.1805 \times 10^{-2}} \quad (179.4029)$$

Step 5. Calculate $f\,L_{\text{max}1}/D$

$$\frac{f L_{\text{max}1}}{D} = \frac{\gamma + 1}{2\gamma} \ln\left(\frac{(\gamma + 1)M_1^2}{\{2 + (\gamma - 1)M_1^2\}}\right) + \frac{1}{\gamma}\left(\frac{1}{M_1^2} - 1\right) \quad (216..2075)$$

SECTION 2 (EXIT)

Step 6. Calculate $\hat{F}_{f2}$

$$\hat{F}_{f2} = \frac{\dot{m}\sqrt{RT_0}}{Ap_2} = \frac{0.11 \times \sqrt{287 \times 300}}{1.3726 \times 10^{-3} \times 1.50 \times 10^5} \quad (0.1568)$$

Step 7. Calculate M_2

$$M_2 = \left(\frac{-\gamma + \sqrt{\gamma^2 + 2\gamma(\gamma - 1)\hat{F}_{f2}^2}}{\gamma(\gamma - 1)}\right)^{\frac{1}{2}} \quad (0.1323)$$

Step 8. Calculate $f\,L_{\text{max}2}/D$

$$\frac{f L_{\text{max}2}}{D} = \frac{\gamma + 1}{2\gamma} \ln\left(\frac{(\gamma + 1)M_2^2}{\{2 + (\gamma - 1)M_2^2\}}\right) + \frac{1}{\gamma}\left(\frac{1}{M_2^2} - 1\right) \quad (36.8046)$$

Step 9. Repeat Steps 1–9, until the following equation is satisfied within a specified tolerance

$$\frac{fL}{D} = \frac{fL_{max1}}{D} - \frac{fL_{max2}}{D}$$

We finally obtain the pipe diameter in Step 3 as $D = 0.04181$ m.

PROBLEM 4.14: NORMAL SHOCK IN A CONVERGENT-DIVERGENT NOZZLE ATTACHED TO A FANNO PIPE

As shown in Figure 4.8, an isentropic convergent-divergent nozzle having an area ratio (exit area/throat area) of 2 discharges air into an insulated pipe of length L and diameter D. The nozzle inlet air has the stagnation pressure of 7×10^5 Pa and the stagnation temperature of 300 K, and the pipe discharges into a space where the static pressure is 2.8×10^5 Pa. Calculate fL/D of the pipe and the mass flow rate per unit area in the pipe for the cases where a normal shock stands (1) at the nozzle throat, (2) at the nozzle exit, and (3) at the pipe exit.

SOLUTION FOR PROBLEM 4.14

As shown in Figure 4.8, C-D nozzle exit diameter equals the downstream pipe diameter $(A_{exit} = A_{pipe})$. As the flow is adiabatic in both the C-D nozzle and the pipe, the total temperature remains constant at $T_0 = 300$ K. For the isentropic flow in the C-D nozzle, the total pressure remains constant at $p_0 = 7 \times 10^5$ Pa. In all three cases, the mass flow rate is fixed by the choked C-D nozzle throat; as a result, the Mach numbers at the pipe exit in all three cases must be subsonic and equal. The flow being subsonic at the pipe exit, the exit static pressure must equal the given ambient pressure $p_{amb} = 2.8 \times 10^5$ Pa.

(a) NORMAL SHOCK IN THE NOZZLE THROAT

At the C-D nozzle throat, the flow is choked with $M = 1$. We compute the mass flow from the conditions at the throat as

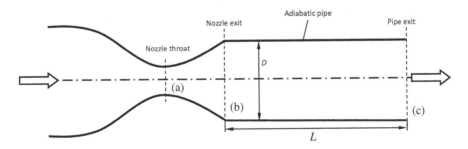

FIGURE 4.8 A convergent-divergent nozzle exhausting into a Fanno pipe (Problem 4.14).

$$\dot{m} = \frac{\hat{F}_{f0}^* A_{\text{throat}} p_0}{\sqrt{RT_0}}$$

$$\frac{\dot{m}}{A_{\text{throat}}} = \frac{\hat{F}_{f0}^* p_0}{\sqrt{RT_0}} = \frac{0.6847 \times 7 \times 10^5}{\sqrt{287 \times 300}} = 1633.416 \frac{\text{kg}}{\text{m}^2 \text{s}}$$

$$\frac{\dot{m}}{A_{\text{pipe}}} = \frac{\dfrac{\dot{m}}{A_{\text{throat}}}}{\dfrac{A_{\text{pipe}}}{A_{\text{throat}}}} = \frac{1633.416}{2} = 816.708 \frac{\text{kg}}{\text{m}^2 \text{s}}$$

The occurrence of a normal shock at the throat, infinitesimally downstream of the throat where the flow has just turned supersonic, implies that the downstream flow becomes subsonic and the diverging section of the C-D nozzle operates as a diffuser. For the nozzle area ratio $\left(A_{\text{exit}} / A_{\text{throat}} \right)$, from the isentropic flow table (Sultanian, 2015) with linear interpolation between the tabular values, we obtain $M_{\text{nozzle\_exit}} = 0.306$, which becomes the Mach number at the pipe inlet.

As the mass flow rate remains constant through the pipe, we can write the following relation at the pipe exit.

$$\frac{\dot{m}}{A_{\text{pipe}}} = \frac{\hat{F}_{f\_pipe\_exit} p_{\text{amb}}}{\sqrt{RT_0}}$$

giving

$$\hat{F}_{f\_pipe\_exit} = \left(\frac{\dot{m}}{A_{\text{pipe}}} \right) \frac{\sqrt{RT_0}}{p_{\text{amb}}} = \frac{816.708 \times \sqrt{287 \times 300}}{2.8 \times 10^5} = 0.856$$

Using the flow functions table (Sultanian, 2015) with linear interpolation between the tabular values, we obtain $M_{\text{pipe\_exit}} = 0.691$ for $\hat{F}_{f\_pipe\_exit} = 0.856$.

Thus, for the Fanno flow in the pipe, the inlet and exit Mach numbers are 0.306 and 0.691, respectively. Using the Fanno flow table (Sultanian, 2015) with linear interpolation between the tabular values, we obtain the following quantities:

$$\left(\frac{f L_{\text{max}}}{D} \right)_{M=0.306} = 5.031$$

and

$$\left(\frac{f L_{\text{max}}}{D} \right)_{M=0.691} = 0.226$$

giving

$$\left(\frac{fL}{D}\right) = \left(\frac{f\,L_{max}}{D}\right)_{M=0.306} - \left(\frac{f\,L_{max}}{D}\right)_{M=0.691} = 5.031 - 0.226 = 4.805$$

(b) Normal Shock at the Nozzle Exit Plane

In this case, both the mass flow rate and Mach number at the pipe exit are the same as we obtained in (a). For the normal shock, the upstream Mach number corresponds to the value at the C-D nozzle exit. For the nozzle area ratio of 2, we obtain the exit supersonic Mach number of 2.197 from the isentropic flow table (Sultanian, 2015) with linear interpolation between the tabular values. This is the normal shock upstream Mach number. Using the normal shock table (Sultanian, 2015) with linear interpolation between the tabular values, we obtain 0.548 as the Mach number downstream of the normal shock, equal to the pipe inlet Mach number.

For the Fanno flow in the pipe, we now have the inlet and exit Mach numbers of 0.548 and 0.691, respectively. Using the Fanno flow table (Sultanian, 2015) with linear interpolation between the tabular values, we obtain the following quantities:

$$\left(\frac{f\,L_{max}}{D}\right)_{M=0.548} = 0.743$$

and

$$\left(\frac{f\,L_{max}}{D}\right)_{M=0.691} = 0.226$$

giving

$$\left(\frac{fL}{D}\right) = \left(\frac{f\,L_{max}}{D}\right)_{M=0.548} - \left(\frac{f\,L_{max}}{D}\right)_{M=0.691} = 0.743 - 0.226 = 0.517$$

(c) Normal Shock at the Pipe Exit Plane

In this case, the mass flow rate and the Mach number at the pipe exit remain the same as in (a) and (b). The Mach number at the pipe inlet equals 2.197, which we computed in (b) at the C-D nozzle exit. For $M_{pipe\_exit} = 0.691$ obtained in (a) and (b), which becomes the Mach number downstream of the normal shock, we obtain from the normal shock table (Sultanian, 2015) with linear interpolation between the tabular values the Mach number upstream of the normal shock as 1.529.

For the Fanno flow in the pipe, we now know the inlet and exit Mach numbers of 2.197 and 1.529, respectively. Using the Fanno flow table (Sultanian, 2015) with linear interpolation between the tabular values, we obtain the following quantities:

$$\left(\frac{fL_{max}}{D}\right)_{M=2.197} = 0.360$$

and

$$\left(\frac{fL_{\text{max}}}{D}\right)_{M=1.529} = 0.147$$

giving

$$\left(\frac{fL}{D}\right) = \left(\frac{fL_{\text{max}}}{D}\right)_{M=2.197} - \left(\frac{fL_{\text{max}}}{D}\right)_{M=1.529} = 0.360 - 0.147 = 0.213$$

PROBLEM 4.15: NORMAL SHOCK ANALYSIS USING FANNO FLOW AND ISENTROPIC FLOW TABLES

Total pressure, static pressure, and static temperature immediately upstream of a normal shock in a compressible air flow are given. Using a Fanno flow table and, if needed, an isentropic flow tables, for example, given in Sultanian (2015), write a step-by-step procedure to determine the static pressure and temperature immediately downstream of the normal shock. For air, assume $\gamma = 1.4$ and $R = 287 \, \text{J}/(\text{kg} \, \text{K})$.

SOLUTION FOR PROBLEM 4.15

In the solution of this problem, we make use of the fact that the impulse pressure remains constant across a normal shock. Denoting the quantities upstream and downstream of the normal shock by subscripts 1 and 2, respectively, we summarize the required step-by-step solution procedure as follows:

Step 1. For the given p_{01}, p_1, and T_1, compute p_{01}/p_1 and use the isentropic flow table to find M_1.

Step 2. For the supersonic M_1 from Step 1, find p_i/p_i^*, p_1/p^*, and T_1/T^* from the Fanno flow table and compute $p^* = p_1/(p_1/p^*)$ and $T^* = T_1/(T_1/T^*)$.

Step 3. For p_i/p_i^* from Step 2, find p_2/p^* and T_2/T^* in the subsonic section of the Fanno flow table.

Step 4. Using p^* and T^* from Step 2 and p_2/p^* and T_2/T^* from Step 3, determine

$$p_2 = p^*\left(\frac{p_2}{p^*}\right)$$

and

$$T_2 = T^*\left(\frac{T_2}{T^*}\right)$$

PROBLEM 4.16:   NORMAL SHOCK ANALYSIS USING RAYLEIGH
FLOW AND ISENTROPIC FLOW TABLES

Total pressure, static pressure, and static temperature immediately upstream of a normal shock in a compressible air flow are given. Using a Rayleigh flow table and, if needed, an isentropic flow table, for example, given in Sultanian (2015), write a step-by-step procedure to determine the static pressure and temperature immediately downstream of the normal shock. For air, assume $\gamma = 1.4$ and $R = 287\,\text{J}/(\text{kg}\,\text{K})$.

SOLUTION FOR PROBLEM 4.16

In the solution of this problem, we make use of the fact that the total temperature remains constant across a normal shock. Denoting the quantities upstream and downstream of the normal shock by subscripts 1 and 2, respectively, we summarize the required step-by-step solution procedure as follows:

Step 1. For the given p_{01}, p_1, and T_1, compute p_{01}/p_1 and use the isentropic flow table to find M_1 and T_{01}/T_1, giving $T_{01} = T_1(T_{01}/T_1)$.

Step 2. For the supersonic M_1 from Step 1, find T_0/T_0^*, p_1/p^*, and T_1/T^* from the Rayleigh flow table and compute $p^* = p_1/(p_1/p^*)$ and $T^* = T_1/(T_1/T^*)$.

Step 3. For T_0/T_0^* from Step 2, find p_2/p^* and T_2/T^* in the subsonic section of the Rayleigh flow table.

Step 4. Using p^* and T^* from Step 2 and p_2/p^* and T_2/T^* from Step 3, compute

$$p_2 = p^*\left(\frac{p_2}{p^*}\right)$$

and

$$T_2 = T^*\left(\frac{T_2}{T^*}\right)$$

PROBLEM 4.17:   SUPERSONIC AIR FLOW OVER A WEDGE
WITH MEASURED WAVE ANGLE

A wedge of angle $2\theta = 30°$ is used to measure the Mach number of an oncoming supersonic air flow. If the observed wave angle is $\beta = 39°$, find M_1, p_2/p_1, T_2/T_1, M_2, and the entropy change across the oblique wave. For air as a perfect gas, assume $\gamma = 1.4$ and $R = 287\,\text{J}/(\text{kg}\,\text{K})$.

SOLUTION FOR PROBLEM 4.17

We rearrange the equation for an oblique shock

$$\frac{\tan\beta}{\tan(\beta-\theta)} = \frac{(\gamma+1)M_{n1}^2}{2+(\gamma-1)M_{n1}^2}$$

to yield the following equation to directly calculate the normal component M_{n1} of the oncoming flow Mach number M_1.

$$M_{n1} = \sqrt{\frac{2\tan\beta}{(\gamma+1)\tan(\beta-\theta)-(\gamma-1)\tan\beta}}$$

$$M_{n1} = \sqrt{\frac{2\tan 39°}{(1.4+1)\tan(39°-15°)-(1.4-1)\tan 39°}} = 1.4748$$

We calculate the Mach number of the oncoming flow as

$$M_1 = \frac{M_{n1}}{\sin\beta} = \frac{1.4748}{\sin 39°} = 2.3435$$

For $M_{n1} = 1.4748$, using linear interpolation between adjacent values, we obtain $p_2/p_1 = 2.3708$, $T_2/T_1 = 1.3035$, and $M_{n2} = 0.7102$ from the normal shock table (Sultanian, 2015). We further obtain

$$M_2 = \frac{M_{n2}}{\sin(\beta-\theta)} = \frac{0.7102}{\sin(39°-15°)} = 1.7462$$

PROBLEM 4.18: SUPERSONIC AIR FLOW OVER A WEDGE WITH MEASURED PRESSURE RATIO

In Problem 4.17, suppose that we measure the pressure ratio p_2/p_1 instead of the wave angle β. Describe a solution method to determine M_1 and the remaining unknowns.

SOLUTION FOR PROBLEM 4.18

In this case, we know the wedge half-angle θ and the static pressure ratio p_2/p_1. We summarize here the solution method to determine M_1, M_2, β, and T_2/T_1.

Step 1. Entering the normal shock table (Sultanian, 2015) with the known value of p_2/p_1, determine M_{n1}, M_{n2}, and using linear interpolation between consecutive tabular values for each quantity.

Step 2. Iteratively solve the following equation, e.g., using "Goal Seek" in MS Excel, to determine the wave angle β:

$$\frac{\tan \beta}{\tan (\beta - \theta)} = \frac{(\gamma + 1) M_{n1}^2}{2 + (\gamma - 1) M_{n1}^2}$$

Step 3. Calculate M_1 as

$$M_1 = \frac{M_{n1}}{\sin \beta}$$

and M_2 as

$$M_2 = \frac{M_{n2}}{\sin (\beta - \theta)}$$

PROBLEM 4.19: PRESSURE, TEMPERATURE, AND AREA CHANGES IN A DUCT WITH CONSTANT MACH NUMBER

Consider the frictionless flow of a calorically perfect gas in a duct with heat transfer. For the case where the duct wall is shaped to keep the Mach number constant, find an expression for p_2/p_1, p_{02}/p_{01}, and A_2/A_1 in terms of γ, M, and T_{02}/T_{01}.

SOLUTION FOR PROBLEM 4.19

TOTAL PRESSURE RATIO $\left(p_{02}/p_{01} \right)$

We use here the following relationship between the change in total pressure due to change in total temperature given by (see Sultanian, 2015):

$$\frac{dp_0}{dT_0} = -\gamma M^2 \left(\frac{p_0}{T_0} \right)$$

Let us rewrite this equation as

$$\frac{dp_0}{p_0} = -\gamma M^2 \left(\frac{dT_0}{T_0} \right)$$

Integrating this equation between duct sections 1 and 2 yields

$$\int_1^2 \frac{dp_0}{p_0} = -\gamma M^2 \int_1^2 \frac{dT_0}{T_0}$$

$$\ln \left(\frac{p_{02}}{p_{01}} \right) = -\gamma M^2 \ln \left(\frac{T_{02}}{T_{01}} \right)$$

$$\frac{p_{02}}{p_{01}} = \left(\frac{T_{02}}{T_{01}} \right)^{-\gamma M^2}$$

STATIC PRESSURE RATIO $\left(p_2 / p_1 \right)$

As the Mach number is constant in the duct, we obtain

$$\frac{p_2}{p_1} = \frac{p_{02}}{p_{01}} = \left(\frac{T_{02}}{T_{01}} \right)^{-\gamma M^2}$$

AREA RATIO $\left(A_2 / A_1 \right)$

As the mass flow rate is constant in the duct, using the total-pressure mass flow function, we obtain

$$\dot{m}_1 = \frac{\hat{F}_{f01} A_1 p_{01}}{\sqrt{RT_{01}}} = \dot{m}_2 = \frac{\hat{F}_{f02} A_2 p_{02}}{\sqrt{RT_{02}}}$$

For constant Mach number in the duct, we have $\hat{F}_{f01} = \hat{F}_{f02}$, which reduces this equation to

$$\frac{A_1 p_{01}}{\sqrt{T_{01}}} = \frac{A_2 p_{02}}{\sqrt{T_{02}}}$$

giving

$$\frac{A_2}{A_1} = \frac{p_{01}}{p_{02}} \sqrt{\frac{T_{02}}{T_{01}}} = \left(\frac{T_{02}}{T_{01}} \right)^{\gamma M^2} \sqrt{\frac{T_{02}}{T_{01}}} = \left(\frac{T_{02}}{T_{01}} \right)^{\left(1 + 2\gamma M^2 \right)\!/2}$$

NOMENCLATURE

A	Area
A^*	Critical throat area with $M = 1$
c_p	Specific heat at constant pressure
c_v	Specific heat at constant volume
C	Speed of sound
C^*	Characteristic speed of sound at T^*
C_V	Velocity coefficient $\left(C_V = V_n / V \right)$
C_f	Shear coefficient or Fanning friction factor
D	Nozzle or pipe diameter
D_h	Mean hydraulic diameter
f	Moody or Darcy friction factor
F	Force magnitude
F_f	Static-pressure mass flow function
$\hat{F}_f$	Dimensionless total-pressure mass flow function
F_{f0}	Total-pressure mass flow function
$\hat{F}_{f0}$	Dimensionless total-pressure mass flow function

g Acceleration due to gravity

h Specific enthalpy; heat transfer coefficient

I_f Static-pressure impulse function

I_{f0} Total-pressure impulse function

L Length

$\dot{m}$ Mass flow rate

M Mach number $\left(M = V/C \right)$

M^* Characteristic Mach number $\left(M^* = V/C^* \right)$

N Normal shock function

N^∞ Asymptotic value of N as $M \to \infty$

p Static pressure

p_i Impulse pressure

p_w Wetted perimeter of a duct

$\dot{Q}$ Heat transfer rate

$\delta \dot{Q}$ Rate of heat transfer into a differential control volume

R Gas constant; pipe radius

Re Reynolds number

s Specific entropy

S_T Stream thrust

t Time

T Static temperature

V Velocity magnitude; through-flow velocity relative a duct

W_j Relative jet velocity

$\dot{W}$ Work transfer rate

$\delta \dot{W}$ Rate of work transfer into a differential control volume

y Cartesian coordinate y

z Cartesian coordinate z

SUBSCRIPTS AND SUPERSCRIPTS

0   Total (stagnation)

1   Location 1; Section 1

2   Location 2; Section 2

cn   Convergent nozzle

dn   Divergent nozzle

max   Maximum

n   Velocity component normal to shock wave

ns   Normal shock

o   Origin

rot   Rotation

sh   Shear

w   Wall

x Component along x-coordinate direction; axial direction

y Component in y-coordinate direction

z Component in z-coordinate direction

β Velocity component along the shock wave

θ Tangential direction

(*)   Properties at $M = 1$; characteristic values; critical value $M = 1/\sqrt{\gamma}$ for an isothermal flow in a constant-area duct with friction

($^-$)   Section-average value

GREEK SYMBOLS

β Wave angle of an oblique shock

δ Parameter in the equation to calculate wave angle from known upstream Mach number and deflection angle: $\delta = 0$ for a strong shock and $\delta = 1$ for a weak shock

η Number of transfer units (NTU)

θ Wedge half-angle or deflection angle

γ Ratio of specific heats $\left(\gamma = c_p/c_v\right)$

λ Parameter in the equation to calculate wave angle from known upstream Mach number and deflection angle

μ Dynamic viscosity; Mach wave angle $\left\{\mu = \sin^{-1}\left(1/M\right)\right\}$

ρ Density

τ_w Wall shear stress

χ Parameter in the equation to calculate wave angle from known upstream Mach number and deflection angle

ϕ Prandtl-Meyer function

Ω Rotational (angular) speed

REFERENCE

Sultanian, B.K. 2015. *Fluid Mechanics: An Intermediate Approach*. Boca Raton, FL: Taylor & Francis.

BIBLIOGRAPHY

Anderson, J.D. 2003. *Modern Compressible Flow with Historical Perspective*, 3rd edition. Boston, MA: McGraw-Hill.

Oosthuizen, P.H. and W.E. Carscallen. 2013. *Introduction to Compressible Fluid Flow*, 2nd edition (Heat Transfer). Boca Rotan, FL: Taylor & Francis.

Shapiro, A.H. 1953. The Dynamics and Thermodynamics of Compressible Fluid Flow, Vols. 1 and 2. New York: Ronald Press.

Sultanian, B.K. 2018: *Gas Turbines: Internal Flow Systems Modeling* (Cambridge Aerospace Series #44). Cambridge: Cambridge University Press.

Sultanian, B.K. 2019. *Logan's Turbomachinery: Flowpath Design and Performance Fundamentals*, 3rd edition. Boca Raton, FL: Taylor & Francis.

5 Potential Flow

REVIEW OF KEY CONCEPTS

The velocity vector field of a three-dimensional potential flow is obtainable from the gradient of a three-dimensional scalar potential function—hence the name potential flow. An irrotational flow features zero curl of its velocity vector field. As the curl of the gradient of a scalar function is identically zero from vector calculus, a potential flow must always be irrotational or vice versa. Being isentropic, potential flows are ideal flows with their total pressure remaining constant everywhere.

In this chapter, we primarily deal with problems on two-dimensional potential flows, which are incompressible and inviscid. However, most of the key concepts presented here are equally applicable to three-dimensional potential flows. For further details on these concepts, see, for example, Sultanian (2015).

VELOCITY POTENTIAL

The gradient of a scalar function $\Phi(x,y,z)$ yields a vector field $\left(V = \nabla\Phi\right)$. Note that V identically satisfies the condition of irrotationality $\nabla \times V = 0$, and for it to represent a flow field, it must satisfy the continuity constraint of zero divergence, that is, $\nabla \cdot V = 0$. Substituting $V = \nabla\Phi$ in $\nabla \cdot V = 0$ yields the Laplace equation $\nabla^2\Phi = 0$. Under these conditions, $\Phi(x,y,z)$ qualifies to be a velocity potential.

STREAM FUNCTION

For a two-dimensional incompressible flow, we define a scalar stream function Ψ to yield the velocity components in the Cartesian coordinate system as

$$V_x = \frac{\partial\Psi}{\partial y} \tag{5.1}$$

$$V_y = -\frac{\partial\Psi}{\partial x} \tag{5.2}$$

Note that V_x and V_y satisfy the continuity equation at every point in the flow field, that is,

$$\frac{\partial V_x}{\partial x} + \frac{\partial V_y}{\partial y} = \frac{\partial^2\Psi}{\partial x \partial y} - \frac{\partial^2\Psi}{\partial x \partial y} = 0$$

The condition of irrotationality for a two-dimensional potential flow in the x-y plane yields

$$\frac{\partial V_y}{\partial x} - \frac{\partial V_x}{\partial y} = 0 \tag{5.3}$$

Substituting for V_x from Equation 5.1 and V_y from Equation 5.2 in this equation yields

$$\frac{\partial^2 \Psi}{\partial x^2} + \frac{\partial^2 \Psi}{\partial y^2} = \nabla^2 \Psi = 0 \tag{5.4}$$

Thus, the stream function defined for a two-dimensional potential flow must also satisfy the Laplace equation (Equation 5.4).

In a two-dimensional flow, let us consider the line of constant stream function. For a differential change in Ψ along such a flow line, we write

$$d\Psi = \frac{\partial \Psi}{\partial x} dx + \frac{\partial \Psi}{\partial y} dy = 0$$

Substituting for $\partial \Psi / \partial x$ from Equation 5.2 and $\partial \Psi / \partial y$ from Equation 5.1 in this equation yields

$$-V_y dx + V_x dy = 0$$

$$\left(\frac{dy}{dx}\right)_\Psi = \frac{V_y}{V_x} \tag{5.5}$$

which shows that the local velocity vector is tangent to a constant-Ψ line. As the velocity vector is also tangent to the streamline, a constant-Ψ line, therefore, represents a streamline.

Figure 5.1 shows a two-dimensional duct of unit depth formed by two streamlines. We can express the volumetric flow rate Q per unit depth through an arbitrary section AB as

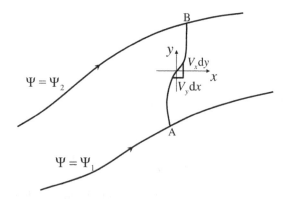

FIGURE 5.1 Volumetric flow rate through a two-dimensional duct formed by two streamlines.

$$Q = \int_A^B V_x dy - \int_A^B V_y dx = \int_A^B (V_x dy - V_y dx)$$

$$Q = \int_A^B \left(\frac{\partial \Psi}{\partial y} dy + \frac{\partial \Psi}{\partial x} dx \right) = \int_A^B d\Psi = \Psi_2 - \Psi_1$$

(5.6)

which shows that Q is the difference in the values of stream functions that the bounding streamlines represent.

In a two-dimensional flow, Ψ is constant along a streamline and Φ is constant along an equipotential line. Let us now find the angle of their intersection. For an equipotential line $(d\Phi = 0)$, we write

$$d\Phi = \frac{\partial \Phi}{\partial x} dx + \frac{\partial \Phi}{\partial y} dy = 0$$

$$V_x dx + V_y dy = 0 \qquad (5.7)$$

$$\left(\frac{dy}{dx} \right)_\Phi = -\frac{V_x}{V_y}$$

From Equations 5.5 and 5.7, at the intersection point of an equipotential line and a streamline, we obtain the relation between their slopes as

$$\left(\frac{dy}{dx} \right)_\Psi \left(\frac{dy}{dx} \right)_\Phi = -1$$

which implies that they are orthogonal to each other.

From the foregoing, we express V_x and V_y in Cartesian coordinates as

$$V_x = \frac{\partial \Phi}{\partial x} = \frac{\partial \Psi}{\partial y}$$

and

$$V_y = \frac{\partial \Phi}{\partial y} = -\frac{\partial \Psi}{\partial x}$$

Similarly, in cylindrical polar coordinates, we express V_r and V_θ as

$$V_r = \frac{\partial \Phi}{\partial r} = \frac{\partial \Psi}{r \partial \theta}$$

and

$$V_\theta = \frac{\partial \Phi}{r \partial \theta} = -\frac{\partial \Psi}{\partial r}$$

CIRCULATION

For a positively oriented closed contour C in a two-dimensional incompressible flow, we define circulation as

$$\Gamma = \oint_C V \cdot \mathrm{d}l \tag{5.8}$$

The circulation Γ measures the total vorticity contained within the closed contour C.

COMPLEX POTENTIAL

For a two-dimensional potential flow, we define a complex potential as

$$F(z) = \Phi(x,y) + i\Psi(x,y) \tag{5.9}$$

where the complex coordinate $z = x + iy$. As both $\Phi(x,y)$ and $\Psi(x,y)$ are conjugate functions, they satisfy the necessary and sufficient conditions for the complex potential $F(z)$ to be analytic.

COMPLEX VELOCITY

In the Cartesian coordinate system with $z = x + iy$, we obtain the complex conjugate velocity, see Sultanian (2015) for details, as

$$\frac{\mathrm{d}F(z)}{\mathrm{d}z} = V_x - iV_y = \overline{W} \tag{5.10}$$

with the complex velocity $W = V_x + iV_y$.

Similarly, in cylindrical polar coordinate system with $z = re^{i\theta}$, we obtain

$$\frac{\mathrm{d}F(z)}{\mathrm{d}z} = \overline{W} = (V_r - iV_\theta)e^{-i\theta} \tag{5.11}$$

with the complex velocity $W = (V_r + iV_\theta)e^{-i\theta}$.

COMPLEX CIRCULATION

For a simply connected closed contour C in a two-dimensional potential flow, the complex circulation $C(z)$ is defined by the equation

$$C(z) = \oint_C \overline{W}\,\mathrm{d}z = \oint_C (V_x - iV_y)(\mathrm{d}x + i\mathrm{d}y)$$

$$C(z) = \oint_C (V_x\mathrm{d}x + V_y\mathrm{d}y) + i\oint_C (V_x\mathrm{d}y - V_y\mathrm{d}x) \tag{5.12}$$

$$C(z) = \oint_C V \cdot \mathrm{d}l + i\oint_C V \cdot \hat{n}\,\mathrm{d}l = \Gamma + iQ$$

In this equation, $\hat{n}$ represents a unit vector normal to the line segment dl. The real part of $C(z)$ is the circulation Γ of the fluid and the imaginary part gives the volumetric flow rate Q (per unit length perpendicular to the plane of the flow) that results from the sources enclosed within the contour of integration. For a singularity-free, simply connected flow region, $C(z) = 0$.

UNIFORM FLOW PARALLEL TO THE X-AXIS

The complex potential function in this case is

$$F(z) = Uz \tag{5.13}$$

which yields

$$\bar{W} = V_x - iV_y = \frac{dF}{dz} = U \tag{5.14}$$

giving $V_x = U$ and $V_y = 0$.

UNIFORM FLOW PARALLEL TO THE Y-AXIS

In this case, we write the complex potential as

$$F(z) = -iVz \tag{5.15}$$

which yields

$$\bar{W} = V_x - iV_y = \frac{dF}{dz} = -iV \tag{5.16}$$

giving $V_x = 0$ and $V_y = V$.

UNIFORM FLOW INCLINED AT AN ANGLE α TO THE X-AXIS

The complex potential function

$$F(z) = Ve^{-i\alpha}z \tag{5.17}$$

yields

$$\bar{W} = V_x - iV_y = \frac{dF}{dz} = Ve^{-i\alpha} = V\cos\alpha - iV\sin\alpha \tag{5.18}$$

giving $V_x = V\cos\alpha$ and $V_y = V\sin\alpha$, which shows that multiplying the complex potential by $e^{-i\alpha}$ rotates the entire flow field counterclockwise by an angle α.

SOURCE AND SINK

From the complex potential function

$$F(z) = \Phi + i\Psi = c\ln z = c\ln\left(re^{i\theta}\right) = c\ln r + ic\theta \qquad (5.19)$$

we obtain $\Phi = c\ln r$ and $\Psi = c\theta$. Thus, in the flow field generated from $F(z) = c\ln z$, the lines of constant r are the equipotential lines and the lines of constant θ are the streamlines. We obtain the complex conjugate velocity from this potential function as

$$\bar{W} = \frac{dF(z)}{dz} = \frac{c}{z} = \frac{c}{r}e^{-i\theta} = \left(\frac{c}{r} - i0\right)e^{-i\theta} = (V_r - iV_\theta)e^{-i\theta} \qquad (5.20)$$

giving $V_r = c/r$ and $V_\theta = 0$. For a positive value of c, the complex potential $F(z) = c\ln z$ represents a source; for a negative value of c, it represents a sink, each located at the origin. For the source or the sink located at $z_0 = x_0 + iy_0$, we express the complex potential by $F(z) = c\ln(z - z_0)$.

From the continuity equation, the total volumetric flow rate m (per unit depth) crossing each equipotential line, which is a circle, must remain constant. Accordingly, we obtain

$$m = \int_0^{2\pi} \frac{c}{r} rd\theta = 2\pi c$$

$$c = \frac{m}{2\pi} \qquad (5.21)$$

We write the complex potential function for a source located at z_0 as

$$F(z) = \frac{m}{2\pi}\ln(z - z_0) \qquad (5.22)$$

where m is the source strength, which is negative for a sink.

VORTEX

By multiplying the complex potential of a source, given by Equation 5.22, by $-i$, we locally rotate the entire flow field counterclockwise by 90° and obtain the complex potential of a vortex given by

$$F(z) = -ic\ln z = -ic\ln(re^{i\theta}) = c\theta - ic\ln r \qquad (5.23)$$

which yields $\Phi = c\theta$ and $\Psi = -c\ln r$. In this flow, the lines of constant radius r are the streamlines and the lines of constant θ are the equipotential lines. This potential function yields

$$\bar{W} = \frac{dF(z)}{dz} = \frac{-ic}{z} = \frac{-ic}{r}e^{-i\theta} = \left(0 - i\frac{c}{r}\right)e^{-i\theta} = (V_r - iV_\theta)e^{-i\theta} \qquad (5.24)$$

which yields $V_r = 0$ and $V_\theta = c/r$. The complex potential $F(z) = -ic \ln z$ represents a vortex located at the origin. For a positive value of c, the counterclockwise rotating vortex is considered positive. For a negative value of c, we obtain a clockwise rotating vortex. The complex potential $F(z) = -ic \ln(z - z_0)$ represents a vortex located at $z_0 = x_0 + iy_0$.

We measure the strength of a vortex by its circulation Γ obtained from

$$\Gamma = \oint_C V \cdot dl$$

$$\Gamma = \int_0^{2\pi} V_\theta r \, d\theta = \int_0^{2\pi} \frac{c}{r} r \, d\theta = 2\pi c$$

$$c = \frac{\Gamma}{2\pi}$$

Thus, we express the complex potential function for a vortex located at z_0 as

$$F(z) = -i \frac{\Gamma}{2\pi} \ln(z - z_0) \tag{5.25}$$

where Γ is the vortex strength, which is positive for a counterclockwise vortex and negative for a clockwise vortex.

DIPOLE

The superposition of a source flow and a sink flow creates a dipole from the complex potential

$$F(z) = \frac{m}{2\pi} \ln(z - \varepsilon) - \frac{m}{2\pi} \ln(z + \varepsilon) \tag{5.26}$$

where we have a source of strength m at $(\varepsilon, 0)$ and a sink of strength m at $(-\varepsilon, 0)$.

DOUBLET

A double is the limiting case of a dipole where the distance between the source and the sink is made vanishingly small while keeping the product of their strength and the distance between them constant. We write the complex potential for a dipole as

$$F(z) = -\frac{m}{2\pi} \ln\left(\frac{z + \varepsilon}{z - \varepsilon}\right) = -\frac{m}{2\pi} \ln\left(\frac{1 + \frac{\varepsilon}{z}}{1 - \frac{\varepsilon}{z}}\right)$$

For $(-1 < \varepsilon/z < 1)$, the power series expansion of the logarithmic term in this equation yields

$$F(z) = -\frac{m}{2\pi}\left[2\frac{\varepsilon}{z} + \frac{2}{3}\left(\frac{\varepsilon^3}{z^3}\right) + \frac{2}{5}\left(\frac{\varepsilon^5}{z^5}\right) + \dots\right]$$

Letting $\varepsilon \to 0$ and $m \to \infty$ in such a way as to yield $m\varepsilon = \pi\mu$, where μ is a constant, we obtain

$$F(z) = -\frac{\mu}{z} \tag{5.27}$$

Thus, a doublet results from the superposition of a strong source and an equally strong sink in proximity. We write the complex potential of a doublet located at $z = z_0$ as

$$F(z) = -\frac{\mu}{(z - z_0)} \tag{5.28}$$

where μ is the strength of the doublet. The principal use of a doublet is in generating more complex and practical potential flows by linear superposition with other simple flows.

FLOW AROUND A CYLINDER WITHOUT CIRCULATION

The superposition of a uniform flow and a doublet generates the potential flow around a cylinder without circulation. We write the resulting complex potential as

$$F(z) = Uz + \frac{\mu}{z} = Ure^{i\theta} + \frac{\mu}{r}e^{-i\theta} = \left(Ur + \frac{\mu}{r}\right)\cos\theta + i\left(Ur - \frac{\mu}{r}\right)\sin\theta \tag{5.29}$$

which yields

$$\Psi = \left(Ur - \frac{\mu}{r}\right)\sin\theta \tag{5.30}$$

whose value on a circle of radius $r = a$ becomes

$$\Psi_a = \left(Ua - \frac{\mu}{a}\right)\sin\theta$$

For $\psi_a = 0$, this equation yields $\mu = Ua^2$. Thus, if we superimpose a uniform flow of velocity U on a doublet of strength Ua^2, we obtain a potential flow over a cylinder of radius $r = a$ whose surface corresponds to zero stream function. Within the cylinder, the sink absorbs the entire mass flow rate generated by the adjacent source of the doublet. The resulting complex potential function and the stream function are

$$F(z) = U\left(z + \frac{a^2}{z}\right) \tag{5.31}$$

$$\Psi = U\left(r - \frac{a^2}{r}\right)\sin\theta \tag{5.32}$$

Stagnation Points

At a stagnation point in a flow, all velocity components become zero. We write

$$\bar{W}(z) = \frac{\mathrm{d}F}{\mathrm{d}z} = U\left(1 - \frac{a^2}{z^2}\right)$$

$$\bar{W}(z) = \left[U\left(1 - \frac{a^2}{r^2}\right)\cos\theta + iU\left(1 + \frac{a^2}{r^2}\right)\sin\theta\right]e^{-i\theta} \tag{5.33}$$

which yields

$$V_r = U\left(1 - \frac{a^2}{r^2}\right)\cos\theta \tag{5.34}$$

$$V_\theta = -U\left(1 + \frac{a^2}{r^2}\right)\sin\theta \tag{5.35}$$

Equation 5.34 yields zero radial velocity everywhere on the cylinder surface ($r = a$) on which, according to Equation 5.35, the tangential velocity becomes zero at $\theta = 0$ and $\theta = \pi$, the stagnation point locations on the cylinder surface.

FLOW AROUND A CYLINDER WITH CIRCULATION

Adding a clockwise rotating vortex to the flow around a cylinder results in the complex potential for the flow around a cylinder with circulation as

$$F(z) = U\left(z + \frac{a^2}{z}\right) + i\frac{\Gamma}{2\pi}\ln z + c_1$$

In this equation, the constant c_1 is added to render $\Psi = 0$ at the cylinder surface ($r = a$). This has no effect on the velocity and pressure distributions in the flow field. Expressing this equation in cylindrical polar coordinates, we obtain

$$F(z) = \left[U\left(r + \frac{a^2}{r}\right)\cos\theta - \frac{\Gamma\theta}{2\pi}\right] + i\left[U\left(r - \frac{a^2}{r}\right)\sin\theta + \frac{\Gamma}{2\pi}\ln r\right] + c_1$$

where the substitution $c_1 = -(i\Gamma/2\pi)(\ln a)$ yields $\Psi = 0$ at $r = a$, giving

$$F(z) = U\left(z + \frac{a^2}{z}\right) + i\frac{\Gamma}{2\pi}\ln\frac{z}{a} \tag{5.36}$$

and

$$\Psi = U\left(r - \frac{a^2}{r}\right)\sin\theta + \frac{\Gamma}{2\pi}\ln r - \frac{\Gamma}{2\pi}\ln a \tag{5.37}$$

Stagnation Points

We write

$$\bar{W}(z) = \frac{dF}{dz} = U\left(1 - \frac{a^2}{z^2}\right) + i\frac{\Gamma}{2\pi z}$$

$$\bar{W}(z) = \left[U\left(1 - \frac{a^2}{r^2}\right)\cos\theta + i\left\{U\left(1 + \frac{a^2}{r^2}\right)\sin\theta + \frac{\Gamma}{2\pi r}\right\}\right]e^{-i\theta} \tag{5.38}$$

which yields

$$V_r = U\left(1 - \frac{a^2}{r^2}\right)\cos\theta \tag{5.39}$$

$$V_\theta = -U\left(1 + \frac{a^2}{r^2}\right)\sin\theta - \frac{\Gamma}{2\pi r} \tag{5.40}$$

Equation 5.39 yields zero radial velocity everywhere on the cylinder surface $(r = a)$. For $V_\theta = 0$ on the cylinder surface, Equation 5.40 yields

$$\sin\theta = -\frac{\Gamma}{4\pi U a} \tag{5.41}$$

PROBLEM 5.1:   FINDING STREAM FUNCTION FROM GIVEN POTENTIAL FUNCTION

Verify that

$$\Phi = 2xy + x^2 - y^2$$

is the potential function for a two-dimensional incompressible flow. Find the corresponding stream function. Plot a few streamlines, indicating their flow directions.

SOLUTION FOR PROBLEM 5.1

Any potential function for an incompressible flow field must satisfy the Laplace equation $\nabla^2\Phi = 0$, which we write for the given potential function as

$$\nabla^2\Phi = \nabla^2\left(2xy+x^2-y^2\right) = \frac{\partial^2}{\partial x^2}\left(2xy+x^2-y^2\right) + \frac{\partial^2}{\partial y^2}\left(2xy+x^2-y^2\right) = 2-2 = 0$$

which proves that the given potential function represents a two-dimensional incompressible potential flow. We obtain the corresponding velocity components as

$$V_x = \frac{\partial}{\partial x}\left(2xy+x^2-y^2\right) = 2y+2x$$

and

$$V_y = \frac{\partial}{\partial y}\left(2xy+x^2-y^2\right) = 2x-2y$$

Using Equation 5.1, we obtain

$$\frac{\partial\Psi}{\partial y} = V_x = 2y+2x$$

whose integration yields

$$\Psi = y^2 + 2xy + f(x)$$

This equation, in conjunction with Equation 5.2, yields

$$-\frac{\partial\Psi}{\partial x} = -2y - f'(x) = V_y = 2x - 2y$$

$$-f'(x) = 2x$$

$$f(x) = -x^2$$

where, without any loss of generality, we have assumed that the constant of integration is zero. Thus, we obtain the stream function for the given velocity potential as

$$\Psi = y^2 + 2xy - x^2$$

To plot the streamlines in the x–y plane, let us rearrange this quadratic equation for y as

$$y^2 + 2xy - \left(x^2 + \Psi\right) = 0$$

with the roots

$$y = -x + \sqrt{2x^2 + \Psi}$$

and

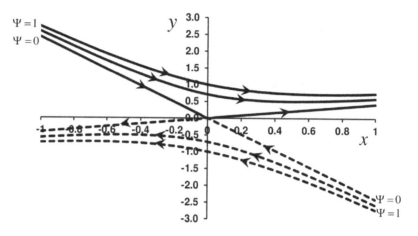

FIGURE 5.2   Streamlines for the given potential function (Problem 5.1).

$$y = -x - \sqrt{2x^2 + \Psi}$$

The streamlines with the flow direction are plotted in Figure 5.2. The solid stream-lines in the upper half correspond to the first root, and those shown by the dashed line in the lower half correspond to the second root.

PROBLEM 5.2:   SUPERPOSITION OF A UNIFORM FLOW
AND A SOURCE (RANKINE HALF-BODY)

Figure 5.3 shows the Rankine half-body flow obtained by superposing a uniform flow and a source with its complex potential given by

$$F(z) = Uz + \frac{m}{2\pi} \ln z$$

a. Find the stream function in cylindrical polar coordinates
b. Show $V_r = V_\theta = 0$ at the stagnation point S

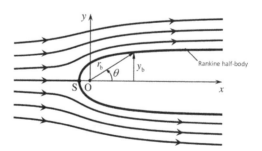

FIGURE 5.3   Superposition of a uniform flow and a source: Rankine half-body (Problem 5.2).

SOLUTION FOR PROBLEM 5.2

(a) STREAM FUNCTION IN CYLINDRICAL POLAR COORDINATES

We express

$$F(z) = Uz + \frac{m}{2\pi}\ln z$$

in cylindrical polar coordinates as

$$F(z) = Ure^{i\theta} + \frac{m}{2\pi}\ln\left(re^{i\theta}\right)$$

$$F(z) = Ur(\cos\theta + i\sin\theta) + \frac{m}{2\pi}\ln r + i\frac{m}{2\pi}\theta$$

$$F(z) = \left(Ur\cos\theta + \frac{m}{2\pi}\ln r\right) + i\left(Ur\sin\theta + \frac{m}{2\pi}\theta\right)$$

which according to Equation 5.9 yields

$$\Psi = Ur\sin\theta + \frac{m}{2\pi}\theta$$

(b) STAGNATION POINT S WITH $V_r = V_\theta = 0$

Figure 5.3 shows a few streamlines generated from the stream function obtained in (a). In this figure, S represents the stagnation point where all velocity components are zero. To verify this, let us compute V_r and V_θ as

$$V_r = \frac{\partial\Psi}{r\,\partial\theta} = U\cos\theta + \frac{m}{2\pi r}$$

$$V_\theta = -\frac{\partial\Psi}{\partial r} = -U\sin\theta$$

For $V_r = V_\theta = 0$ at the stagnation point S, we obtain $\theta_S = \pi$ and $r_S = m/(2\pi U)$. These values yield $\Psi_S = m/2$ for the streamline that corresponds to the Rankine half-body, defined by the equation

$$r_b = \frac{m}{2U\sin\theta}\left(1 - \frac{\theta}{\pi}\right)$$

yielding

$$2y_b = 2r_b\sin\theta = \frac{m}{U}\left(1 - \frac{\theta}{\pi}\right)$$

whose maximum value of m/U occurs asymptotically as θ approaches zero.

(c) Static Pressure Distribution of the Rankine Half-Body Surface

With the free stream static pressure of p_∞, far away from the Rankine half-body where the velocity is U, we obtain the total pressure as

$$p_0 = p_\infty + \frac{1}{2}\rho U^2$$

As the total pressure remains constant in a potential flow, the static pressure at any point in this flow is obtained by the following equation (Bernoulli equation)

$$p = p_\infty + \frac{1}{2}\rho U^2 - \frac{1}{2}\rho\left(V_r^2 + V_\theta^2\right)$$

which yields at a point (r_b, θ) on the Rankine half-body

$$p = p_\infty + \frac{1}{2}\rho U^2 - \frac{1}{2}\rho\left(U\cos\theta + \frac{m}{2\pi r_b}\right)^2 - \frac{1}{2}\rho(U\sin\theta)^2$$

$$\frac{p - p_\infty}{\frac{1}{2}\rho U^2} = -\left(\frac{m\cos\theta}{\pi r_b U} + \frac{m^2}{4\pi^2 r_b^2 U^2}\right)$$

Substituting for r_b and simplifying the resulting expression, we finally obtain

$$\frac{p - p_\infty}{\frac{1}{2}\rho U^2} = -\frac{\sin 2\theta}{(\pi - \theta)} - \frac{\sin^2\theta}{(\pi - \theta)^2}$$

which is plotted in Figure 5.4. This equation is independent of the source strength m. At the stagnation point, which corresponds to $\theta = 180°$, the static pressure equals the

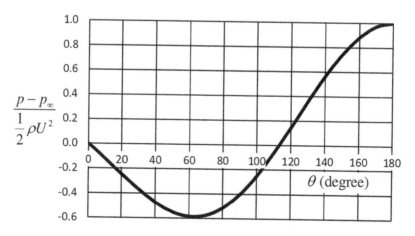

FIGURE 5.4 Static pressure distribution over a Rankine half-body (Problem 5.2).

stagnation pressure. With decreasing θ, the flow accelerates along the Rankine half-body with decrease in static pressure. At $\theta = 113.2°$, the velocity equals the incoming uniform velocity U and the static pressure equals the free stream value p_∞. At $\theta \cong 63°$, the velocity on the surface reaches its maximum value with the minimum static pressure. As θ approaches zero, the static pressure and velocity attain their free stream value.

PROBLEM 5.3:   ADDING A SOURCE TO A GIVEN STREAM FUNCTION

A two-dimensional potential flow with stagnation point at the origin is represented by the stream function $\Psi_1 = Axy$, where A is a constant. When we add a source of strength m to this flow at the origin, the stagnation point moves up by a distance h along the y-axis. Find the value of h in terms of A and m. For $A = 4$ and $m = 1$, plot a few streamlines in the first quadrant of the x-y plane. Include the streamline that contains the stagnation point.

SOLUTION FOR PROBLEM 5.3

We can write the stream function $\Psi_1 = Axy$ in polar coordinates as

$$\Psi_1 = Axy = A(r\cos\theta)(r\sin\theta) = \frac{Ar^2}{2}\sin 2\theta$$

We express the stream function of a point source of strength m located at the origin as $\Psi_2 = m\theta/2\pi$. Combining these stream functions, we obtain

$$\Psi = \Psi_1 + \Psi_2 = \frac{Ar^2}{2}\sin 2\theta + \frac{m}{2\pi}\theta$$

giving

$$V_r = \frac{\partial\Psi}{r\partial\theta} = Ar\cos 2\theta + \frac{m}{2\pi r}$$

$$V_\theta = -\frac{\partial\Psi}{\partial r} = -Ar\sin 2\theta$$

At the stagnation point, we have $V_\theta = V_r = 0$, giving

$$Ar\sin 2\theta = 0$$

$$\theta = \pi/2$$

$$Ar\cos 2\theta + \frac{m}{2\pi r} = 0$$

which, with the substitution $\theta = \pi/2$ and $r = h$, yields

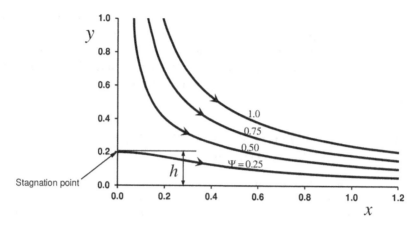

FIGURE 5.5   Streamlines for the given stream function (Problem 5.3).

$$h = \sqrt{\frac{m}{2\pi A}}$$

and

$$\Psi = \frac{m}{4}$$

For $A = 4$ and $m = 1$, we obtain $h = 0.199$ and $\Psi = 0.25$. Figure 5.5 shows the stream-lines that correspond to the stream function $\Psi = 2r^2 \sin 2\theta + \theta/2\pi$, indicating that the stagnation point lies at a distance $h = 0.199$ from the origin on the streamline with $\Psi = 0.25$.

PROBLEM 5.4:   STREAMLINES IN A SPIRAL VORTEX

The velocity potential for a spiral vortex flow is given by

$$\Phi = \Phi_{\text{vortex}} + \Phi_{\text{sink}} = \left(\frac{\Gamma}{2\pi}\right)\theta - \left(\frac{m}{2\pi}\right)\ln r$$

where Γ and m are constants representing the vortex strength and sink strength, respectively. Find the corresponding stream function and plot a few streamlines. Also show that for a spiral vortex, the angle between the velocity vector and the radial direction is constant throughout the flow field.

SOLUTION FOR PROBLEM 5.4

We write the complex potential for a vortex as

$$F_{\text{vortex}} = \Phi_{\text{vortex}} + i\Psi_{\text{vortex}} = \left(\frac{\Gamma}{2\pi}\right)\theta - i\left(\frac{\Gamma}{2\pi}\right)\ln r$$

and for a sink as

$$F_{sink} = \Phi_{sink} + i\Psi_{sink} = -\left(\frac{m}{2\pi}\right)\ln r - i\left(\frac{m}{2\pi}\right)\theta$$

The superposition of F_{vortex} and F_{sink} yields

$$F = F_{vortex} + F_{sink} = \Phi + i\Psi = \left[\left(\frac{\Gamma}{2\pi}\right)\theta - \left(\frac{m}{2\pi}\right)\ln r\right] - i\left[\left(\frac{\Gamma}{2\pi}\right)\ln r + \left(\frac{m}{2\pi}\right)\theta\right]$$

From which we obtain the stream function for the spiral vortex as

$$\Psi = -\left(\frac{\Gamma}{2\pi}\right)\ln r - \left(\frac{m}{2\pi}\right)\theta$$

For $m = 1$ and $\Gamma = 2.5$, Figure 5.6 shows three streamlines with $\Psi = 0.50$, 0.75, and 1.0. These streamlines spiral down to the sink at the origin.

The velocity potential yields

$$V_r = \frac{\partial\Phi}{\partial r} = -\left(\frac{m}{2\pi}\right)\frac{1}{r} = -\frac{m}{2\pi r}$$

$$V_\theta = \frac{1}{r}\frac{\partial\Phi}{\partial\theta} = \frac{1}{r}\left(\frac{\Gamma}{2\pi}\right) = \frac{\Gamma}{2\pi r}$$

which we write in vector notation as

$$V = V_r\hat{e}_r + V_\theta\hat{e}_\theta = -\frac{m}{2\pi r}\hat{e}_r + \frac{\Gamma}{2\pi r}\hat{e}_\theta$$

We obtain the angle between V and the unit vector in the radial direction $(\hat{e}_r)$ as

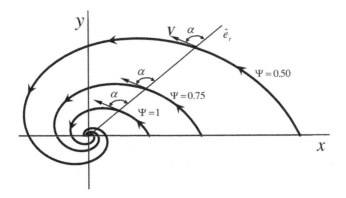

FIGURE 5.6   Streamlines in the given spiral vortex (Problem 5.4).

$$V \cdot \hat{e}_r = V \cos\alpha$$

$$\cos\alpha = \frac{V \cdot \hat{e}_r}{V} = \frac{-m}{\sqrt{m^2 + \Gamma^2}} = \frac{-1}{\sqrt{1 + \left(\dfrac{\Gamma}{m}\right)^2}}$$

Therefore, for the given values of m and Γ, we obtain $\alpha = 111.8°$, which is constant throughout the flow field.

PROBLEM 5.5:   FINDING STREAM FUNCTION FOR A GIVEN POTENTIAL FUNCTION

Consider an incompressible flow with velocity potential given by $\Phi = y^2 - x^2$. Does this function satisfy the Laplace equation? For $\Psi = 3$, sketch the streamlines with their flow directions.

SOLUTION FOR PROBLEM 5.5

We write the Laplace equation for the given potential function as

$$\frac{\partial^2 \Phi}{\partial x^2} + \frac{\partial^2 \Phi}{\partial y^2} = 0$$

$$-2 + 2 = 0$$

which shows that the given potential function satisfies the Laplace equation.
   We obtain V_x as

$$V_x = \frac{\partial \Psi}{\partial y} = \frac{\partial \Phi}{\partial x} = -2x$$

which upon integration yields

$$\Psi = -2xy + f(x)$$

Similarly, we obtain V_y as

$$V_y = -\frac{\partial \Psi}{\partial x} = 2y + f'(x) = \frac{\partial \Phi}{\partial y} = 2y$$

which yields $f'(x) = 0$ and which upon integration yields $f(x) = C$. Assuming $C = 0$, we obtain $\Psi = -2xy$, which, for $\Psi = 3$, gives the streamline equation as

$$y = -\frac{3}{2x}$$

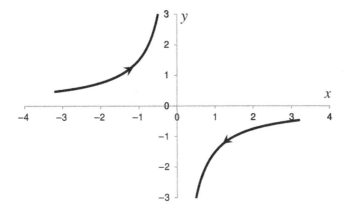

FIGURE 5.7 Streamlines with flow directions (Problem 5.5).

We have $V_x = -2x$ and $V_y = 2y = -3/x$ on each streamline. These velocities are negative for positive x and positive for negative x. Figure 5.7 shows two streamlines along with the flow direction.

PROBLEM 5.6: POTENTIAL FLOW OVER A SINUSOIDAL WAVY WALL

The following complex potential function represents the flow over a sinusoidally wavy wall:

$$F(z) = V_0 \left[z + y_0 e^{2i\pi x/\lambda} \right]$$

where $z = x + iy$, y_0 is the wave-amplitude, and λ is the wavelength. Find the stream function Ψ and the variation of y for the streamline $\Psi = 0$.

SOLUTION FOR PROBLEM 5.6

Let us expand the complex potential function in real and imaginary parts.

$$F(z) = V_0 \left[x + iy + y_0 \left(e^{2i\pi x/\lambda} \right) e^{-2\pi y/\lambda} \right]$$

$$F(z) = V_0 \left[x + iy + y_0 e^{-2\pi y/\lambda} \left\{ \cos\left(\frac{2\pi x}{\lambda} \right) \right\} + iy_0 e^{-2\pi y/\lambda} \left\{ \sin\left(\frac{2\pi x}{\lambda} \right) \right\} \right]$$

$$F(z) = \Phi + i\Psi = V_0 \left[x + y_0 e^{-2\pi y/\lambda} \left\{ \cos\left(\frac{2\pi x}{\lambda} \right) \right\} \right] + iV_0 \left[y + y_0 e^{-2\pi y/\lambda} \left\{ \sin\left(\frac{2\pi x}{\lambda} \right) \right\} \right]$$

from which we obtain the stream function

$$\Psi = V_0 \left[y + y_0 e^{-2\pi y/\lambda} \left\{ \sin\left(\frac{2\pi x}{\lambda} \right) \right\} \right]$$

which for $\Psi = 0$ yields

$$y + y_0 e^{-2\pi y/\lambda} \left\{ \sin\left(\frac{2\pi x}{\lambda}\right) \right\} = 0$$

$$y = -y_0 e^{-2\pi y/\lambda} \left\{ \sin\left(\frac{2\pi x}{\lambda}\right) \right\}$$

For $2\pi y/\lambda \ll 1$, we have $e^{-2\pi y/\lambda} \approx 1$, yielding

$$y = -y_0 \sin\left(\frac{2\pi x}{\lambda}\right)$$

NOMENCLATURE

a	Cylinder radius
C	Complex circulation; closed contour
$\hat{e}_r$	Unit vector in the radial direction
$\hat{e}_\theta$	Unit vector in the θ direction
F	Complex potential function; force
i	Imaginary number $\left(i = \sqrt{-1}\right)$
$\hat{k}$	Unit vector along z-axis
$d\mathbf{l}$	Differential length vector along a closed contour
m	Source strength
$\hat{n}$	Unit vector normal to $d\mathbf{l}$
p	Pressure; static pressure
Q	Volumetric flow rate
r	Coordinate r of cylindrical polar coordinate system
S	Stagnation point
U	Uniform velocity in x-coordinate direction
V	Uniform velocity in y-coordinate direction
$\mathbf{V}$	Velocity vector
W	Complex velocity
$\overline{W}$	Complex conjugate velocity
x	Cartesian coordinate x
y	Cartesian coordinate y
z	Complex coordinate $\left(z = x + iy\right)$
$\overline{z}$	Complex conjugate coordinate $\left(\overline{z} = x - iy\right)$

SUBSCRIPTS AND SUPERSCRIPTS

0	Total (stagnation)
b	Point on the Rankine-half body
r	Component in r coordinate direction
S	Stagnation point

x	Component in x-coordinate direction
y	Component in y-coordinate direction
z	Belonging to z-plane (physical plane)
θ	Component in θ coordinate direction
∞	Infinity (far away)

GREEK SYMBOLS

α	Angle between the velocity vector and the x-axis; angle of attack
Γ	Circulation; vortex strength
ε	Distance of source or sink from the origin
θ	Cylindrical polar coordinate θ
ρ	Density
Φ	Velocity potential function
Ψ	Stream function

REFERENCES

Sultanian, B.K. 2015. *Fluid Mechanics: An Intermediate Approach*. Boca Raton, FL: Taylor & Francis.

BIBLIOGRAPHY

Currie, I.G. 2013. *Fundamental Mechanics of Fluids*, 4th edition. Boca Raton, FL: CRC Press.
Kirchhoff, R.H. 1985. *Potential Flows: Computer Graphic Solutions*, 1st edition. New York: Marcel Dekker.
Milne-Thomson, L.M. 1968. *Theoretical Hydrodynamics*. New York: Dover Publications.
Zdravkovich, M.M. 1997. *Flow around Circular Cylinders Volume 1: Fundamentals*. Oxford: Oxford University Press.

6 Navier-Stokes Equations: Exact Solutions

REVIEW OF KEY CONCEPTS

Navier-Stokes equations are the time-dependent partial differential equations that govern the force and linear momentum balance in a Newtonian fluid flow, be it laminar or turbulent, incompressible or compressible. The major difficulty in solving the N-S equations lies in the nonlinear inertia (convection) terms, which involve the product of velocity with its spatial gradient. Only for a limited number of laminar flow problems, the N-S equations are known to have analytical solutions. These equations are invariably solved along with the continuity equation—the first law of fluid flow. Sultanian (2015) presents a few exact solutions for fully developed two-dimensional laminar flows. Additional solutions may, for example, be found in Bird, Stewart, and Lightfoot (2006) and Riley and Drazin (2006).

Detailed derivations of the Navier-Stokes equations may be found in many graduate-level textbooks, for example, Sultanian (2015). We present here key concepts on surface shear stresses and gravitational body force on a fluid element, including the summary of the continuity and N-S equations in both Cartesian and cylindrical polar coordinate systems for an incompressible flow with constant viscosity. We use these equations in the solutions of various problems presented in this chapter.

SURFACE FORCES DUE TO STRESSES

Figure 6.1 shows all nine components of the 3×3 stress tensor τ_{ij} on an infinitesimal fluid element in a flow. The first subscript of τ_{ij} denotes the direction of the normal to the element face on which the stress is acting, and the second subscript indicates the stress direction. For example, τ_{xy} acts in the y-direction on the element face whose normal is in the x-direction. For $i \neq j$, τ_{ij} represents shear stresses; for $i = j$, τ_{ii} represents normal stresses. The stress tensor being symmetric $\left(\tau_{ij} = \tau_{ji} \right)$, we only have six different stresses—three normal and three shear stresses. For positive τ_{ij}, i and j must be either both positive or both negative relative to the coordinate system used. At a wall-fluid interface, according to this convention, τ_{ij} for each face has the same sign—their face-normal vectors are in the opposite directions, and so are the forces they produce.

Figure 6.2 shows an infinitesimal control volume measuring δx, δy, and δz along the Cartesian-coordinate directions. Included here are the variations of τ_{yx}, τ_{yy}, and τ_{yz} in the y-direction from the point $\left(x, y, z \right)$ located at the center of the control volume. Let us evaluate the contributions of these stresses (on the positive

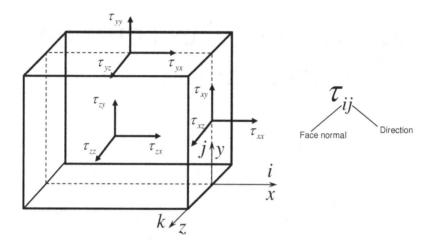

FIGURE 6.1   Surface forces acting on a fluid element.

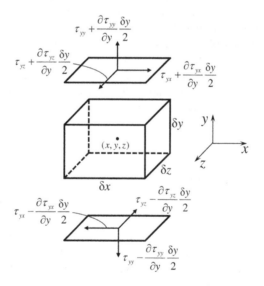

FIGURE 6.2   Gradients of stresses at a point in fluid flow.

and negative y faces) to the total surface force acting on the control volume in each coordinate direction. For τ_{yx}, for example, we write

$$\delta F_{s\_yx} = \left(\tau_{yx} + \frac{\partial \tau_{yx}}{\partial y} \frac{\delta y}{2} \right) \delta x \delta z - \left(\tau_{yx} - \frac{\partial \tau_{yx}}{\partial y} \frac{\delta y}{2} \right) \delta x \delta z$$

$$\frac{\delta F_{s\_yx}}{\delta x \delta y \delta z} = \frac{\partial \tau_{yx}}{\partial y}$$

(6.1)

which shows that $\partial\tau_{yx}/\partial y$ contributes to the x-direction surface force (per unit volume) acting on the control volume. Considering similar variations of stresses in x- and z-directions, we obtain $\partial\tau_{xx}/\partial x$ and $\partial\tau_{zx}/\partial z$ as additional contributions to this force. Thus, we can write the net surface force per unit volume acting on the control volume in each coordinate direction as

$$\delta f_{sx} = \frac{\partial\tau_{xx}}{\partial x} + \frac{\partial\tau_{yx}}{\partial y} + \frac{\partial\tau_{zx}}{\partial z} \qquad (6.2)$$

$$\delta f_{sy} = \frac{\partial\tau_{xy}}{\partial x} + \frac{\partial\tau_{yy}}{\partial y} + \frac{\partial\tau_{zy}}{\partial z} \qquad (6.3)$$

$$\delta f_{sz} = \frac{\partial\tau_{xz}}{\partial x} + \frac{\partial\tau_{yz}}{\partial y} + \frac{\partial\tau_{zz}}{\partial z} \qquad (6.4)$$

Equations 6.2–6.4 show that not the stresses but their gradients act on the infinitesimal control volume as surface forces per unit volume.

BODY FORCE DUE TO GRAVITY

We assume that the body force due to gravity acts vertically. Figure 6.3 shows an arbitrary orientation of the Cartesian coordinate axes, and the height h for each point in the flow is measured from a fixed datum. We write the x component of the gravitational body force acting on the control volume as

$$\delta F_{bx} = -\rho g\left(\delta x\delta y\delta z\right)\sin\theta$$

$$\frac{\delta F_{bx}}{\delta x\delta y\delta z} = -\rho g\sin\theta$$

which with

$$\sin\theta = \frac{\left(h + \dfrac{\partial h}{\partial x}\delta x\right) - h}{\delta x} = \frac{\partial h}{\partial x}$$

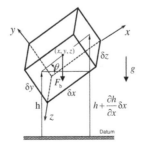

FIGURE 6.3   Body force due to gravity.

becomes

$$\frac{\delta F_{bx}}{\delta x \delta y \delta z} = \delta f_{bx} = -\rho g \frac{\partial h}{\partial x}$$

where δf_{bx} is the x-direction body force per unit volume. We can similarly obtain body forces due to gravity in y- and z-directions, writing all three in the compact tensor notation as

$$\delta f_{bi} = -\rho g \frac{\partial h}{\partial x_i} \qquad i = 1, 2, 3 \tag{6.5}$$

SUMMARY OF NAVIER-STOKES AND CONTINUITY EQUATIONS— CONSTANT VISCOSITY AND CONSTANT DENSITY

Cartesian Coordinates

Continuity Equation

$$\frac{\partial V_x}{\partial x} + \frac{\partial V_y}{\partial y} + \frac{\partial V_z}{\partial z} = 0 \tag{6.6}$$

Navier-Stokes Equations

$$\rho \frac{DV_x}{Dt} = -\frac{\partial \hat{p}}{\partial x} + \mu \nabla^2 V_x \tag{6.7}$$

$$\rho \frac{DV_y}{Dt} = -\frac{\partial \hat{p}}{\partial y} + \mu \nabla^2 V_y \tag{6.8}$$

$$\rho \frac{DV_z}{Dt} = -\frac{\partial \hat{p}}{\partial z} + \mu \nabla^2 V_z \tag{6.9}$$

where $\hat{p} = p + h \partial g$,

$$\frac{D}{Dt} = \frac{\partial}{\partial t} + V_x \frac{\partial}{\partial x} + V_y \frac{\partial}{\partial y} + V_z \frac{\partial}{\partial z}$$

and

$$\nabla^2 = \frac{\partial^2}{\partial x^2} + \frac{\partial^2}{\partial y^2} + \frac{\partial^2}{\partial z^2}$$

Cylindrical Polar Coordinates

Continuity Equation

$$\frac{1}{r}\frac{\partial}{\partial r}(rV_r)+\frac{1}{r}\frac{\partial V_\theta}{\partial \theta}+\frac{\partial V_x}{\partial x}=0 \tag{6.10}$$

Navier-Stokes Equations

$$\rho\left(\frac{DV_r}{Dt}-\frac{V_\theta^2}{r}\right)=-\frac{\partial \hat{p}}{\partial r}+\mu\left(\nabla^2 V_r-\frac{V_r}{r^2}-\frac{2}{r^2}\frac{\partial V_\theta}{\partial \theta}\right) \tag{6.11}$$

$$\rho\left(\frac{DV_\theta}{Dt}+\frac{V_rV_\theta}{r}\right)=-\frac{1}{r}\frac{\partial \hat{p}}{\partial \theta}+\mu\left(\nabla^2 V_\theta-\frac{V_\theta}{r^2}+\frac{2}{r^2}\frac{\partial V_r}{\partial \theta}\right) \tag{6.12}$$

$$\rho\frac{DV_x}{Dt}=-\frac{\partial \hat{p}}{\partial x}+\mu\nabla^2 V_x \tag{6.13}$$

where

$$\frac{D}{Dt}=\frac{\partial}{\partial t}+V_r\frac{\partial}{\partial r}+\frac{V_\theta}{r}\frac{\partial}{\partial \theta}+V_x\frac{\partial}{\partial x}$$

and

$$\nabla^2=\frac{\partial^2}{\partial r^2}+\frac{1}{r}\frac{\partial}{\partial r}+\frac{1}{r^2}\frac{\partial^2}{\partial \theta^2}+\frac{\partial^2}{\partial x^2}$$

PROBLEM 6.1:   A FLUID FILM FALLING UNDER GRAVITY ALONG A VERTICAL FLAT PLATE

As shown in Figure 6.4, a thin fluid film falls under gravity along a flat plate. The static pressure is constant throughout the flow, and there is no wind shear at the edge of the film. Modeling the fluid film as a two-dimensional, incompressible, laminar flow, find:

a. the *w*-velocity distribution,
b. the ratio of the maximum and average velocities,
c. the distribution of shear stress in the film and its magnitude and direction at the wall, and
d. the wall shear stress using the force-momentum balance on the control volume shown in the figure.

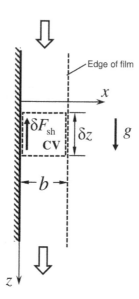

FIGURE 6.4   A fluid film falling under gravity along a flat plate (Problem 6.1).

SOLUTION FOR PROBLEM 6.1

In this flow along the vertical plate, the body force due to gravity acts downward (positive z-direction). In the absence of wind shear at the edge of the film, the wall shear force acts upward (negative z-direction). The flow geometry favors the use of Cartesian coordinates.

ASSUMPTIONS

In this problem, we make the following assumptions:

- both ρ and μ are constant,
- the flow is laminar and steady $\left(\partial/\partial t = 0\right)$ with static pressure gradient $\partial p/\partial x = 0$,
- the flow is fully developed in the flow direction $\left(\partial/\partial z = 0\right)$, and
- the flow is two-dimensional $\left(v = 0\right)$.

Because the flow is fully developed in the z-direction, w can be a function of x only, that is, $w = w(x)$.

CONTINUITY EQUATION

Under the assumption of steady, incompressible flow, the continuity equation in Cartesian coordinates reduces to

$$\frac{\partial u}{\partial x} + \frac{\partial v}{\partial y} + \frac{\partial w}{\partial z} = 0$$

which, for the present two-dimensional, fully developed flow, further reduces to

$$\frac{\partial u}{\partial x} = \frac{du}{dx} = 0$$

Upon integration, this equation yields

$$u = C$$

where C is a constant. As $u = 0$ at $x = 0$, we conclude that $u = 0$ everywhere in the film.

PRESSURE GRADIENT UNDER GRAVITATIONAL BODY FORCE

$$\hat{p} = p + \rho g h$$

$$\frac{d\hat{p}}{dz} = \frac{d}{dz}(p + \rho g h) = \frac{dp}{dz} + \rho g \frac{dh}{dz} = \frac{dp}{dz} - \rho g$$

where we have used $dh/dz = -1$. As $dp/dz = 0$, this equation yields

$$\frac{d\hat{p}}{dz} = -\rho g$$

REDUCED NAVIER-STOKES EQUATION

Under our assumptions for this problem, the Navier-Stokes equations reduce to the following governing equation for the velocity component in the z-direction.

$$\mu \frac{d^2 w}{dx^2} = \frac{d\hat{p}}{dz} = -\rho g$$

$$\frac{d^2 w}{dx^2} = -\frac{g}{v}$$

which is a second-order nonhomogeneous linear ordinary differential equation (ODE) where the kinematic viscosity $v = \mu/\rho$.

BOUNDARY CONDITIONS

For a unique solution, the second-order ODE requires two boundary conditions. These boundary conditions are as follows:

$$w = 0 \quad @ \, x = 0$$

and

$$dw/dx = 0 \, @ \, x = b$$

SOLUTIONS AND DISCUSSION

(a) Distribution of Velocity w

Integrating the ODE and applying the boundary conditions yield

$$w = -\frac{g}{2v}x^2 + \frac{bg}{v}x$$

which yields the maximum velocity $w_{max} = bg/2v$ at $x = b$. Using $\tilde{w} = w/w_{max}$ and $\xi = x/b$, we express the $\tilde{w}$-velocity distribution as

$$\tilde{w} = 2\xi - \xi^2$$

As shown in Figure 6.5, this velocity distribution is semi-parabolic for $0 \leq \xi = \leq 1$. Treating the edge of the film as the line of symmetry, the solution also represents the velocity profile of the flow falling under gravity between two vertical plates separated by $2b$ with no imposed static pressure gradient. The dash line in this figure represents the second half of the parabola.

(b) Ratio of Maximum-to-Average Velocity

We calculate the average velocity $\bar{w}$ as

$$\bar{w} = \frac{\int_0^b \left(-\frac{g}{2v}x^2 + \frac{bg}{v}x \right)dx}{b} = \frac{gb^2}{3v}$$

giving

$$\frac{w_{max}}{\bar{w}} = \frac{3}{2}$$

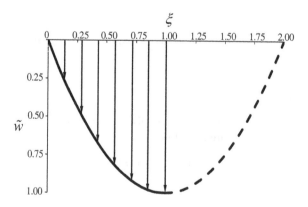

FIGURE 6.5 Velocity profile of the fluid film flow along a vertical plate (Problem 6.1).

(c) Shear Stress Distribution and Wall Shear Stress

We evaluate the local shear stress in the flow as

$$\tau_{xz} = \mu\frac{dw}{dx} = \mu\frac{d}{dx}\left(-\frac{g}{2v}x^2 + \frac{bg}{v}x\right) = -\rho gx + \rho gb$$

which shows that the shear stress becomes zero at $x = b$. At the wall $(x = 0)$, we obtain $(\tau_{xz})_{x=0} = \rho gb$, which is positive. As shown in Figure 6.1, for a positive shear stress at a point, both the face normal and shear stress directions should be either positive or negative relative to the chosen coordinate axes. As the fluid face normal at $x = 0$ is in the negative x-direction, the positive value of $(\tau_{xz})_{x=0} = \rho gb$ implies that it is along the negative z-direction (vertically up). As the shear force acting on the wall must be equal and opposite to the shear force acting on the fluid surface in contact, we obtain $\tau_w = \rho gb$ in the positive z-direction (vertically down).

(d) Wall Shear Stress from Control Volume Analysis

For this fully developed flow, the net efflux of z momentum flux over the control volume shown in Figure 6.4 is zero, and as a result, the shear force balances the gravitational body force, yielding

$$\tau_w \delta z\,\delta y = \rho gb\,\delta z\,\delta y$$

$$\tau_w = \rho gb$$

which is identical to the value from the shear stress distribution obtained in (c). This example shows the strength of the control volume analysis if one is interested in knowing only the shear stress at the wall.

PROBLEM 6.2:   A POROUS PLATE MOVING AT CONSTANT VELOCITY THROUGH A STAGNANT FLUID

Consider a steady fully developed incompressible laminar flow over a porous plate of infinite length, as shown in Figure 6.6. The plate moves in the x-direction at a constant velocity V_{x0}. The fluid is removed through the porous plate at a constant velocity V_{y0}. Clearly stating all relevant assumptions, find:

a. the velocity profile over the plate and
b. the shear stress at the wall, validating it against the result from a control volume analysis.

SOLUTION FOR PROBLEM 6.2

In this case, a porous plate of infinite dimension is pulled along the x-direction at a constant velocity V_{x0}. Due to no-slip condition, the fluid in contact with the plate also moves with the same velocity. At a large distance from the plate in the y direction, the effect of viscosity becomes negligible, and the fluid has zero velocity in

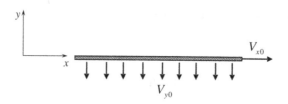

FIGURE 6.6 A porous plate moving at constant velocity through a stagnant fluid (Problem 6.2).

the x-direction. The fluid flows through the porous plate in the negative y-direction at a uniform velocity V_{y0}, which must become uniform in the entire flow field to satisfy the continuity equation. The flow geometry of this example favors the use of Cartesian coordinates.

Assumptions

In this problem, we make the following assumptions:

- both ρ and μ are constant,
- the flow is laminar and steady $(\partial/\partial t = 0)$ with zero static pressure gradient $(\partial p/\partial x = 0)$,
- the flow is fully developed in the flow direction $(\partial/\partial x = 0)$, and
- the flow is two-dimensional $(\partial/\partial z = 0 \text{ and } V_z = 0)$.

Because the flow is fully developed in the x-direction, V_x is a function of y only—i.e., $V_x = V_x(y)$.

Continuity Equation

Under the foregoing assumptions, the continuity equation

$$\frac{\partial V_x}{\partial x} + \frac{\partial V_y}{\partial y} + \frac{\partial V_z}{\partial z} = 0$$

reduces to

$$\frac{\partial V_y}{\partial y} = 0$$

whose integration yields

$$V_y = C_1$$

where C_1 is the integration constant. Because $V_y = -V_{y0}$ at the wall, we conclude that $V_y = -V_{y0}$ everywhere in the flow.

REDUCED NAVIER-STOKES EQUATIONS

The x-momentum equation reduces to

$$-\rho V_{y0}\frac{dV_x}{dy} = \mu\frac{d^2V_x}{dy^2}$$

$$\frac{d^2V_x}{dy^2} + \frac{V_{y0}}{\nu}\frac{dV_x}{dy} = 0$$

where the kinematic viscosity $\nu = \mu/\rho$. We solve this homogenous second-order linear ODE for the following boundary conditions.

BOUNDARY CONDITIONS

The boundary conditions in this case are $V_x = V_{x0}$ at $y = 0$ and $V_x = 0$ at $y = \infty$.

RESULTS AND DISCUSSION

(a) Velocity Profile $V_x = V_x(y)$

Integrating the ODE once yields

$$\frac{dV_x}{dy} + \frac{V_{y0}V_x}{\nu} = C_2$$

where C_2 is the integration constant. We can write the general solution of this nonhomogeneous first-order ODE as

$$V_x = C_3\,V_{xh} + V_{xp}$$

where V_{xh} is the solution for the corresponding homogeneous ODE and V_{xp} is a particular solution, obtained as

$$V_{xh} = e^{-\frac{V_{y0}y}{\nu}}$$

and

$$V_{xp} = \frac{\nu}{V_{y0}}C_2$$

We can write the general solution of the ODE as

$$V_x = C_3V_{xh} + V_{xp} = C_3e^{-\frac{V_{y0}y}{\nu}} + \frac{\nu}{V_{y0}}C_2$$

Using the boundary condition $V_x = 0$ at $y = \infty$ yields $C_2 = 0$. Using the boundary condition $V_x = V_{x0}$ at $y = 0$ yields $C_3 = V_{x0}$, finally giving the required velocity profile as

$$V_x = V_{x0}e^{-\frac{V_{y0}y}{\nu}}$$

(b) Wall Shear Stress

From the foregoing velocity distribution, we obtain the shear stress distribution as

$$\tau_{yx} = \mu\frac{dV_x}{dy} = \mu\frac{d}{dy}\left(V_{x0}e^{-\frac{V_{y0}y}{\nu}}\right) = -\rho V_{y0}V_{x0}e^{-\frac{V_{y0}y}{\nu}}$$

from which we obtain the wall shear stress as

$$\tau_w = \tau_{yx\_(y=0)} = -\rho V_{y0}V_{x0}e^{-\frac{V_{y0}y}{\nu}}\bigg|_{y=0} = -\rho V_{x0}\,V_{y0}$$

The wall has its face normal in the positive y-direction. The negative sign on the right-hand side of this result implies that τ_w acts in the positive x-direction.

On the fluid control volume, shown in Figure 6.7, we obtain the total shear force acting in the positive x-direction as

$$F_{sh} = (-\tau_w)L = \rho V_{x0}V_{y0}L$$

Because the flow is fully developed in the x-direction, for a unit length perpendicular to the plane of the control volume shown in this figure, the x-momentum flux entering AB equals that exiting CD. Considering the remaining x-momentum fluxes through AC and BD, the force-momentum balance on the control volume ABDC in the x-direction yields

$$F_{sh} = \left(\rho V_{x0}V_{y0}L\right)_{BD} - (0)_{AC} = \rho\,V_{x0}V_{y0}\,L$$

which is identical to the value we obtained in the foregoing from the exact solution of the N-S equation.

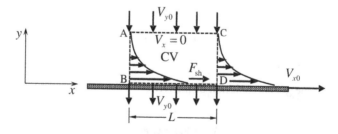

FIGURE 6.7   Control volume analysis of the flow over a moving porous plate (Problem 6.2).

PROBLEM 6.3:   ANNULAR FLOW BETWEEN THE PIPE
AND THE WIRE MOVING AT CONSTANT
VELOCITY ALONG THE PIPE AXIS

Consider a fully developed, steady, incompressible, laminar flow induced when a wire of radius R_1 is drawn at a constant velocity W down the centerline of a pipe of radius R_2, as shown in Figure 6.8. Develop the appropriate form of the Navier-Stokes equations governing the flow with no applied pressure gradient. Find:

a. an expression for the velocity distribution $V_z(r)$ in the annular region between the pipe and the wire and
b. an expression for the force needed to draw a wire of radius R_1 and length L in this flow.

SOLUTION FOR PROBLEM 6.3

In this case, a wire of radius R_1 and infinite length is pulled in the z-direction at a constant velocity W. Due to the no-slip condition, the fluid in contact with the wire moves with the same velocity. This sets the flow in the annular region between the pipe and the wire even with zero imposed pressure gradient. The fluid velocity is zero at the pipe wall. The flow geometry of this case favors the use of cylindrical polar coordinates.

ASSUMPTIONS

In this problem, we make the following assumptions:

- both ρ and μ are constant,
- the flow is laminar and steady $(\partial/\partial t = 0)$ with zero static pressure gradient $(\partial p/\partial z = 0)$,
- the flow is fully developed in the flow direction $(\partial/\partial z = 0)$, and
- the flow is axisymmetric $(V_\theta = \partial/\partial\theta = 0)$.

As the flow is fully developed in the z direction, V_z can be a function of r only, that is, $V_z = V_z(r)$.

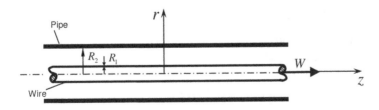

CONTINUITY EQUATION

For the steady incompressible flow, the continuity equation in cylindrical polar coordinates reads

$$\frac{1}{r}\frac{\partial}{\partial r}(rV_r)+\frac{1}{r}\frac{\partial V_\theta}{\partial \theta}+\frac{\partial V_z}{\partial z}=0$$

For $\dfrac{\partial V_z}{\partial z}=\dfrac{\partial V_\theta}{\partial \theta}=0$ from our assumptions, this equation reduces to

$$\frac{\partial(rV_r)}{\partial r}=0$$

whose integration yields $rV_r=C_1$, where C_1 is the integration constant. As $V_r=0$ both at the pipe wall $(r=R_2)$ and at the wire surface $(r=R_1)$, it must be zero everywhere in the flow, giving $C_1=0$.

REDUCED NAVIER-STOKES EQUATIONS

In the absence of static pressure gradient in the flow direction, the Navier-Stokes equations reduce to the following governing equation for the velocity component v_z

$$\frac{1}{r}\frac{d}{dr}\left(r\frac{dV_z}{dr}\right)=0$$

which is a homogeneous second-order ODE.

BOUNDARY CONDITIONS

Two boundary conditions needed for a unique solution of the ODE in this case are

$$V_z=W \quad @r=R_1$$

and

$$V_z=0 \quad @r=R_2$$

RESULTS AND DISCUSSION

(a) Velocity Distribution

Integrating the ODE once yields

$$V_z=C_1\ln r+C_2$$

where C_1 and C_2 are the integration constants to be determined from the boundary conditions. Applying the boundary condition at the wire surface $(r=R_1)$, we obtain

$$W = C_1 \ln R_1 + C_2$$

Similarly, applying the boundary condition at the pipe wall $(r = R_2)$ yields

$$0 = C_1 \ln R_2 + C_2$$

These two equations yield

$$C_1 = -\frac{W}{\ln(R_2/R_1)}$$

and

$$C_2 = \frac{W \ln R_2}{\ln(R_2/R_1)}$$

Thus, we finally obtain the radial distribution of the axial velocity in the annulus between the pipe and the wire as

$$V_z = -\frac{W \ln(r/R_2)}{\ln(R_2/R_1)}$$

(b) Force Necessary to Draw a Wire of Radius R_1 and length L

We obtain the shear stress distribution from the velocity distribution in (a) as

$$\tau_{rz} = \mu \frac{dV_z}{dr} = -\frac{1}{r}\left(\frac{\mu W}{\ln(R_2/R_1)}\right)$$

which shows that the shear stress on the fluid surface in contact with wire surface is negative. As the fluid surface normal is in the negative r-direction, the direction of the shear stress on the fluid surface must be in the positive z-direction. Accordingly, the direction of the shear stress on the wire surface must be in the negative z-direction, giving

$$\tau_{wire} = -\frac{1}{R_1}\left(\frac{\mu W}{\ln(R_2/R_1)}\right)$$

Thus, the required force in the positive z-direction to draw a wire of length L is calculated as

$$F_{wire} = -\tau_{wire}(2\pi R_1 L)$$

$$F_{wire} = \frac{2\pi W \mu L}{\ln(R_2/R_1)}$$

This result shows that the force needed to pull the wire is directly proportional to the wire velocity and the fluid viscosity. The result is obviously not valid for $R_1 = 0$ in which case we will have no wire. The result is also not valid for $R_1 = R_2$, which implies zero gap between the wire and the pipe with no fluid flow. For $0 < R_1 < R_2$, F_{wire} increases with the wire radius, initially at a low rate of increase for a thin wire but becoming very large as the wire radius approaches the pipe radius (thin annulus).

PROBLEM 6.4:   FLOW BETWEEN INFINITE PARALLEL PLATES

Figure 6.9 shows a two-dimensional (*x-y* plane) fully developed incompressible laminar flow between two infinite parallel plates separated by $2H$. Flow properties do not vary in the *z*-direction. The flow between the plates is entirely driven by an imposed static pressure gradient. Find:

   a. the fully developed velocity $V_x(y)$,
   b. the average velocity and its relation to the maximum velocity,
   c. the shear stress distribution between the plates,
   d. the expression for Darcy friction factor in terms of Reynolds number, and
   e. the expression to compute the static pressure loss over length L in the flow direction.

SOLUTION FOR PROBLEM 6.4

Assumptions

In this problem, we make the following assumptions:

   • both ρ and μ are constant,
   • the flow is laminar and steady $(\partial/\partial t = 0)$,
   • the flow is fully developed in the flow direction $(\partial/\partial x = 0)$, and
   • the flow is two-dimensional $(\partial/\partial z = 0 \text{ and } V_z = 0)$.

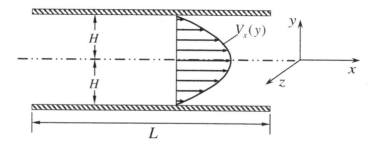

FIGURE 6.9   Flow between infinite parallel plates (Problem 6.4).

CONTINUITY EQUATION

Under the foregoing assumptions, the continuity equation

$$\frac{\partial V_x}{\partial x} + \frac{\partial V_y}{\partial y} + \frac{\partial V_z}{\partial z} = 0$$

reduces to

$$\frac{\partial V_y}{\partial y} = 0$$

whose integration yields

$$V_y = C_1$$

where C_1 is the integration constant. As $V_y = 0$ at the wall, it must be zero everywhere in the flow, giving $C_1 = 0$.

REDUCED NAVIER-STOKES EQUATIONS

Under the assumptions made in this case, the z-momentum equation is identically satisfied. Because $V_y = 0$, the y-momentum equation yields

$$\frac{\partial p}{\partial y} = 0$$

That is, the static pressure remains uniform in the y-direction, keeping the streamlines parallel to the x-axis. Finally, the x-momentum equation reduces to

$$\mu \frac{d^2 V_x}{dx^2} = \frac{dp}{dx}$$

Because the left-hand side of this equation is a function of y only and the right-hand side is a function of x only, each must be a constant, giving

$$\mu \frac{d^2 V_x}{dy^2} = \frac{dp}{dx} = C_2$$

which is a second order ODE, which we solve for the following boundary conditions.

BOUNDARY CONDITIONS

The no-slip condition for each plate yields $V_x = 0$ at $y = \pm H$.

Results and Discussion

(a) Fully Developed Velocity Profile: $V_x(y)$

Integrating the ODE twice yields the general solution

$$V_x = \frac{1}{2\mu}\frac{dp}{dx}y^2 + C_3 y + C_4$$

where the integration constants C_3 and C_4 are obtained from the boundary conditions as

$$C_3 = 0 \text{ and } C_4 = -\frac{1}{2\mu}\frac{dp}{dx}H^2$$

giving

$$V_x = -\frac{1}{2\mu}\frac{dp}{dx}\left(H^2 - y^2\right)$$

with the maximum value of V_x occurring at $y = 0$ (symmetry plane). For positive V_x, dp/dx must be negative, that is, the static pressure decreases in the flow direction to overcome the shearing action of viscosity. With

$$V_{x\_max} = -\frac{H^2}{2\mu}\frac{dp}{dx}$$

we express the fully developed parabolic velocity profile between the parallel plates as

$$\frac{V_x}{V_{x\_max}} = 1 - \left(\frac{y}{H}\right)^2$$

(b) Average Velocity and Its Relation to the Maximum Velocity

$$\overline{V}_x = \frac{V_{x\_max}\displaystyle\int_{-H}^{H}\left(\frac{V_x}{V_{x\_max}}\right)dy}{2H} = \frac{V_{x\_max}}{2}\int_{-1}^{1}\left\{1 - \left(\frac{y}{H}\right)^2\right\}d\left(\frac{y}{H}\right)$$

$$\overline{V}_x = \frac{2}{3}V_{x\_max}$$

which we can also write as

$$\overline{V}_x = -\frac{H^2}{3\mu}\frac{dp}{dx}$$

(c) Shear Stress Distribution between the Plates

$$\tau_{yx} = \mu \frac{dV_x}{dy}$$

$$\tau_{yx} = -\frac{2\mu V_{x\_max} y}{H^2} = -\frac{3\mu \overline{V}_x y}{H^2}$$

which shows that the shear stress varies linearly from 0 at $y = 0$ to the maximum absolute value at $y = \pm H$ (walls). We obtain the shear stress at $y = H$ as

$$\tau_{yx\_(y=H)} = -\frac{3\overline{V}_x \mu}{H}$$

which is negative, being in negative x-direction on the face whose normal is in the positive y-direction. Similarly, at $y = -H$, we obtain

$$\tau_{yx\_(y=-H)} = \frac{3\overline{V}_x \mu}{H}$$

which is positive, being in the negative x-direction on the face whose normal is in the negative y-direction.

(d) Darcy Friction Factor

Rearranging the relation for the average velocity obtained in the foregoing, we write

$$-\frac{dp}{dx} = \frac{3\mu \overline{V}_x}{H^2}$$

For the flow through the gap of $2H$ between the plates, the hydraulic mean diameter works out to be $4H$. Defining the Reynolds number for this flow as

$$Re = \frac{\rho \overline{V}_x (4H)}{\mu}$$

we express the static pressure gradient as

$$-\frac{dp}{dx} = \frac{96}{Re} \left(\frac{1}{4H} \right) \left(\frac{1}{2} \rho \overline{V}_x^2 \right)$$

which yields the Darcy friction factor as

$$f = \frac{96}{Re}$$

(e) Pressure Loss over Length L

Integrating the pressure gradient expression over L yields the required pressure loss

$$\Delta p_{loss} = \frac{96}{Re}\frac{L}{4H}\left(\frac{1}{2}\rho \bar{V}_x^{\,2}\right)$$

PROBLEM 6.5: HAGEN-POISEUILLE FLOW

The Hagen-Poiseuille flow, shown in Figure 6.10, is a two-dimensional fully developed incompressible laminar flow in a circular pipe of radius R. This flow is a cylindrical analog to the plane flow between two parallel plates we considered in Problem 6.4. Using the cylindrical polar coordinate system shown in the figure, find:

    a. the fully developed velocity $V_x(r)$,
    b. the average velocity and its relation to the maximum velocity,
    c. the shear stress distribution,
    d. shear coefficient in terms of Reynolds number, and
    e. the expression to compute the static pressure loss over length L and the Darcy friction factor in terms of Reynolds number.

SOLUTION FOR PROBLEM 6.5

Assumptions

In this example, we make the following assumptions:

- both ρ and μ are constant,
- the flow is laminar and steady $(\partial/\partial t = 0)$,
- the flow is fully developed in the flow direction $(\partial/\partial x = 0)$, and
- the flow is axisymmetric $(\partial/\partial\theta = 0)$ and $V_\theta = 0$.

Continuity Equation

Under the foregoing assumptions, the continuity equation

$$\frac{1}{r}\frac{\partial}{\partial r}(rV_r) + \frac{1}{r}\frac{\partial V_\theta}{\partial\theta} + \frac{\partial V_x}{\partial x} = 0$$

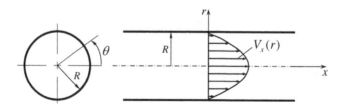

FIGURE 6.10   Hagen-Poiseuille flow (Problem 6.5).

reduces to

$$\frac{\partial(rV_r)}{\partial r} = 0$$

whose integration yields $rV_r = C_1$, where C_1 is the integration constant. As $V_r = 0$ at the wall $(r = R)$, it must be zero everywhere in the flow with $C_1 = 0$.

REDUCED NAVIER-STOKES EQUATIONS

With $\partial/\partial\theta = V_\theta = 0$, the momentum equation in the θ direction is identically satisfied. With $V_r = 0$, the r-momentum equation reduces to

$$\frac{\partial p}{\partial r} = 0$$

which implies that p is a function of x (flow direction) only.

$$\left(\frac{\partial^2 V_x}{\partial r^2} + \frac{1}{r}\frac{\partial V_x}{\partial r} \right) = \frac{1}{\mu}\frac{dp}{dx}$$

Because the left-hand side of this equation is a function of r only and the right-hand side a function of x only, each side must be equal to a constant, giving

$$\frac{1}{r}\frac{d}{dr}\left(r\frac{dV_x}{dr} \right) = \frac{1}{\mu}\frac{dp}{dx} = C_2$$

which is a second-order ODE, which we solve for the following boundary conditions.

BOUNDARY CONDITIONS

The no-slip condition at the pipe wall yields $V_x = 0$ at $r = R$. From the flow symmetry about the pipe axis, we obtain $dV_x/dr = 0$ at $r = 0$.

RESULTS AND DISCUSSION

(a) Velocity Profile $V_x(r)$

Integrating the ODE once yields

$$\frac{dV_x}{dr} = \left(\frac{1}{2\mu}\frac{dp}{dx} \right)r + C_3$$

where C_3 is the integration constant, which becomes zero from the boundary condition $dV_x/dr = 0$ at $r = 0$, giving

$$\frac{dV_x}{dr} = \left(\frac{1}{2\mu}\frac{dp}{dx} \right)r$$

whose integration yields

$$V_x = \left(\frac{1}{4\mu}\frac{dp}{dx}\right)r^2 + C_4$$

where C_4 is another integration constant. Applying the boundary condition $V_x = 0$ at $r = R$ to this equation, we obtain

$$C_4 = -\left(\frac{1}{4\mu}\frac{dp}{dx}\right)R^2$$

$$V_x = \left(-\frac{1}{4\mu}\frac{dp}{dx}\right)\left(R^2 - r^2\right)$$

which is the required solution for $V_x(r)$, showing that the radial variation of the axial velocity is parabolic in a fully developed laminar pipe flow. In this equation, for positive V_x, dp/dx must be negative—i.e., the static pressure decreases in the flow direction.

The maximum velocity at the pipe axis $(r = 0)$ becomes

$$V_{x\_max} = -\frac{R^2}{4\mu}\frac{dp}{dx}$$

giving

$$\frac{V_x}{V_{x\_max}} = 1 - \left(\frac{r}{R}\right)^2$$

(b) Average Velocity and Its Relation to the Maximum Velocity

As the flow is steady, the mass flow rate at each cross-section in the pipe (including the entrance region not shown in Figure 6.10) must remain constant. We compute this flow rate as

$$\dot{m} = \int_A \rho V_x dA = \rho V_{x\_max} \int_0^R \left(1 - \frac{r^2}{R^2}\right)2\pi r dr$$

where A is the flow area. This integration yields

$$\dot{m} = \pi R^2 \rho \bar{V}_x = \frac{\pi R^2 \rho V_{x\_max}}{2}$$

$$\bar{V}_x = \frac{V_{x\_max}}{2} = -\frac{R^2}{8\mu}\frac{dp}{dx}$$

where $\bar{V}_x$ is the section-average velocity.

We can also write

$$\dot{m} = -\frac{\pi R^4 \rho}{8\mu}\left(\frac{dp}{dx}\right)$$

which reveals that, in a Hagen-Poiseuille flow, the mass flow rate is linearly dependent on the static pressure gradient.

(c) Shear Stress Distribution

$$\tau_{rx} = \mu\frac{dV_x}{dr} = -\frac{2\mu V_{x\_max}r}{R^2} = -\frac{4\mu\bar{V}_x r}{R^2}$$

which shows that the shear stress varies linearly from zero at the pipe axis to its maximum absolute value at the wall $(r = R)$. The negative sign is consistent with our convention that the shear stress is in the negative x-direction on the face whose normal is in the positive r-direction and vice versa.

(d) Shear Coefficient

We obtain the wall shear stress as

$$\tau_w = \tau_{rx\_(r=R)} = -\frac{4\mu\bar{V}_x}{R}$$

Because the normal to the pipe wall in contact with the fluid is in the negative r direction, the negative value of τ_w indicates that it acts in the positive x-direction. We can write the magnitude of τ_w as

$$|\tau_w| = \frac{16}{Re}\left(\frac{1}{2}\rho\bar{V}_x^2\right) = C_f\left(\frac{1}{2}\rho\bar{V}_x^2\right)$$

where $Re = 2\rho\bar{V}_x R/\mu$ is the flow Reynolds number in the pipe and C_f is the shear coefficient for which this equation yields

$$C_f = \frac{16}{Re}$$

(e) Static Pressure Loss and Darcy Friction Factor

For the fully developed flow in a pipe of length L and diameter D, we obtain the following relationship between the wall shear stress and the static pressure drop

$$\Delta p_{loss} = 4|\tau_w|\left(\frac{L}{D}\right)$$

which with the substitution for $|\tau_w|$ yields

$$\Delta p_{\text{loss}} = 4C_{\text{f}}\left(\frac{L}{D}\right)\left(\frac{1}{2}\rho\bar{V}_x^2\right)$$

which, using the Darcy friction factor f, we can write as

$$\Delta p_{\text{loss}} = f\left(\frac{L}{D}\right)\left(\frac{1}{2}\rho\bar{V}_x^2\right)$$

yielding $f = 4C_{\text{f}} = 64/Re$. Note that $f = 4C_{\text{f}}$ is also valid for a fully developed turbulent pipe flow.

PROBLEM 6.6:   FLOW IN AN ANNULUS BETWEEN CONCENTRIC PIPES

Figure 6.11 shows an axisymmetric fully developed incompressible laminar flow through an annulus between two concentric pipes of radii R_1 and R_2. The flow in the annulus is entirely driven by an imposed static pressure gradient with no contribution from the gravitational body force. Find:

a. the fully developed velocity $V_x(r)$,
b. the radial location of the maximum velocity,
c. the maximum velocity,
d. the average velocity,
e. the shear stress distribution within the annulus,
f. the expression for Darcy friction factor in terms of Reynolds number, and
g. the loss in static pressure over a length L in the flow direction.

SOLUTION FOR PROBLEM 6.6

Under the assumptions identical to those in Problem 6.5 for the Hagen-Poiseuille flow, we obtain $V_r = 0$ and the reduced Navier-Stokes equation as the following second-order ODE for $V_x(r)$

$$\frac{1}{r}\frac{d}{dr}\left(r\frac{dV_x}{dr}\right) = \frac{1}{\mu}\frac{dp}{dx} = C_1$$

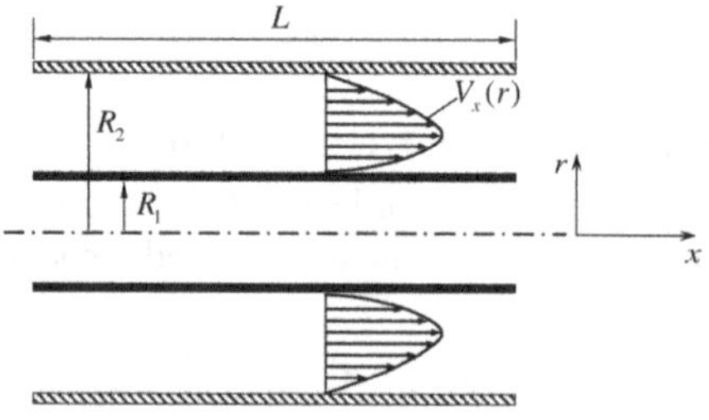

FIGURE 6.11 Fully developed laminar flow in an annulus between concentric pipes (Problem 6.6).

where C_1 is a constant.

BOUNDARY CONDITIONS

In this flow, the no-slip condition on the annulus inner and outer walls yields $V_x(R_1) = V_x(R_2) = 0$.

RESULTS AND DISCUSSIONS

(a) Velocity Profile $V_x(r)$

Integrating the ODE once yields

$$\frac{dV_x}{dr} = \frac{C_1}{2}r + \frac{C_2}{r}$$

where C_2 is the integration constant. Integrating this equation again, we obtain

$$V_x = \frac{C_1}{4}r^2 + C_2 \ln r + C_3$$

where C_3 is another integration constant. Applying the boundary conditions at the annulus inner and outer walls yield the following equations:

$$\frac{C_1}{4}R_1^2 + C_2 \ln R_1 + C_3 = 0$$

$$\frac{C_1}{4}R_2^2 + C_2 \ln R_2 + C_3 = 0$$

from which we obtain C_2 and C_3 as

$$C_2 = -\frac{C_1}{4}\frac{\left(R_2^2 - R_1^2\right)}{\ln\left(R_2/R_1\right)}$$

$$C_3 = -\frac{C_1}{4}R_2^2 + \frac{C_1}{4}\ln R_2 \frac{\left(R_2^2 - R_1^2\right)}{\ln\left(R_2/R_1\right)}$$

which we substitute in the foregoing solution for $V_x(r)$ to yield

$$V_x = -\frac{C_1}{4}\left[R_2^2 - r^2 - \left(R_2^2 - R_1^2\right)\left(\frac{\ln R_2 - \ln r}{\ln R_2 - \ln R_1}\right)\right]$$

$$V_x = -\frac{1}{4\mu}\frac{dp}{dx}\left[R_2^2 - r^2 - \left(R_2^2 - R_1^2\right)\left(\frac{\ln R_2 - \ln r}{\ln R_2 - \ln R_1}\right)\right]$$

With $\zeta = r/R_2$, $\beta = R_1/R_2$,

$$K_1 = -\frac{R_2^2}{4\mu}\frac{dp}{dx}$$

and

$$K_2 = \frac{1-\beta^2}{\ln(1/\beta)}$$

we can alternatively write V_x as

$$V_x = K_1\left(1 - \zeta^2 + K_2\ln\zeta\right)$$

(b) Radial Location of the Maximum Velocity

At the point of maximum velocity, we have $dV_x/dr = 0$, giving

$$\frac{dV_x}{dr} = \frac{dV_x}{R_2 d\zeta} = \frac{K_1 d}{R_2 d\zeta}\left(1 - \zeta^2 + K_2\ln\zeta\right) = 0$$

from which we obtain

$$\zeta_{max} = \sqrt{\frac{K_2}{2}} = \sqrt{\frac{1-\beta^2}{2\ln(1/\beta)}}$$

(c) Maximum Velocity

Substituting $\zeta = \zeta_{max}$ in the expression for the velocity profile, we obtain the maximum velocity as

$$V_{x\_max} = K_1\left(1 - \zeta_{max}^2 + K_2\ln\zeta_{max}\right)$$

$$V_{x\_max} = -\frac{R_2^2}{4\mu}\frac{dp}{dx}\left[1 - \frac{1-\beta^2}{2\ln(1/\beta)}\left\{1 - \ln\left(\frac{1-\beta^2}{2\ln(1/\beta)}\right)\right\}\right]$$

(d) Average Velocity

$$\bar{V}_x = \frac{\int_0^{2\pi}\int_{R_1}^{R_2}V_x r\,dr\,d\theta}{\int_0^{2\pi}\int_{R_1}^{R_2}r\,dr\,d\theta} = \frac{\int_{R_1}^{R_2}V_x r\,dr}{\int_{R_1}^{R_2}r\,dr} = \frac{2\int_{R_1}^{R_2}V_x r\,dr}{\left(R_2^2 - R_1^2\right)} = \frac{2\int_{\beta}^{1}V_x\zeta\,d\zeta}{\left(1-\beta^2\right)}$$

$$\bar{V}_x = \frac{2\int_{\beta}^{1}K_1\left(1-\zeta^2 + K_2\ln\zeta\right)\zeta\,d\zeta}{\left(1-\beta^2\right)} = \frac{2K_1}{\left(1-\beta^2\right)}\int_{\beta}^{1}\left(\zeta - \zeta^3 + K_2\zeta\ln\zeta\right)d\zeta$$

$$\bar{V}_x = -\frac{R_2^2}{8\mu}\frac{dp}{dx}\left[\left(1+\beta^2\right) - \frac{1-\beta^2}{\ln(1/\beta)}\right]$$

(e) Shear-Stress Distribution

$$\tau_{rx} = \mu \frac{dV_x}{dr} = \frac{\mu}{R_2} \frac{dV_x}{d\zeta}$$

$$\tau_{rx} = \frac{\mu K_1}{R_2} \frac{d}{d\zeta}\left(1 - \zeta^2 + K_2 \ln\zeta\right)$$

$$\tau_{rx} = -\frac{R_2}{2} \frac{dp}{dx}\left[\frac{1-\beta^2}{2\zeta \ln(1/\beta)} - \zeta\right]$$

(f) Darcy Friction Factor

For the annulus, the hydraulic mean diameter equals $2(R_2 - R_1)$. Using the flow Reynolds number

$$Re = \frac{2(R_2 - R_1)\rho \bar{V}_x}{\mu}$$

and

$$\lambda = R_2^2\left[\left(1+\beta^2\right) - \frac{1-\beta^2}{\ln(1/\beta)}\right]$$

which is constant for a given annulus, we write the average velocity as

$$\bar{V}_x = -\left(\frac{1}{8\mu} \frac{dp}{dx}\right)\lambda$$

from which we obtain

$$-\frac{dp}{dx} = \frac{8\mu \bar{V}_x}{\lambda}$$

$$-\frac{dp}{dx} = f \frac{1}{2(R_2 - R_1)}\left(\frac{1}{2}\rho \bar{V}_x^2\right)$$

where in terms of Re and λ yields the Darcy friction factor f as

$$f = \frac{64}{Re} \frac{(R_2 - R_1)^2}{\lambda}$$

(g) Static Pressure Loss over the Annulus Length L

Integrating the constant pressure gradient term over the annulus length L, we obtain

$$\Delta p_{loss} = \frac{32}{Re} \frac{(R_2 - R_1)L}{\lambda}\left(\frac{1}{2}\rho \bar{V}_x^2\right)$$

PROBLEM 6.7: LIMITING CASES OF FULLY DEVELOPED LAMINAR FLOW IN AN ANNULUS BETWEEN CONCENTRIC PIPES

Using the fully developed velocity profile

$$V_x = -\frac{1}{4\mu}\frac{dp}{dx}\left[R_2^2 - r^2 - \left(R_2^2 - R_1^2\right)\left(\frac{\ln R_2 - \ln r}{\ln R_2 - \ln R_1}\right)\right]$$

in an annulus we obtained in Problem 6.6 show that

  a. for $R_1 = 0$ and $R_2 = R$, the velocity profile reduces to that in a Hagen-Poiseuille flow in a pipe of radius R obtained in Problem 6.5 and
  b. for $R_2 \approx R_1$, the solution reduces to that between two infinite parallel plates, which are 2H apart, obtained in Problem 6.4.

SOLUTION FOR PROBLEM 6.7

(a) Velocity profile for $R_1 = 0$ and $R_2 = R$

Let us rewrite the given velocity profile as

$$V_x = -\frac{1}{4\mu}\frac{dp}{dx}\left[R^2 - r^2 - \left(R^2 - R_1^2\right)\frac{\ln\left(R/r\right)}{\ln\left(R/R_1\right)}\right]$$

For $R_1 = 0$, $\ln\left(R/R_1\right) = \infty$, the second term within brackets of this equation becomes zero, reducing it to

$$V_x = -\frac{1}{4\mu}\frac{dp}{dx}\left(R^2 - r^2\right)$$

which is identical to the velocity profile we obtained in Problem 6.5 for a fully developed laminar flow in a pipe of radius R (Hagen-Poiseuille flow).

(b) Velocity Profile for $R_2 \approx R_1$

Using the coordinates shown in Figure 6.12 and the relations $R_m = (R_1 + R_2)/2$, $R_1 = R_m - H$, $R_2 = R_m + H$, $r = R_m + y$, $\xi = y/R_m$, and $\kappa = H/R_m$, we write various terms in the equation for the given velocity profile in an annulus as follows:

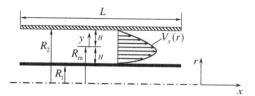

FIGURE 6.12 Additional coordinates in the annular flow between concentric pipes (Problem 6.7).

$$R_2^2 - r^2 = (R_2 - r)(R_2 + r) = (H - y)(2R_m + H + y)$$

$$R_2^2 - R_1^2 = (R_2 - R_1)(R_2 + R_1) = (2H)(2R_m) = 4HR_m$$

$$\ln r/R_m = \ln(1 + \xi) = \xi - \frac{\xi^2}{2} + \frac{\xi^3}{3} - \frac{\xi^4}{4} + \cdots \quad (\text{for } \xi \ll 1)$$

$$\ln R_1/R_m = \ln(1 - \kappa) = -\kappa - \frac{\kappa^2}{2} - \frac{\kappa^3}{3} - \frac{\kappa^4}{4} - \cdots \quad (\text{for } \kappa \ll 1)$$

$$\ln R_2/R_m = \ln(1 + \kappa) = \kappa - \frac{\kappa^2}{2} + \frac{\kappa^3}{3} - \frac{\kappa^4}{4} + \cdots \quad (\text{for } \kappa \ll 1)$$

Neglecting cubic and higher-order terms in these Taylor-series expansions of the logarithmic expressions, we obtain

$$\ln R_2 - \ln r = \ln(R_2/R_m) - \ln(r/R_m) = \ln(1 + \kappa) - \ln(1 + \xi)$$

$$\ln R_2 - \ln r = \kappa - \frac{\kappa^2}{2} - \xi + \frac{\xi^2}{2} = \frac{1}{2}(\kappa - \xi)(2 - \kappa - \xi)$$

$$\ln R_2 - \ln R_1 = \ln(R_2/R_m) - \ln(R_1/R_m) = \ln(1 + \kappa) - \ln(1 - \kappa)$$

$$\ln R_2 - \ln R_1 = \kappa - \frac{\kappa^2}{2} + \kappa + \frac{\kappa^2}{2} = 2\kappa$$

Substituting the foregoing expressions in the given solution for V_x yields

$$V_x = -\frac{1}{4\mu} \frac{dp}{dx} \left[(H - y)(2R_m + H + y) - 4HR_m \left(\frac{(\kappa - \xi)(2 - \kappa - \xi)}{4\kappa} \right) \right]$$

which upon resubstituting $\xi = y/R_m$ and $\kappa = H/R_m$, after further simplifications, results in

$$V_x = -\frac{1}{4\mu} \frac{dp}{dx} (H^2 - y^2)$$

which is identical to the velocity profile we obtained in Problem 6.4 for a fully developed laminar flow between two parallel plates separated by $2H$.

NOMENCLATURE

b	Film thickness
C_1, C_2, C_3, C_4	Integration constants
C_f	Shear coefficient
D	Pipe diameter

f	Darcy friction factor
F	Force
g	Acceleration due to gravity
h	Height measured vertically up (for positive h) from a fixed datum
H	Half the distance between two infinite parallel plates
L	Length
$\dot{m}$	Mass flow rate
p	Pressure
$\hat{p}$	Modified static pressure that includes local hydrostatic pressure
r	Cylindrical polar coordinate r; radius
R	Pipe radius, gas constant
R_1	Inner radius
R_2	Outer radius
Re	Reynolds number
t	Time
u	Velocity in x-direction
v	Velocity in y-direction
V	Velocity
$\bar{V}$	Section-average velocity
w	Velocity in z-direction
$\tilde{w}$	Dimensionless velocity $\tilde{w} = w/w_{max}$
W	Weight, wire velocity
x	Cartesian coordinate x
y	Cartesian coordinate y
z	Cartesian coordinate z

SUBSCRIPTS AND SUPERSCRIPTS

g	Due to gravity
h	Solution of the homogeneous part of an ODE
i	Inner cylinder
max	Maximum value
o	Outer cylinder
p	Particular solution of a nonhomogeneous ODE
r	Component in r-coordinate direction
w	Wall
x	Component in x-coordinate direction
y	Component in y-coordinate direction
z	Component in z-coordinate direction
θ	Component in θ-coordinate direction

GREEK SYMBOLS

β	Radius ratio $\left(\zeta = R_1/R_2 \right)$
Γ	Torque

δf_b	Body force per unit volume acting on an infinitesimal control volume
δf_s	Surface force per unit volume acting on an infinitesimal control volume
δF_{bx}	Body force acting in the x-direction on an infinitesimal control volume
δF_{s_yx}	Contribution of $\partial \tau_{yx}/\partial y$ to the surface force per unit volume
ζ	Dimensionless radial distance $\left(\zeta = r/R_2 \right)$
θ	Coordinate θ in cylindrical polar coordinate system
μ	Dynamic viscosity
v	Kinematic viscosity $\left(v = \mu/\rho \right)$
ξ	Dimensionless distance $\left(\xi = x/b \right)$
ρ	Density
σ_{ii}	Normal stress in tensor notation
τ_{ij}	Stress acting in the j-direction on the face, whose normal is in the i direction, of a differential control volume $\left(\tau_{ij} = \tau_{ji} \right)$

REFERENCES

Bird, R.B., W.E. Stewart, and E.N. Lightfoot. 2006. *Transport Phenomena*, 2nd edition. New York: John Wiley & Sons.

Riley, N. and P.G. Drazin. 2006. *The Navier-Stokes Equations: A Classification of Flows and Exact Solutions*. London: Cambridge University Press.

Sultanian, B.K. 2015. *Fluid Mechanics: An Intermediate Approach*. Boca Raton, FL: Taylor & Francis.

BIBLIOGRAPHY

Doering, C.R. and J.D. Gibbon. 1993. *Applied Analysis of the Navier-Stokes Equation*. Cambridge: Cambridge University Press.

Schlichting, H. 1979. *Boundary Layer Theory*, 7th edition. New York: McGraw-Hill.

Sultanian, B.K. 2018. *Gas Turbines: Internal Flow Systems Modeling* (Cambridge Aerospace Series). Cambridge: Cambridge University Press.

Sultanian, B.K. 2019. *Logan's Turbomachinery: Gaspath Design and Performance Fundamentals*. Boca Raton, FL: Taylor & Francis.

7 Boundary Layer Flow

REVIEW OF KEY CONCEPTS

The problems and solutions presented in this chapter pertain to laminar boundary layer flows of an incompressible fluid with constant viscosity. Schlichting (1979) presents comprehensive analyses of various boundary layers, including turbulent boundary layers.

DIFFERENTIAL BOUNDARY LAYER EQUATIONS

The continuity and momentum equations (Navier-Stokes equations) for a constant-viscosity two-dimensional incompressible laminar flow, shown in Figure 7.1, presented in Chapter 6 are as follows:

Continuity Equation

$$\frac{\partial u}{\partial x} + \frac{\partial v}{\partial y} = 0 \tag{7.1}$$

Momentum Equation in x-Direction

$$\frac{\partial u}{\partial t} + u\frac{\partial u}{\partial x} + v\frac{\partial u}{\partial y} = -\frac{1}{\rho}\frac{\partial p}{\partial x} + v\left(\frac{\partial^2 u}{\partial x^2} + \frac{\partial^2 u}{\partial y^2}\right) \tag{7.2}$$

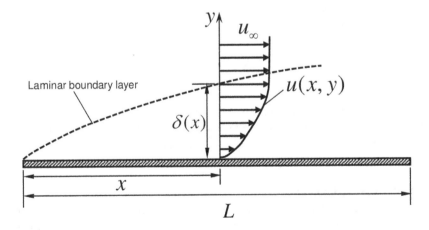

FIGURE 7.1 Laminar boundary layer flow over a solid surface.

Momentum Equation in y-Direction

$$\frac{\partial v}{\partial t} + u\frac{\partial v}{\partial x} + v\frac{\partial v}{\partial y} = -\frac{1}{\rho}\frac{\partial p}{\partial y} + v\left(\frac{\partial^2 v}{\partial x^2} + \frac{\partial^2 v}{\partial y^2}\right) \tag{7.3}$$

As per the derivation details given in Sultanian (2015), we obtain the following boundary layer equation:

$$\frac{\partial u}{\partial t} + u\frac{\partial u}{\partial x} + v\frac{\partial u}{\partial y} = u_\infty\frac{du_\infty}{dx} + \frac{1}{\rho}\frac{\partial \tau_{yx}}{\partial y} \tag{7.4}$$

which is a second-order partial differential equation and, along with the continuity equation, (Equation 7.1), is sufficient to compute the distribution of u and v in the boundary layer for the given initial and boundary conditions.

VON KARMAN MOMENTUM INTEGRAL EQUATION

At each value of x, integrating Equation 7.4 in the y-direction over the boundary layer thickness yields the following ordinary differential equation with quantities varying only in the x-direction, which is the predominant flow direction:

$$\frac{d}{dx}\left(u_\infty^2\delta_2\right) + u_\infty\delta_1\frac{du_\infty}{dx} = \frac{\tau_0}{\rho} \tag{7.5}$$

which is known as the Von Karman momentum integral equation for a steady incompressible boundary layer flow in the x–y plane. We can also express this equation as

$$\frac{d\delta_2}{dx} + (2+H)\frac{\delta_2}{u_\infty}\frac{du_\infty}{dx} = \frac{\tau_0}{\rho u_\infty^2} = \frac{C_f}{2} \tag{7.6}$$

where the shape factor $H = \delta_1/\delta_2$, and the shear coefficient (also called Fanning friction factor) C_f is defined as

$$C_f = \frac{\tau_0}{\frac{1}{2}\rho u_\infty^2} \tag{7.7}$$

Sultanian (2015) derives Equation 7.5 in a couple of ways: first, by direct integration of Equation 7.4; second, using a control volume analysis. We present here the integral quantities, boundary layer thickness (δ), displacement thickness (δ_1), and momentum thickness (δ_2), used in Equations 7.5 and 7.6.

Boundary Layer Thickness (δ)

The boundary layer thickness δ is the characteristic length scale measuring the extent from the wall in which the viscous effects are considered significant. Although, theoretically, the boundary layer extends up to infinity in the cross-flow direction,

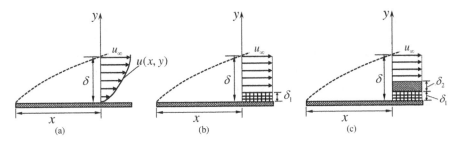

FIGURE 7.2 (a) Actual boundary layer flow showing $u(x,y)$ and the boundary layer thickness δ, (b) hypothetical uniform velocity profile with δ_1, and (c) hypothetical uniform velocity profile with δ_1 and δ_2.

we will neglect the effect of fluid viscosity beyond δ where the velocity in the flow direction reaches 99% of the local free stream velocity. As shown in Figure 7.2a, for all practical purposes, δ defines the edge of the boundary layer where $u = u_\infty(x)$. The shear stress has its maximum value at the wall, and we assume it to be zero at the edge of the boundary layer.

Displacement Thickness (δ_1)

At each x location in a boundary layer, the velocity $u(x,y)$ varies in the y-direction from zero at the wall to $u_\infty(x)$ at $y = \delta$. To model this velocity profile in terms of a uniform velocity u_∞ to yield the same mass flow rate, as shown in Figure 7.2b, we need a step function of zero velocity up to a distance δ_1 and uniform velocity u_∞ over $(\delta - \delta_1)$. Both the actual velocity profile $u(x,y)$ shown in Figure 7.2a and the hypothetical profile in Figure 7.2b yield equal mass flow rates. For a constant density and unit depth in the z-direction, we can write the mass flow for each velocity profile as

$$\dot{m} = \rho \int_0^\delta u\,dy = \rho \int_{\delta_1}^\delta u_\infty\,dy = \rho \int_0^\delta u_\infty\,dy - \rho \int_0^{\delta_1} u_\infty\,dy = \rho \int_0^\delta u_\infty\,dy - \rho\delta_1 u_\infty$$

giving

$$\delta_1 = \int_0^\delta \left(1 - \frac{u}{u_\infty}\right) dy \tag{7.8}$$

which defines the displacement thickness, as shown in Figure 7.2b. In other words, if we displace the wall by δ_1, we can compute the same mass flow rate using the uniform velocity profile $u = u_\infty$ over the rest of the boundary layer thickness. Alternatively, for a given velocity profile $u(x,y)$ with known δ and δ_1, we can calculate the associated mass flow rate per unit depth as

$$\dot{m} = \rho(\delta - \delta_1)u_\infty \tag{7.9}$$

Momentum Thickness (δ_2)

While it simulates the mass flow rate of the actual local velocity profile in the boundary layer, the hypothetical step-function velocity profile shown in Figure 7.2b fails to simulate its x-momentum flow rate. For this, we need to further modify our step-function profile with an additional step, called momentum thickness δ_2, as shown in Figure 7.2c. Thus, we can write

$$\dot{M}_x = \rho \int_0^\delta u^2 \, dy = \rho \int_{\delta_1+\delta_2}^\delta u_\infty^2 \, dy = \rho \int_0^\delta u_\infty^2 \, dy - \rho \int_0^{\delta_1} u_\infty^2 \, dy - \rho \int_0^{\delta_2} u_\infty^2 \, dy$$

$$\int_0^\delta u^2 \, dy = \int_0^\delta u_\infty^2 \, dy - u_\infty^2 \delta_1 - u_\infty^2 \delta_2$$

Substituting in this equation for δ_1 from Equation 7.8 yields

$$\int_0^\delta u^2 \, dy = \int_0^\delta u_\infty^2 \, dy - u_\infty^2 \int_0^\delta \left(1 - \frac{u}{u_\infty}\right) dy - u_\infty^2 \delta_2$$

$$\int_0^\delta u^2 \, dy = u_\infty^2 \int_0^\delta \frac{u}{u_\infty} \, dy - u_\infty^2 \delta_2 \tag{7.10}$$

$$\delta_2 = \int_0^\delta \frac{u}{u_\infty}\left(1 - \frac{u}{u_\infty}\right) dy$$

For a given velocity profile $u(x,y)$ with known δ, δ_1, and δ_2, we can calculate the associated x-momentum flow rate per unit depth as

$$\dot{M}_x = \rho(\delta - \delta_1 - \delta_2)u_\infty^2 \tag{7.11}$$

PROBLEM 7.1:   TWO-DIMENSIONAL INCOMPRESSIBLE LAMINAR FLOW ON A FLAT PLATE

For the stream-wise velocity in a two-dimensional incompressible laminar boundary layer flow on a flat plate, consider the following self-similar dimensionless profile

$$U = C_1 + C_2\eta + C_3\eta^3$$

where $U = u/u_\infty$ and $\eta = y/\delta(x)$. This velocity profile must satisfy the boundary conditions: (1) at $\eta = 0$, $U = 0$; (2) at $\eta = 1$, $U = 1$; and (3) at $\eta = 1$, $dU/d\eta = 0$. Find:

   a. the boundary layer thickness $\delta(x)$,
   b. the displacement thickness $\delta_1(x)$,
   c. the momentum thickness $\delta_2(x)$, and
   d. the total drag force acting on the flat plate of length L and width b.

SOLUTION FOR PROBLEM 7.1

We use the given boundary conditions to evaluate C_1, C_2, and C_3 in the cubic polynomial assumed for the velocity profile. The first boundary condition yields $C_1 = 0$. From the remaining two boundary conditions we obtain $C_2 + C_3 = 1$ and $C_2 + 3C_3 = 0$, which together yield $C_2 = 3/2$ and $C_3 = -1/2$. The cubic polynomial for the velocity profile becomes

$$U = \frac{3}{2}\eta - \frac{1}{2}\eta^3$$

(a) THE BOUNDARY LAYER THICKNESS $\delta(x)$

The momentum integral equation (Equation 7.5) for the flat plate boundary layer flow with constant u_∞ reduces to

$$\frac{d\delta_2}{dx} = \frac{\tau_0}{\rho u_\infty^2}$$

Wall Shear Stress τ_0

$$\tau_0 = \mu\left(\frac{\partial u}{\partial y}\right)_{y=0} = \frac{\mu u_\infty}{\delta}\left(\frac{\partial U}{\partial \eta}\right)_{\eta=0} = \frac{3}{2}\frac{\mu u_\infty}{\delta}$$

Momentum Thickness δ_2

$$\frac{\delta_2}{\delta} = \int_0^1 U(1-U)d\eta$$

$$= \int_0^1 \left(\frac{3}{2}\eta - \frac{1}{2}\eta^3\right)\left(1 - \frac{3}{2}\eta + \frac{1}{2}\eta^3\right)d\eta$$

$$= \int_0^1 \left(\frac{3}{2}\eta - \frac{9}{4}\eta^2 - \frac{1}{2}\eta^3 + \frac{3}{2}\eta^4 - \frac{1}{4}\eta^6\right)d\eta$$

$$= \left(\frac{3}{4} - \frac{9}{12} - \frac{1}{8} + \frac{3}{10} - \frac{1}{28}\right)$$

$$\frac{\delta_2}{\delta} = \frac{39}{280}$$

Substituting for τ_0 and δ_2 in the momentum integral equation yields

$$\frac{39}{280}\frac{d\delta}{dx} = \frac{3\mu}{2\delta\rho u_\infty}$$

$$\frac{\delta\,d\delta}{dx} = \frac{140\mu}{13\rho u_\infty}$$

Separating variables in this equation and integrating we obtain

$$\delta = \sqrt{\frac{280\mu x}{13\rho u_\infty}} + C$$

As at $x = 0$, the boundary layer thickness is zero, we obtain $C = 0$, giving

$$\delta = \sqrt{\frac{280\mu x}{13\rho u_\infty}} = 4.641\sqrt{\frac{vx}{u_\infty}}$$

This approximate result for the boundary layer thickness is about 7% lower than the Blasius exact solution given by

$$\delta = 5\sqrt{\frac{vx}{u_\infty}}$$

(b) THE DISPLACEMENT THICKNESS $\delta_1(x)$

$$\frac{\delta_1}{\delta} = \int_0^1 (1 - U)\,d\eta = \int_0^1 \left(1 - \frac{3}{2}\eta + \frac{1}{2}\eta^3\right)d\eta$$

$$\frac{\delta_1}{\delta} = \left(1 - \frac{1}{2} + \frac{1}{8}\right) = \frac{3}{8}$$

Substituting for δ from (a), we obtain

$$\delta_1 = 1.740\sqrt{\frac{vx}{u_\infty}}$$

which is about 1% higher than the Blasius exact solution given by $\delta_1 = 1.7208\sqrt{vx/u_\infty}$ — very impressive for an approximate solution method.

(c) THE MOMENTUM THICKNESS $\delta_1(x)$

In (a), we obtained $\delta_2/\delta = 39/280$. Substituting for δ yields

$$\delta_2 = 0.646\sqrt{\frac{vx}{u_\infty}}$$

which is about 3% lower than the Blasius exact solution given by $\delta_2 = 0.664\sqrt{vx/u_\infty}$.

(d) TOTAL DRAG FORCE ON THE FLAT PLATE

Substituting for δ in the solution for the local wall shear stress τ_0 obtained in (a) yields

TABLE 7.1

Results of Laminar Boundary Layer over a Flat Plate from Momentum Integral Methods along with the Exact Solution

Velocity Profile $U = u/u_\infty$; $\eta = y/\delta(x)$	$\delta_1 \sqrt{\dfrac{u_\infty}{vx}}$	$\delta_2 \sqrt{\dfrac{u_\infty}{vx}}$	$\dfrac{\tau_0}{\rho u_\infty^2} \sqrt{\dfrac{u_\infty x}{v}}$	$\bar{C}_f \sqrt{\dfrac{u_\infty L}{v}}$	$H = \delta_1/\delta_2$
$U = \eta$	1.732	0.578	0.289	1.155	3.00
$U = \dfrac{3}{2}\eta - \dfrac{1}{2}\eta^3$	1.740	0.646	0.323	1.292	2.70
$U = 2\eta - 2\eta^3 + \eta^4$	1.752	0.686	0.343	1.372	2.55
$U = \sin\left(\dfrac{\pi\eta}{2}\right)$	1.741	0.654	0.327	1.310	2.66
Exact solution	1.721	0.664	0.332	1.328	2.59

$$\tau_0(x) = 0.323 \rho u_\infty^2 \sqrt{\frac{v}{u_\infty x}}$$

$$F_D = b \int_0^L \tau_0(x)\,dx = 0.323 b \rho u_\infty^2 \sqrt{\frac{v}{u_\infty}} \int_0^L \sqrt{\frac{1}{x}}\,dx$$

$$F_D = 0.646(bL)\rho u_\infty^2 \sqrt{\frac{v}{u_\infty L}}$$

which is about 3% below the value computed from the Blasius exact solution for the local wall shear coefficient given by $C_f = 0.664/\sqrt{Re_x}$. This solution demonstrates the power of the momentum integral method to quickly compute the key parameters of a laminar boundary layer on a flat plate using a cubic velocity profile. The results of the solution for this problem are summarized in Table 7.1 along with the Blasius exact solution.

PROBLEM 7.2: TOTAL DRAG ON A FLAT PLATE IN AN ONCOMING UNIFORM WATER FLOW

Consider a $0.5m$ by $1.0m$ thin flat plate placed in an oncoming flow of water with uniform velocity $u_\infty = 10\,m/s$. Using the exact solution of Blasius boundary layer equation, compute the total drag on both top and bottom surfaces of the plate for both orientations (long and short sides parallel to the flow). Assuming a thin plate, neglect any contribution of the form drag.

SOLUTION FOR PROBLEM 7.2

In this solution, we assume for water $\rho = 1000 \ kg/m^3$ and $v = 1.0 \times 10^{-6} \ m^2/s$.
   Total area of plate top and bottom surfaces: $A = 2 \times 0.5 \times 1.0 = 1.0 \ m^2$.

The Blasius exact solution (Table 7.1) yields the average shear coefficient over a flat plate of length L (along the flow direction) as

$$\bar{C}_f = \frac{1.328}{\sqrt{\dfrac{u_\infty L}{v}}} = \frac{1.328}{\sqrt{Re_L}}$$

We compute the dynamic pressure of the free stream velocity $u_\infty = 10\,\text{m/s}$ as

$$\frac{1}{2}\rho u_\infty^2 = \frac{1}{2} \times 1000 \times (10)^2 = 5 \times 10^4 \text{ Pa}$$

LONG SIDE PARALLEL TO FLOW

$$Re_L = \frac{u_\infty L}{v} = \frac{10 \times 1}{1.0 \times 10^{-6}} = 10^7$$

$$\bar{C}_f = \frac{1.328}{\sqrt{10^7}} = 4.200 \times 10^{-4}$$

which yields the total drag force as

$$F_D = \bar{C}_f \left(\frac{1}{2}\rho u_\infty^2 \right) A = 4.200 \times 10^{-4} \times 5 \times 10^4 \times 1.0 = 21 \text{ N}$$

$$F_D = 21 \text{ N}$$

SHORT-SIDE PARALLEL TO FLOW

$$Re_L = \frac{u_\infty L}{v} = \frac{10 \times 0.5}{1.0 \times 10^{-6}} = 5 \times 10^6$$

$$\bar{C}_f = \frac{1.328}{\sqrt{5 \times 10^5}} = 5.939 \times 10^{-4}$$

which yields the total drag force as

$$F_D = \bar{C}_f \left(\frac{1}{2}\rho u_\infty^2 \right) A = 5.939 \times 10^{-4} \times 5 \times 10^4 \times 1.0 = 29.659 \text{ N}$$

$$F_D = 29.659 \text{ N}$$

Therefore, the total drag force (neglecting any form drag) on the plate top and bottom surfaces is 21 N when the long side is parallel to the flow and 29.659 N when the short side is parallel to the flow. These results assume that the Blasius exact solution is valid over the entire plate.

As the flow Reynolds in both cases is higher than $Re_L = 3 \times 10^6$, which is the limit for the laminar boundary layer transition to the turbulent boundary layer with a higher drag coefficient, the drag force on the flat plate with its long side parallel to the flow will be higher than 29.659 N, and that for the short side parallel to the flow will be higher than 21 N.

PROBLEM 7.3: MOMENTUM INTEGRAL METHOD: LAMINAR BOUNDARY LAYER ON A FLAT PLATE

For a laminar boundary layer on a flat plate, using momentum integral method, compute various quantities given in Table 7.1 for the following approximate velocity profile:

$$U = \frac{u}{u_\infty} = 1 - e^{-\alpha\eta}$$

where α is a constant. Compare your results with the exact solutions given in the table.

SOLUTION FOR PROBLEM 7.3

Let us first evaluate the boundary layer thickness $\delta(x)$ for the given velocity profile. The momentum integral equation (Equation 7.5) for the flat plate boundary layer flow with constant u_∞ reduces to

$$\frac{d\delta_2}{dx} = \frac{\tau_0}{\rho u_\infty^2}$$

WALL SHEAR STRESS τ_0

$$\tau_0 = \mu\left(\frac{\partial u}{\partial y}\right)_{y=0} = \frac{\mu u_\infty}{\delta}\left(\frac{\partial U}{\partial \eta}\right)_{\eta=0} = \frac{\alpha\mu u_\infty}{\delta}\left(e^{-\alpha\eta}\right)_{\eta=0} = \frac{\alpha\mu u_\infty}{\delta}$$

MOMENTUM THICKNESS δ_2

$$\frac{\delta_2}{\delta} = \int_0^1 U(1-U)d\eta$$

$$= \int_0^1 e^{-\alpha\eta}\left(1 - e^{-\alpha\eta}\right)d\eta$$

$$= \int_0^1 \left(e^{-\alpha\eta} - e^{-2\alpha\eta}\right)d\eta$$

$$= \left[\frac{e^{-2\alpha\eta}}{2\alpha} - \frac{e^{-\alpha\eta}}{\alpha}\right]_0^1$$

$$= \frac{e^{-2\alpha}}{2\alpha} - \frac{e^{-\alpha}}{\alpha} + \frac{1}{2\alpha}$$

$$= \frac{\left(1 - e^{-\alpha}\right)^2}{2\alpha}$$

Substituting for τ_0 and δ_2 in the momentum integral equation for the flat plate yields

$$\frac{\left(1 - e^{-\alpha}\right)^2}{2\alpha} \frac{d\delta}{dx} = \frac{\alpha\mu}{\delta\rho u_\infty}$$

$$\frac{d\delta}{dx} = \frac{2\alpha^2}{\left(1 - e^{-\alpha}\right)^2} \frac{v}{\delta u_\infty}$$

Separating the variables in this equation and integrating, we obtain

$$\delta = \sqrt{\frac{4\alpha^2}{\left(1 - e^{-\alpha}\right)^2} \frac{vx}{u_\infty} + C}$$

As the boundary layer thickness is zero at $x = 0$, we obtain $C = 0$, giving

$$\delta = \frac{2\alpha}{1 - e^{-\alpha}} \sqrt{\frac{vx}{u_\infty}}$$

Thus, the momentum thickness becomes

$$\delta_2 = \frac{\left(1 - e^{-\alpha}\right)^2}{2\alpha} \times \frac{2\alpha}{1 - e^{-\alpha}} \sqrt{\frac{vx}{u_\infty}}$$

$$\delta_2 \sqrt{\frac{u_\infty}{vx}} = 1 - e^{-\alpha}$$

DISPLACEMENT THICKNESS δ_1

$$\frac{\delta_1}{\delta} = \int_0^1 (1 - U)\,d\eta = \int_0^1 e^{-\alpha\eta}\,d\eta$$

$$\frac{\delta_1}{\delta} = \left[-\frac{e^{-\alpha\eta}}{\alpha}\right]_0^1 = \frac{1 - e^{-\alpha}}{\alpha}$$

Substituting for δ in this equation yields

$$\delta_1 = \left(\frac{1-e^{-\alpha}}{\alpha}\right)\frac{2\alpha}{1-e^{-\alpha}}\sqrt{\frac{vx}{u_\infty}}$$

$$\delta_1\sqrt{\frac{u_\infty}{vx}} = 2$$

Substituting for δ in the foregoing equation for the local wall shear stress τ_0, we obtain

$$\frac{\tau_0}{\rho u_\infty^2}\sqrt{\frac{u_\infty x}{v}} = \frac{1-e^{-\alpha}}{2}$$

For evaluating the average shear coefficient for a plate of length L and width b, we first determine the total frictional drag force on the plate as

$$F_D = b\int_0^L \tau_0(x)\,dx = \left(\frac{1-e^{-\alpha}}{2}\right)b\rho u_\infty^2\sqrt{\frac{v}{u_\infty}}\int_0^L\sqrt{\frac{1}{x}}\,dx$$

$$F_D = \left(1-e^{-\alpha}\right)(bL)\rho u_\infty^2\sqrt{\frac{v}{u_\infty L}}$$

which yields

$$\bar{C}_f = \frac{2F_D}{(bL)\rho u_\infty^2} = 2\left(1-e^{-\alpha}\right)\sqrt{\frac{v}{u_\infty L}}$$

$$\bar{C}_f\sqrt{\frac{u_\infty L}{v}} = 2\left(1-e^{-\alpha}\right)$$

Finally, we compute the shape factor as

$$H = \frac{\delta_1}{\delta_2} = \frac{2}{1-e^{-\alpha}}$$

For the assumed velocity profile $\left(U = u/u_\infty = 1-e^{-\alpha}\right)$, except for the displacement thickness δ_1, all other boundary layer quantities (reported in Table 7.1) are functions of the constant α. For $\alpha = 1.091$, the momentum thickness δ_2 in this case is equal to the value obtained from the exact solution. The numerical values of all quantities for $\alpha = 1.091$ along with the values in parentheses for the exact solution are summarized as follows:

$$\delta_1 \sqrt{\frac{u_\infty}{vx}} = 2 \quad (1.721)$$

$$\delta_2 \sqrt{\frac{u_\infty}{vx}} = 1 - e^{-\alpha} = 0.664 \quad (0.664)$$

$$\frac{\tau_0}{\rho u_\infty^2} \sqrt{\frac{u_\infty x}{v}} = \frac{1 - e^{-\alpha}}{2} = 0.332 \quad (0.332)$$

$$\bar{C}_f \sqrt{\frac{u_\infty L}{v}} = 2(1 - e^{-\alpha}) = 1.328 \quad (1.328)$$

$$H = \frac{2}{1 - e^{-\alpha}} = 3.012 \quad (2.590)$$

PROBLEM 7.4:   DRAG FORCE DISTRIBUTION IN A FLAT PLATE BOUNDARY LAYER

A flat plate is placed in an incompressible laminar flow with uniform velocity u_∞. Assuming a laminar boundary layer on the entire plate, at what fraction of the length from the leading edge would the drag force on the front portion be equal to half of the total drag force?

SOLUTION FOR PROBLEM 7.4

From the exact solution summarized in Table 7.1, we have

$$\bar{C}_f \sqrt{\frac{u_\infty L}{v}} = 1.328$$

Using the above relation, we can compute the total drag force on a flat plate of width b and length L_1 along the flow direction from the leading edge as

$$F_{D1} = \bar{C}_f \left(\frac{1}{2} \rho u_\infty^2\right)(L_1 b) = 1.328 \left(\frac{1}{2} \rho u_\infty^2\right)(L_1 b) \sqrt{\frac{v}{u_\infty L_1}}$$

$$F_{D1} = 0.664 b \rho u_\infty^{1.5} v^{0.5} \sqrt{L_1}$$

If the total length of the plate is L and the corresponding drag force is F_D, we can write

$$\frac{F_{D1}}{F_D} = \frac{1}{2} = \frac{0.664 b \rho u_\infty^{1.5} v^{0.5} \sqrt{L_1}}{0.664 b \rho u_\infty^{1.5} v^{0.5} \sqrt{L}}$$

which yields

$$\frac{L_1}{L} = \frac{1}{4} = 0.25$$

Therefore, the drag force on the initial one-quarter (from the leading edge) of the flat plate equals the drag force over the remaining three-fourth (from the trailing edge) of the plate.

PROBLEM 7.5: TURBULENT BOUNDARY LAYER OVER A FLAT PLATE

For an incompressible turbulent boundary layer over a flat plate, the velocity profile may be approximated by the one-seventh power law $U = u/u_\infty = \eta^{1/7}$. From empirical data, the wall shear stress correlates to the following equation:

$$\tau_0 = 0.0228 \rho u_\infty^2 \left(\frac{\nu}{u_\infty \delta(x)} \right)^{1/4}$$

Using the Von Karman momentum integral equation, find the expressions for $\delta(x)$, $\delta_1(x)$, $\delta_2(x)$, $C_f(x)$, and $\bar{C}_f(L)$.

SOLUTION FOR PROBLEM 7.5

Let us first evaluate the boundary layer thickness $\delta(x)$ for the given velocity profile. The momentum integral equation (Equation 7.5) for the flat plate boundary layer flow with constant u_∞ reduces to

$$\frac{d\delta_2}{dx} = \frac{\tau_0}{\rho u_\infty^2}$$

where the wall shear stress is given by

$$\tau_0 = 0.0228 \rho u_\infty^2 \left(\frac{\nu}{u_\infty \delta(x)} \right)^{1/4}$$

and we evaluate the momentum thickness δ_2 for the one-seventh power law profile as

$$\frac{\delta_2}{\delta} = \int_0^1 U(1-U)d\eta$$

$$= \int_0^1 \eta^{1/7}\left(1 - \eta^{1/7}\right)d\eta$$

$$= \int_0^1 \left(\eta^{1/7} - \eta^{2/7} \right)d\eta$$

$$= \left[\frac{7}{8}\eta^{8/7} - \frac{7}{9}\eta^{9/7} \right]_0^1$$

$$\frac{\delta_2}{\delta} = \frac{7}{72}$$

Substituting for τ_0 and δ_2 in the momentum integral equation yields

$$\frac{7}{72}\frac{d\delta}{dx} = 0.0228\left(\frac{v}{u_\infty \delta}\right)^{1/4}$$

$$\delta^{1/4}\frac{d\delta}{dx} = 0.2338\left(\frac{v}{u_\infty}\right)^{1/4}$$

Integrating this equation yields

$$\frac{4}{5}\delta^{5/4} = 0.2338\left(\frac{v}{u_\infty}\right)^{1/4}x + C$$

As the boundary layer thickness is zero at $x = 0$, we obtain $C = 0$, giving

$$\delta = 0.3737\left(\frac{v}{u_\infty}\right)^{1/5}x^{4/5}$$

$$\frac{\delta}{x} = \frac{0.3737}{Re_x^{0.2}}$$

where $Re_x = (u_\infty x)/v$. Thus, the momentum thickness becomes

$$\frac{\delta_2}{x} = \frac{7}{72}\times\frac{0.3737}{Re_x^{0.2}}$$

$$\frac{\delta_2}{x} = \frac{0.0363}{Re_x^{0.2}}$$

DISPLACEMENT THICKNESS δ_1

$$\frac{\delta_1}{\delta} = \int_0^1 (1-U)d\eta = \int_0^1 \left(1-\eta^{1/7}\right)d\eta$$

$$\frac{\delta_1}{\delta} = \left[\eta - \frac{7}{8}\eta^{8/7}\right]_0^1$$

$$\frac{\delta_1}{\delta} = \frac{1}{8}$$

Substituting for δ in this equation, we obtain

$$\frac{\delta_1}{x} = \frac{1}{8}\times\frac{0.3737}{Re_x^{0.2}}$$

$$\frac{\delta_1}{x} = \frac{0.0467}{Re_x^{0.2}}$$

Substituting for δ in the equation for the local wall shear stress τ_0 yields

$$\tau_0 = 0.0228\rho u_\infty^2 \left(\frac{\nu}{u_\infty \delta(x)}\right)^{1/4}$$

$$\tau_0 = 0.0228\rho u_\infty^2 \left(\frac{Re_x^{0.2}}{0.3737 Re_x}\right)^{1/4}$$

$$\tau_0 = 0.0292 \frac{\rho u_\infty^2}{Re_x^{0.2}}$$

which further yields

$$C_f = \frac{\tau_0}{\frac{1}{2}\rho u_\infty^2} = \frac{0.0583}{Re_x^{0.2}}$$

$$C_f = \frac{0.0583}{Re_x^{0.2}}$$

The average shear coefficient $\overline{C}_f$ for a plate of length L is obtained as

$$\overline{C}_f = \frac{\int_0^L C_f \, dx}{L}$$

$$\overline{C}_f = \frac{\int_0^L \left(\frac{0.0583}{Re_x^{0.2}}\right) dx}{L}$$

$$\overline{C}_f = \frac{0.0729}{Re_L^{0.2}}$$

Various boundary layer quantities computed for the one-seventh power law velocity profile are summarized as follows:

$$\frac{\delta}{x} = \frac{0.3737}{Re_x^{0.2}}, \frac{\delta_1}{x} = \frac{0.0467}{Re_x^{0.2}}, \frac{\delta_2}{x} = \frac{0.0363}{Re_x^{0.2}}, C_f = \frac{0.0583}{Re_x^{0.2}}, \text{ and } \overline{C}_f = \frac{0.0729}{Re_L^{0.2}}$$

PROBLEM 7.6:   TRANSVERSE VELOCITY IN THE LAMINAR BOUNDARY LAYER OVER A FLAT PLATE

The u-velocity profile in a laminar boundary layer over an impermeable flat plate is given by

$$\frac{u}{U} = 2\frac{y}{\delta} - \left(\frac{y}{\delta}\right)^2$$

where U is the free-stream velocity and δ is the local boundary layer thickness given by $\delta = K\sqrt{x}$. Find an expression for the velocity component v that is perpendicular to u.

SOLUTION FOR PROBLEM 7.6

From the given u-velocity profile, we obtain

$$\frac{u}{U} = 2\frac{y}{\delta} - \left(\frac{y}{\delta}\right)^2$$

$$\frac{\partial u}{\partial x} = U\left[-2\frac{y}{\delta^2} + 2\frac{y^2}{\delta^3}\right]\left(\frac{K}{2\sqrt{x}}\right)$$

Substituting this in the two-dimensional continuity equation

$$\frac{\partial u}{\partial x} + \frac{\partial v}{\partial y} = 0$$

yields

$$\frac{\partial v}{\partial y} = -\frac{\partial u}{\partial x} = U\left[2\frac{y}{\delta^2} - 2\frac{y^2}{\delta^3}\right]\left(\frac{K}{2\sqrt{x}}\right) = \frac{KU}{\sqrt{x}}\left[\frac{y}{\delta^2} - \frac{y^2}{\delta^3}\right]$$

$$v = \frac{KU}{\sqrt{x}}\left[\frac{y^2}{2\delta^2} - \frac{y^3}{3\delta^3}\right] + C$$

where C is the constant of integration. As $v = 0$ at $y = 0$ (impermeable flat plate), we obtain $C = 0$, giving

$$v = \frac{KU}{\sqrt{x}}\left[\frac{y^2}{2\delta^2} - \frac{y^3}{3\delta^3}\right]$$

NOMENCLATURE

C_f Shear coefficient
$f(\eta)$ Dimensionless stream function used in the similarity solution, similarity variable
f' First derivative of f with respect to η
f'' Second derivative of f with respect to η
f''' Third derivative of f with respect to η
H Shape factor $\left(H = \delta_1/\delta_2\right)$
L Characteristic length scale
L_e Entrance length in a pipe flow

Re	Reynolds number
u, v	Velocity components in the x- and y-directions, respectively
U	Dimensional velocity in the x-direction $\left(U = u/u_\infty\right)$
x	Cartesian coordinate x
y	Cartesian coordinate y

SUBSCRIPTS AND SUPERSCRIPTS

$(\bar{})$	Average value
∞	Infinity (free stream)

GREEK SYMBOLS

δ	Boundary layer thickness
δ_1	Displacement thickness
δ_2	Momentum thickness
η	Dimensionless similarity variable $\left(\eta = y/\delta\right)$
μ	Dynamic viscosity
ν	Kinematic viscosity $\left(\nu = \mu/\rho\right)$
π	Ratio of the circumference of a circle to its diameter
ρ	Density
τ_0	Wall shear stress

REFERENCES

Schlichting, H. 1979. *Boundary Layer Theory*, 7th edition. New York: McGraw-Hill.
Sultanian, B.K. 2015. *Fluid Mechanics: An Intermediate Approach*. Boca Raton, FL: Taylor & Francis.

BIBLIOGRAPHY

Anderson, J.D. 2005. Ludwig Prandtl's boundary layer. *Physics Today*. 58(12):42–48.
Carnahan, B., H.A. Luther, and J.O. Wilkes. 1969. *Applied Numerical Methods*. New York: John Wiley & Sons.
Falkner, V.M. and S.W. Skan. 1931. Some approximate solutions of the boundary layer equations. *Philosophical Magazine*. 12:865–896.
Howarth, L. 1938. On the solution of the laminar boundary layer equations. *Proceedings of the Royal Society of London*. 164:547–579.

8 Centrifugal Pumps and Fans

REVIEW OF KEY CONCEPTS

We concisely present here some key concepts of centrifugal pumps and fans. More details on each topic are given, for example, in Sultanian (2019).

IMPELLER IDEAL HEAD

With $V_{\theta 1} = 0$ at the impeller inlet, we use the Euler's turbomachinery equation (see Appendix A) to compute the ideal or virtual head H_i produced by the impeller as

$$H_i = \frac{U_2 V_{\theta 2}}{g} \tag{8.1}$$

Note that H_i is higher than that found in practice, the hydraulic efficiency being always less than 100%.

IMPELLER PRESSURE RISE

As $U_2 V_{\theta 2}$ is the mechanical energy per unit mass transferred from the pump impeller to the fluid, we can express the increase in total pressure, which is the fluid total mechanical energy per unit volume, over the impeller as

$$p_{02} - p_{01} = \rho U_2 V_{\theta 2}$$

Substituting $p_{01} = p_1 + \rho V_1^2/2$ and $p_{02} = p_2 + \rho V_2^2/2$ in this equation yields

$$p_2 - p_1 = \rho U_2 V_{\theta 2} - \rho \left(\frac{V_2^2 - V_1^2}{2} \right) \tag{8.2}$$

Using the cosine rule for the velocity triangle at the impeller outlet, we obtain

$$U_2 V_{\theta 2} = \frac{U_2^2 + V_2^2 - W_2^2}{2}$$

whose substitution in Equation 8.2 along with $V_1^2 = W_1^2 - U_1^2$ yields

$$p_2 - p_1 = \frac{\rho \left(U_2^2 - U_1^2 + W_1^2 - W_2^2 \right)}{2} \tag{8.3}$$

SLIP COEFFICIENT

The ratio of the actual $V_{\theta2'}$ to ideal $V_{\theta2}$ is usually known as the slip coefficient μ_s obtained from the equation

$$\mu_s = \frac{V_{\theta2'}}{V_{\theta2}} = 1 - \frac{\pi U_2 \sin\beta_2}{V_{\theta2} n_b} \tag{8.4}$$

where n_b is the number of blades and β_2 is the exit angle for the relative velocity W_2.

IMPELLER ACTUAL HEAD

For a finite number of blades, we must include the effect of slip and use the actual tangential velocity component $V_{\theta2'}$, replacing $V_{\theta2}$, which corresponds to perfect guidance by the blades. In this case, the actual fluid angle is $\beta_{2'}$, and we calculate the energy transfer per unit mass as $V_{\theta2'} U_2$ and the corresponding input head H_{in} as

$$H_{in} = \frac{U_2 V_{\theta2'}}{g} \tag{8.5}$$

EFFICIENCY

All frictional and secondary flow losses in a turbomachine result in the conversion of mechanical energy into thermal (internal) energy. We can write the steady flow mechanical energy equation from suction to discharge as

$$gH = g\left(H_{in} - H_{loss}\right) = \frac{V_d^2 - V_s^2}{2} + \left(z_d - z_s\right)g + \frac{p_d - p_s}{\rho} \tag{8.6}$$

where the subscripts s and d refer to properties at the suction and discharge flanges of the pump casing, and H is the output head. Typically, for computing H, we need to consider only the last term on the right-hand side of this equation, giving

$$H = \frac{p_d - p_s}{\rho g} \tag{8.7}$$

We write the overall pump efficiency as

$$\eta = \eta_m \eta_v \eta_h \tag{8.8}$$

where the mechanical efficiency η_m is defined as

$$\eta_m = \frac{\left(m + m_{leak}\right) g H_{in}}{P} = \frac{\left(Q + Q_{leak}\right) \rho g H_{in}}{P} \tag{8.9}$$

the volumetric efficiency η_v as

$$\eta_v = \frac{m}{m + m_{leak}} = \frac{Q}{Q + Q_{leak}} \tag{8.10}$$

and the hydraulic efficiency η_h as

$$\eta_h = \frac{H}{H_{in}} = \frac{H_{in} - H_{loss}}{H_{in}} \tag{8.11}$$

Karassik et al. (2007) provides the following empirical correlation for η_h:

$$\eta_h = 1 - \frac{0.8}{Q^{1/4}} \tag{8.12}$$

where Q is the capacity in gallons per minute.

In practice, when we measure the power P of the motor that drives the pump from the dynamometer test and total head H by evaluating the mechanical energy terms at the suction and discharge sides of the pump, we express the overall pump efficiency as

$$\eta = \frac{mgH}{P} \tag{8.13}$$

Karassik et al. (2007) provides the volumetric efficiency data whose correlation yields

$$\eta_v = 1 - \frac{C}{Q^n} \tag{8.14}$$

where C and n are constants, which depend on the dimensional specific speed N_s. Some values of these constants are presented in Table 8.1.

PUMP CHARACTERISTIC CURVES

Characteristic curves for a given pump are determined by test, and they consist primarily of a plot of head H as a function of the volume flow rate Q. Expressing Equation 8.1 in terms of Q, we obtain

$$H_i = U_2 \frac{U_2 - Q \cot \beta_2 / A_2}{g} \tag{8.15}$$

TABLE 8.1
Constants for Equation 8.14

$N_s \left[(\text{rpm})(\text{gpm})^{0.5} / \text{ft}^{0.75} \right]$	C	n
500	1.0	0.50
1000	0.35	0.38
2000	0.091	0.24
3000	0.033	0.128

Dividing this equation by the square of twice the tip speed ND_2 yields

$$\frac{gH_i}{N^2 D_2^2} = 0.5\left(0.5 - \frac{D_2 Q \cot \beta_2}{\pi b N D_2^3}\right) \tag{8.16}$$

which indicates a functional relationship between the head coefficient, defined as

$$\Psi_2 = \frac{gH}{N^2 D_2^2} \tag{8.17}$$

and the flow coefficient, defined as

$$\Phi_2 = \frac{Q}{N D_2^3} \tag{8.18}$$

For a given pump, assuming similar flows with constant Ψ and Φ, Equation 8.17 yields

$$\frac{H_1}{N_1^2} = \frac{H_2}{N_2^2} \tag{8.19}$$

and Equation 8.18 yields

$$\frac{Q_1}{N_1} = \frac{Q_2}{N_2} \tag{8.20}$$

Equations 8.19 and 8.20 express the pump (or fan) laws and together yield

$$\frac{H_1}{Q_1^2} = \frac{H_2}{Q_2^2} \tag{8.21}$$

which states that H is proportional to Q^2.

CAVITATION

The cavitation in a pump occurs when the fluid pressure falls below the vapor pressure with the formation and collapse of vapor bubbles. Outward flow in the impeller passage, which is accompanied by pressure rise, results in a collapse of the bubble. The head-capacity $(H - Q)$ curve falls off at the flow corresponding to the onset of cavitation.

To avoid cavitation, the net positive suction head (NPSH), defined as the atmospheric head plus the distance of liquid level above the pump centerline minus the friction head in suction piping minus the liquid vapor pressure (in terms of head) at the operating temperature, should be higher than $(NPSH)_c$, which is obtained from

$$S_c = \frac{N Q^{\frac{1}{2}}}{[g(NPSH)_c]^{\frac{3}{4}}} \tag{8.22}$$

where S_c is the critical specific speed. Shepherd (1956) recommends $S_c = 3$ for single-suction water pumps and $S_c = 4$ for double-suction water pumps, forming useful rules of thumb to prevent cavitation.

PRELIMINARY DESIGN OF PUMPS

The design specifications of a pump typically include head, capacity, and speed. We outline here a stepwise preliminary design process to determine the pump impeller dimensions.

Step 1. Compute the specific speed N_s and determine the overall pump efficiency η from the available test data plotted in the form of η as a function of N_s with Q as the parameter.

Step 2. Calculate the brake power from Equation 8.13 and determine the shaft torque as

$$\Gamma = \frac{P}{N} \tag{8.23}$$

The double-suction impeller shown in Figure 8.1, taking half of the flow on each, is used to maintain low fluid speeds at the impeller eye and to avoid abrupt turning of the fluid when the shroud diameters are large. The shroud diameter should not exceed half of the impeller diameter. For determining overall, hydraulic, or volumetric efficiency, we treat each side of the double-suction impeller as a single-suction impeller. Thus, the specific speed that is used to determine efficiency is based on $Q/2$, rather than Q. On the other hand, for determining the pump power or tip blade width, we use the full flow rate Q.

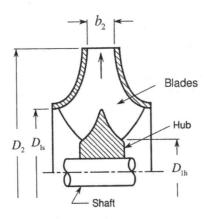

FIGURE 8.1 A double-suction centrifugal impeller.

Step 3. Compute the hydraulic efficiency η_h from Equation 8.12 and, substituting a value of $V_{\theta 2'}/U_2$ in the range of 0.5–0.55, compute the impeller tip speed U_2 from the equation

$$U_2 = \sqrt{\frac{gH}{\eta_h\left(V_{\theta 2'}/U_2\right)}} \tag{8.24}$$

Step 4. Calculate the impeller diameter from $D_2 = 2U_2/N$.

Step 5. Using the recommendation of Karassik et al. (2007), select the flow coefficient in the range $\left(N_s/21{,}600\right) < \varphi_2 < \left(N_s/15{,}900\right) + 0.019$, where N_s is the dimensional specific speed in $(\text{rpm})(\text{gpm})^{0.5}/\text{ft}^{0.75}$ and $\varphi_2 = W_{m2}/U_2$, and hence calculate $W_{m2} = \varphi_2 U_2$.

Step 6. Use Equation 8.14 and Table 8.1 to compute η_v, and hence compute $\left(Q + Q_{\text{leak}}\right)$ from
Equation 8.10 and width b_2 at the blade tip from

$$b_2 = \frac{Q + Q_{\text{leak}}}{\pi D_2 \, W_{m2}} \tag{8.25}$$

Step 7. Using the recommendation of Karassik et al. (2007), calculate the shroud diameter as

$$D_{1s} = 4.54\left(\frac{Q + Q_{\text{leak}}}{kN \tan \beta_{1s}}\right)^{1/3} \tag{8.26}$$

where

$$k = 1 - \left(\frac{D_{1h}}{D_{1s}}\right)^2 \tag{8.27}$$

and the hub diameter D_{1h} and the shroud diameter D_{1s} are in inches, N in rpm, and Q in gpm. Note that Equation 8.26 is for single-suction impellers. For using this equation for double-suction impellers, we must replace $Q + Q_{\text{leak}}$ by $\left(Q + Q_{\text{leak}}\right)/2$. In Equation 8.27, D_{1h}/D_{1s} is in the range of 0–0.5. In Equation 8.26, Karassik et al. (2007) recommend β_{1s} to be in the range of $10° - 25°$.

Step 8. Determine the blade angle β_2 using the following iterative process:

Step A. As recommended by Karassik et al. (2007), select a blade angle in the range of $17° - 25°$.

Step B. As recommended by Pfleiderer (1949) and Church (1972), calculate the optimum number of blades from

$$n_b = 6.5\left(\frac{D_2 + D_{1s}}{D_2 - D_{1s}}\right)\sin\left(\frac{\beta_{1s} + \beta_2}{2}\right) \tag{8.28}$$

The optimum number of blades for pumps lies in the range of 5–12.
Step C. Calculate $V_{\theta 2}/U_2$ from the equation

$$\frac{V_{\theta 2}}{U_2} = \frac{V_{\theta 2'}}{U_2} + \frac{\pi \sin \beta_2}{n_b} \tag{8.29}$$

which we have obtained from Equation 8.4.
Step D. Compute β_2 as

$$\beta_2 = \tan^{-1} \left| \frac{W_{m2}/U_2}{1 - V_{\theta 2}/U_2} \right| \tag{8.30}$$

Step E. Repeat Steps from A to D until the value of β_2 assumed in Step
A agrees within an acceptable error with its value computed in Step D.

CENTRIFUGAL FANS

Pumps handle liquids, and fans handle gases flowing at low Mach numbers
$(M \leq 0.3)$. Although the gases are compressible, their flow at low Mach numbers
may be treated as incompressible. Therefore, but for the fluid used, fan design
parallels that of a pump. However, compared with a pump, a centrifugal fan
requires a smaller radius ratio R_2/R_1. Although it has a volute but has no diffuser
to enhance the pressure rise. As shown in Figure 8.2, the flow passages between
impeller blades are quite short.

The analysis and design of the impeller proceed as with the centrifugal pump.
The small changes of gas density are ignored, and the incompressible equations are
applied as with pumps. Performance curves are qualitatively the same as for pumps,
except that the units of head are customarily given in centimeters of water, and those
of capacity are typically in cubic meter per minute. Similarity laws for fans are
identical to those for pumps.

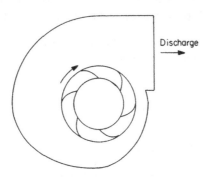

FIGURE 8.2 Centrifugal fan.

PROBLEM 8.1:   DIMENSIONLESS AND DIMENSIONAL SPECIFIC SPEEDS OF A CENTRIFUGAL WATER PUMP

Calculate the dimensionless and dimensional specific speeds for a centrifugal water pump having design-point performance $Q = 545\,m^3/h$, $H = 21\,m$, and $N = 870$ rpm.

SOLUTION FOR PROBLEM 8.1

For computing the dimensionless specific speed of the centrifugal water pump, we first convert each quantity to consistent units, giving

$$Q = 545\,m^3/h = \frac{545}{3600} = 0.1514\,m^3/s$$

and

$$N = 870 \text{ rpm} = \frac{870 \times \pi}{30} = 91.106\,rad/s$$

We compute the dimensional specific speed as

$$N_s = \frac{NQ^{\frac{1}{2}}}{H^{\frac{3}{4}}} = \frac{870 \times (545)^{\frac{1}{2}}}{(21)^{\frac{3}{4}}} = 2070.394 \; \frac{(\text{rpm})(m^3/h)^{\frac{1}{2}}}{m^{\frac{3}{4}}}$$

We compute the dimensionless specific speed as

$$\hat{N}_s = \frac{NQ^{\frac{1}{2}}}{(gH)^{\frac{3}{4}}} = \frac{91.106 \times (0.1514)^{\frac{1}{2}}}{(9.81 \times 21)^{\frac{3}{4}}} = 0.652$$

PROBLEM 8.2:   USING ROTHALPY TO COMPUTE PRESSURE RISE IN A CENTRIFUGAL IMPELLER

Using the concept of rothalpy presented in Appendix A and assuming an isentropic flow, derive the following equation to compute pressure rise in a centrifugal impeller:

$$p_2 - p_1 = \frac{\rho\left(U_2^2 - U_1^2 + W_1^2 - W_2^2\right)}{2}$$

SOLUTION FOR PROBLEM 8.2

For an isentropic flow in the impeller, the rothalpy remains constant between any two points, giving

$$I_1 = c_p T_{0R1} - \frac{U_1^2}{2} = I_2 = c_p T_{0R2} - \frac{U_2^2}{2}$$

which we can write as

$$h_{0R2} - h_{0R1} = \frac{U_2^2}{2} - \frac{U_1^2}{2}$$

$$h_2 - h_1 + \frac{W_2^2}{2} - \frac{W_1^2}{2} = \frac{U_2^2}{2} - \frac{U_1^2}{2}$$

$$h_2 - h_1 = \frac{U_2^2}{2} - \frac{U_1^2}{2} + \frac{W_1^2}{2} - \frac{W_2^2}{2}$$

As we have $dh = dp/\rho$ for an isentropic flow, we write this equation as

$$\frac{p_2 - p_1}{\rho} = \frac{U_2^2}{2} - \frac{U_1^2}{2} + \frac{W_1^2}{2} - \frac{W_2^2}{2}$$

giving

$$p_2 - p_1 = \frac{\rho\left(U_2^2 - U_1^2 + W_1^2 - W_2^2\right)}{2}$$

PROBLEM 8.3:   HYDRAULIC, VOLUMETRIC, AND MECHANICAL EFFICIENCIES OF A CENTRIFUGAL PUMP

A centrifugal water pump, when discharging $227\,\mathrm{m^3/h}$ (~1000 gpm) at the dimensional specific speed of $1859(\mathrm{rpm})\left(\mathrm{m^3/h}\right)^{0.5}/\mathrm{m^{0.75}}$ (~1600(rpm)(gpm)$^{0.5}/\mathrm{ft^{0.75}}$), has the overall efficiency of 0.83. Determine its hydraulic, volumetric, and mechanical efficiencies.

SOLUTION FOR PROBLEM 8.3

Using Equation 8.12, we calculate the hydraulic efficiency as

$$\eta_h = 1 - \frac{0.8}{Q^{1/4}} = 1 - \frac{0.8}{(1000)^{1/4}} = 0.858$$

We determine $C = 0.195$ and $n = 0.296$ for Equation 8.14 by interpolating between values in Table 8.1 and calculate the volumetric efficiency as

$$\eta_v = 1 - \frac{0.195}{Q^{0.296}} = 1 - \frac{0.195}{(1000)^{0.296}} = 0.975$$

Finally, we determine the mechanical efficiency from Equation 4.20 as

$$\eta_m = \frac{\eta}{\eta_h \eta_v} = \frac{0.83}{0.858 \times 0.975} = 0.99$$

PROBLEM 8.4: VARIOUS HEADS OF A SINGLE-SUCTION CENTRIFUGAL PUMP

A single-suction centrifugal pump with six blades, while running at 885 rpm, discharges water at the rate of $2270\,\mathrm{m^3/h}$ (~10,000 gpm). Determine the ideal, input, and output heads, if the impeller diameter is 0.9652 m (~38 in), and the blade angle at the tip is 21.6°.

SOLUTION FOR PROBLEM 8.4

For the rotational speed

$$N = \frac{885 \times \pi}{30} = 92.68 \text{ rad/s}$$

we determine the tip speed as

$$U_2 = \frac{ND_2}{2} = \frac{92.68 \times 0.9652}{2} = 44.726 \,\mathrm{m/s}$$

CALCULATION OF W_{m2}

As we are not given the impeller tip width in this problem, let us choose a value for the flow coefficient φ_2 in the range $(N_s/21,600) < \varphi_2 < (N_s/15,900) + 0.019$ where N_s is the dimensional specific speed in $(\mathrm{rpm})(\mathrm{gpm})^{0.5}/\mathrm{ft}^{0.75}$ and $\varphi_2 = W_{m2}/U_2$. Assuming $N_s = 1000 (\mathrm{rpm})(\mathrm{gpm})^{0.5}/\mathrm{ft}^{0.75}$ and using the upper limit for φ_2, we obtain

$$\varphi_2 = 1000/15,900 + 0.019 = 0.082$$

Hence,

$$W_{m2} = 44.726 \times 0.082 = 3.668 \text{ m/s}$$

CALCULATION OF $V_{\theta 2}$

From the pump outlet velocity triangle, we obtain

$$V_{\theta 2} = U_2 - W_{m2}\cot \beta_2 = 44.726 - 3.668 \times \cot 21.6° = 35.463 \text{ m/s}$$

CALCULATION OF H_i

We obtain from Equation 8.1

$$H_i = \frac{U_2 V_{\theta 2}}{g} = \frac{44.726 \times 35.463}{9.81} = 161.757\,\text{m}$$

CALCULATION OF μ_s AND $V_{\theta 2'}$

With $n_b = 6$, we obtain from Equation 8.4

$$\mu_s = 1 - \frac{3.14159 \times 44.726 \times \sin 21.6°}{35.479 \times 6} = 0.757$$

Hence,

$$V_{\theta 2'} = \mu_s V_{\theta 2} = 0.757 \times 35.479 = 26.858 \text{ m/s}$$

CALCULATION OF H_{in}

From Equation 8.5, we obtain

$$H_{in} = \frac{44.726 \times 26.858}{9.81} = 122.452 \text{ m}$$

CALCULATION OF η_h AND H

From Equation 8.12, we obtain

$$\eta_h = 1 - \frac{0.8}{(10,000)^{\frac{1}{4}}} = 0.92$$

giving from Equation 8.11

$$H = 122.452 \times 0.92 = 112.656 \text{ m}$$

To check the assumed specific speed, we calculate

$$N_s = \frac{NQ^{\frac{1}{2}}}{H^{\frac{3}{4}}} = \frac{(885)(10,000)^{\frac{1}{2}}}{(369.6)^{\frac{3}{4}}} = 1050\,(\text{rpm})(\text{gpm})^{0.5}/\text{ft}^{0.75}$$

Readers may verify that assuming $N_s = 1063\,(\text{rpm})(\text{gpm})^{0.5}/\text{ft}^{0.75}$ instead of $N_s = 1000\,(\text{rpm})(\text{gpm})^{0.5}/\text{ft}^{0.75}$ will finally yield $N_s = 1063\,(\text{rpm})(\text{gpm})^{0.5}/\text{ft}^{0.75}$ with $H_i = 159.697\,\text{m}$, $H_{in} = 120.392\,\text{m}$, and $H = 110.760\,\text{m}$. In this solution, we have also assumed the upper limit for φ_2, which equals 0.082. Using the lower limit of $\varphi_2 = 0.0434$ will yield the following converged solution:

$$N_s = 938(\text{rpm})(\text{gpm})^{0.5}/\text{ft}^{0.75}, \ H_i = 181.548\,\text{m}, \ H_{in} = 142.244\,\text{m, and, } H = 130.863\,\text{m}.$$

PROBLEM 8.5:   ELECTRIC POWER NEEDED TO RUN A CENTRIFUGAL PUMP

Calculate the electric power needed to drive a centrifugal pump running at 1160 rpm with a discharge of 704 m³/h (~3100 gpm). The quadratic polynomial representing the $H - Q$ curve for the pump at this rpm is

$$H = -4.015 \times 10^{-5} Q^2 + 0.0257Q + 32.447$$

where H is in meters and Q is in m³/h. The mechanical efficiency of the pump is 95%. Also determine the dimensionless specific speed of the pump.

SOLUTION FOR PROBLEM 8.5

ROTATIONAL SPEED

$$N = \frac{1160 \times \pi}{30} = 121.475 \text{ rad/s}$$

HEAD

From the given $H - Q$ characteristic, we obtain

$$H = -4.015 \times 10^{-5} \times (704)^2 + 0.0257 \times (704) + 32.447 = 30.641\,\text{m}$$

HYDRAULIC EFFICIENCY

From Equation 8.12, we obtain

$$\eta_h = 1 - \frac{0.8}{(3100)^{1/4}} = 0.893$$

VOLUMETRIC EFFICIENCY

For using Table 8.1, we calculate the dimensional specific speed as

$$N_s = \frac{NQ^{1/2}}{H^{3/4}} = \frac{(1160)(3100)^{1/2}}{(30.641 \times 3.2808)^{3/4}} = 2034.224(\text{rpm})(\text{gpm})^{0.5}/\text{ft}^{0.75}$$

which yields $C = 0.089$ and $n = 0.236$ by interpolating between values in Table 8.1. We calculate the volumetric efficiency from Equation 8.14 as

$$\eta_v = 1 - \frac{0.089}{Q^{0.296}} = 1 - \frac{0.089}{(3100)^{0.236}} = 0.987$$

OVERALL EFFICIENCY

$$\eta = \eta_h \eta_v \eta_m = 0.893 \times 0.987 \times 0.95 = 0.837$$

ELECTRIC POWER FOR THE PUMP MOTOR

From Equation 8.13, we obtain

$$P = \frac{Q(\rho g H)}{\eta} = \frac{704 \times 1000 \times 9.81 \times 30.641}{3600 \times 0.837 \times 1000} = 70.229 \, \text{kW}$$

DIMENSIONLESS SPECIFIC SPEED

$$\hat{N}_s = \frac{NQ^{\frac{1}{2}}}{(gH)^{\frac{3}{4}}} = \frac{121.475 \times (704/3600)^{\frac{1}{2}}}{(9.81 \times 30.641)^{\frac{3}{4}}} = 0.744$$

PROBLEM 8.6: SUCTION SPECIFIC SPEED OF A CENTRIFUGAL PUMP

A single-suction centrifugal pump operating at $T_{amb} = 27^o\text{C}$ and $P_{amb} = 1.0135\,\text{bar}$ delivers $Q = 704\,\text{m}^3/\text{h}$ (~3100 gpm) and $H = 30.48\,\text{m}$ (~100 ft) at $N = 1160\,\text{rpm}$. The pipe connecting the pump suction to the supply reservoir has a diameter D_{su} of 0.2032 m (~8 in), a length L_{su} of 3.048 m (~10 ft), and the absolute roughness of 50 μm for its wall. The pipe lifts water from a reservoir, which is 3.048 m (~10 ft) below the centerline of the pump with its free surface exposed to the ambient. At 27°C, the vapor pressure of water is $p_{vap} = 3560\,\text{Pa}$ and its kinematic viscosity equals $0.852 \times 10^{-6}\,\text{m}^2/\text{s}$. Determine the suction specific speed. Will the pump cavitate at the given operating conditions?

SOLUTION FOR PROBLEM 8.6

ROTATIONAL SPEED

$$N = \frac{1160 \times \pi}{30} = 121.475 \, \text{rad/s}$$

DISCHARGE

$$Q = 704 \, \text{m}^3/\text{h} = \frac{704}{3600} = 0.1956 \, \text{m}^3/\text{s}$$

SUCTION PIPE CROSS-SECTION AREA

$$A_{su} = \frac{\pi \times D_{su}^2}{4} = \frac{\pi \times (0.2032)^2}{4} = 0.03243\,\mathrm{m}^2$$

AVERAGE FLOW VELOCITY IN THE SUCTION PIPE

$$V_{su} = \frac{Q}{A} = \frac{0.1956}{0.03243} = 6.030\,\mathrm{m/s}$$

FLOW REYNOLDS NUMBER IN THE SUCTION PIPE

$$Re = \frac{V_{su} D_{su}}{\nu} = \frac{6.030 \times 0.2032}{0.852 \times 10^{-6}} = 1.438 \times 10^6$$

FRICTION FACTOR

Using the approximate equation proposed by Swamee and Jain (1976), we obtain

$$\frac{1}{\sqrt{f}} = -2.0\,\log_{10}\left(\frac{e/D_{su}}{3.7} + \frac{5.74}{Re^{0.9}}\right) = -2.0\,\log_{10}\left(\frac{50 \times 10^{-6}/0.2032}{3.7} + \frac{5.74}{\left(1.438 \times 10^6\right)^{0.9}}\right)$$

$$f = 0.0150$$

FRICTION HEAD LOSS IN THE SUCTION PIPE

$$h_f = \frac{f\,L_{su} V_{su}^2}{2g D_{su}} = \frac{0.0150 \times 3.048 \times (6.030)^2}{2 \times 9.81 \times 0.2032} = 0.417\,\mathrm{m}$$

NET POSITIVE SUCTION HEAD

We calculate the *NPSH* for the given centrifugal pump from the equation

$$NPSH = \frac{p_{atm}}{\rho g} + z_{su} - h_f - \frac{p_{vap}}{\rho g} = \frac{1.0135 \times 10^5}{9.81 \times 1000} - 3.048 - 0.417 - \frac{3560}{9.81 \times 1000}$$

$$NPSH = 6.50\,\mathrm{m}$$

SUCTION SPECIFIC SPEED (S)

$$S = \frac{N Q^{\frac{1}{2}}}{\left[g(NPSH)\right]^{\frac{3}{4}}} = \frac{121.475 \times (0.1956)^{0.5}}{(9.81 \times 6.50)^{0.75}} = 2.38$$

As this computed suction specific speed is less than its recommended critical value of $S_c = 3$, the pump will not cavitate.

PROBLEM 8.7: PRELIMINARY DESIGN OF A DOUBLE-SUCTION PUMP IMPELLER

The specifications of a double-suction centrifugal water pump correspond to $Q = 545\,\text{m}^3/\text{h}$ (~2400 gpm), $H = 21\,\text{m}$ (~68.9 ft), and $N = 870\,\text{rpm}$. Find $D_2, b_2, D_{1s}, D_{1h}, \beta_2, \beta_{1s}$, and n_b for the impeller.

SOLUTION FOR PROBLEM 8.7

ROTATIONAL SPEED

$$N = \frac{870 \times \pi}{30} = 91.106\,\text{rad/s}$$

DISCHARGE

$$Q = 545\,\text{m}^3/\text{h} = \frac{545}{3600} = 0.1514\,\text{m}^3/\text{s}$$

DIMENSIONAL SPECIFIC SPEED

For the double-suction pump, we calculate dimensional speed using $Q/2$ as

$$N_s = \frac{N(Q/2)^{\frac{1}{2}}}{H^{\frac{3}{4}}} = \frac{(870)(1200)^{\frac{1}{2}}}{(68.9)^{\frac{3}{4}}} = 1260.137\,(\text{rpm})(\text{gpm})^{0.5}/\text{ft}^{0.75}$$

HYDRAULIC EFFICIENCY

For the double-suction pump, we calculate the hydraulic efficiency by using $Q/2$ in place of Q in Equation 8.12

$$\eta_h = 1 - \frac{0.8}{(Q/2)^{\frac{1}{4}}} = 1 - \frac{0.8}{(1200)^{\frac{1}{4}}} = 0.864$$

VOLUMETRIC EFFICIENCY

For the double-suction pump, we calculate the volumetric efficiency by using $Q/2$ in place of Q in Equation 8.14 and determine $C = 0.283$ and $n = 0.344$ from Table 8.1 by interpolation, giving

$$\eta_v = 1 - \frac{0.283}{(1200)^{0.344}} = 0.975$$

For the current design, for the impeller, we choose $D_{1h}/D_{1s} = 0.5$ and $\beta_{1s} = 17°$. With these values fixed, we use the following stepwise preliminary design process:

Step 1. Assuming $V_{\theta 2'}/U_2 = 0.5$, we compute the impeller tip speed U_2 from Equation 8.24

$$U_2 = \sqrt{\frac{gH}{\eta_h \left(V_{\theta 2'}/U_2\right)}} = \sqrt{\frac{9.81 \times 21}{0.864 \times 0.5}} = 21.837 \, \text{m/s}$$

and calculate the impeller diameter

$$D_2 = \frac{2U_2}{N} = \frac{2 \times 21.837}{91.106} = 0.479 \, \text{m}$$

Step 2. Using the recommendation of Karassik et al. (2007), we select the flow coefficient in the range $\left(N_s/21,600\right) < \varphi_2 < \left(N_s/15,900\right) + 0.019$, where N_s is the dimensional specific speed in $(\text{rpm})(\text{gpm})^{0.5}/\text{ft}^{0.75}$ and $\varphi_2 = W_{m2}/U_2$. In this design, we choose the upper limit for φ_2 given by

$$\varphi_2 = \frac{N_s}{15,900} + 0.019 = \frac{1260.137}{15,900} + 0.019 = 0.0983$$

giving

$$W_{m2} = \varphi_2 U_2 = 0.0983 \times 21.837 = 2.146 \, \text{m/s}$$

Step 3. Using the computed hydraulic efficiency $\eta_v = 0.975$, we compute

$$\left(Q + Q_{\text{leak}}\right) = \frac{Q}{\eta_v} = \frac{0.1514}{0.975} = 0.155 \, \text{m}^3/\text{s}$$

and the width b_2 at the blade tip as

$$b_2 = \frac{Q + Q_{\text{leak}}}{\pi D_2 W_{m2}} = \frac{0.155}{3.1415 \times 0.479 \times 2.146} = 0.048 \, \text{m}$$

Step 4. Using the recommendation of Karassik et al. (2007), we calculate the shroud diameter from Equation 8.24 by replacing $\left(Q + Q_{\text{leak}}\right)$ by $\left(Q + Q_{\text{leak}}\right)/2$ and using $k = 1 - \left(D_{1h}/D_{1s}\right)^2 = 1 - (0.5)^2 = 0.75$ as

$$D_{1s} = 4.54 \left(\frac{Q + Q_{\text{leak}}}{kN \tan \beta_{1s}}\right)^{\frac{1}{3}} = 4.54 \left(\frac{1200/0.975}{0.75 \times 870 \times \tan 17°}\right)^{\frac{1}{3}} = 8.325 \, \text{in}$$

$$D_{1s} = 0.211 \, \text{m}$$

giving

$$D_{1h} = 0.211 \times 0.5 = 0.106 \, m$$

Step 5. We determine the blade angle β_2 using the following iterative process:

Step A. Select $\beta_2 = 17° \left(\beta_2 = 17.57° \right)$

Step B. As recommended by Pfleiderer (1949) and Church (1972), calculate the optimum number of blades from Equation 8.28

$$n_b = 6.5 \left(\frac{D_2 + D_{1s}}{D_2 - D_{1s}} \right) \sin \left(\frac{\beta_{1s} + \beta_2}{2} \right) = 6.5 \left(\frac{0.479 + 0.211}{0.479 - 0.211} \right) \sin \left(\frac{17° + 17°}{2} \right) = 4.9$$

$$\left(n_b = 4.98 \right)$$

giving

$$n_b = 5$$

Step C. Calculate $V_{\theta 2}/U_2$ from Equation 8.29

$$\frac{V_{\theta 2}}{U_2} = \frac{V_{\theta 2'}}{U_2} + \frac{\pi \sin \beta_2}{n_b} = 0.5 + \frac{3.1415 \times \sin (17°)}{5} = 0.6837$$

$$\left(\frac{V_{\theta 2}}{U_2} = 0.6896 \right)$$

Step D. Compute β_2 from Equation 8.30

$$\beta_2 = \tan^{-1} \left| \frac{W_{m2}/U_2}{1 - V_{\theta 2}/U_2} \right| = \tan^{-1} \left| \frac{0.0983}{1 - 0.6837} \right| = 17.26°$$

$$\left(\beta_2 = 17.57° \right)$$

Step E. Repeat Steps from A to D until the value of β_2 assumed in Step A agrees within an acceptable error with its value computed in Step D. The converged values in Steps A–D are given within parentheses.

Thus, the pump preliminary design comprises $D_2 = 0.479 \, m$, $b_2 = 0.048 \, m$, $D_{1s} = 0.211 \, m$, $D_{1h} = 0.106 \, m$, $\beta_2 = 17.57°$, $\beta_{1s} = 17°$, and $n_b = 5$. Note that these results are sensitive to the choice of $V_{\theta 2'}/U_2$, which we have assumed to be 0.5. As an exercise, readers may verify that for $V_{\theta 2'}/U_2 = 0.51$, we obtain $D_2 = 0.475 \, m$, $b_2 = 0.049 \, m$, $D_{1s} = 0.211 \, m$, $D_{1h} = 0.106 \, m$, $\beta_2 = 19°$, $\beta_{1s} = 17°$, and $n_b = 5$, and for $V_{\theta 2'}/U_2 = 0.52$, we obtain $D_2 = 0.470 \, m$, $b_2 = 0.050 \, m$, $D_{1s} = 0.211 \, m$, $D_{1h} = 0.106 \, m$, $\beta_2 = 21.4°$, $\beta_{1s} = 17°$, and $n_b = 5$.

NOMENCLATURE

A_2	Flow area at impeller exit
b_2	Width of blade at $r = r_2$ in pump impeller
c_p	Specific heat at constant pressure
C	Constant in Equation 8.14
D_{1s}	Shroud diameter
D_{1h}	Hub diameter
D_2	Impeller tip diameter
D_s	Specific diameter
D_{su}	Diameter of suction pipe
e	Absolute roughness
f	Friction factor
g	Gravitational acceleration
h	Specific static enthalpy
h_f	Friction head loss in suction pipe
h_0	Specific total enthalpy
h_{0R}	Specific total enthalpy in rotor reference frame
H	Output head
H_i	Ideal head
H_{in}	Input head
I	Rothalpy
k	Function of hub-tip ratio (Equation 8.27)
L_{su}	Length of suction pipe
$\dot{m}$	Mass flow rate of fluid at discharge flange of pump
$\dot{m}_{leak}$	Mass flow rate of fluid leaked around the outside of the impeller from the high to low pressure regions
n	Constant in Equation 8.14
n_b	Number of blades in the impeller
N	Rotor speed
N_s	Dimensional specific speed
$\hat{N}_s$	Dimensionless specific speed
$NPSH$	Net positive suction head
p	Static pressure
p_{atm}	Atmospheric pressure
p_d	Static pressure of fluid at discharge flange of pump
p_0	Total pressure
p_{su}	Static pressure of fluid at suction flange of pump
p_{vap}	Vapor pressure of fluid
P	Power to impeller shaft
Q	Capacity, i.e., volumetric flow rate delivered by pump at discharge flange
Q_{leak}	Volume flow rate of leaked fluid
r	Radius
Re	Reynolds number

S Suction specific speed
S_c Critical suction specific speed
T Static temperature
T_0 Total temperature
T_{0R} Total temperature in the rotor reference frame
U_1 Impeller speed at blade leading edge
U_2 Impeller speed at blade tip
V_d Average velocity at the pump discharge flange
V_s Average velocity at the pump suction flange
V_{su} Average velocity in suction pipe
V_1 Absolute velocity at leading edge of impeller blade
$V_{\theta 1}$ Tangential component of V_1
V_{m1} Meridional component of V_1
V_2 Absolute velocity leaving the impeller with an infinite number of blades
$V_{2'}$ Absolute velocity leaving the impeller with a finite number of blades
V_{m2} Meridional component of V_2 or $V_{2'}$
$V_{\theta 2}$ Tangential component of V_2
W_1 Relative velocity at leading edge of impeller blade
W_2 Relative velocity leaving the impeller with an infinite number of blades
$W_{2'}$ Relative velocity leaving the impeller with a finite number of blades
         Meridional component of W_2, $W_{2'}$, V_2, or $V_{2'}$
W_{1s} Relative velocity at shroud
z_d Elevation of the center of the discharge flange above (positive) or below (negative) the centerline of the pump shaft
z_r Elevation of the free surface of the supply reservoir above (positive) or below (negative) the centerline of the pump shaft
z_s Elevation of the center of the suction flange above (positive) or below (negative) the centerline of the pump shaft

GREEK SYMBOLS

β_1 Angle between W_1 and U_1
β_{1s} Angle between W_{1s} and U_{1s}
β_2 Angle between W_2 and U_2; the blade angle at the blade trailing edge
$\beta_{2'}$ Angle between $W_{2'}$ and U_2; actual fluid angle
Γ Torque
η Overall pump efficiency
η_h Hydraulic efficiency
η_m Mechanical efficiency
η_v Volumetric efficiency
μ_s Slip coefficient
ρ Fluid density
φ Flow coefficient
φ_2 Flow coefficient at impeller exit $= W_{m2}/U_2$

REFERENCES

Church, A.H. 1972. *Centrifugal Pumps and Blowers*. Huntington, NY: Krieger.
Karassik, I.J., et al. 2007. *Pump Handbook*, 4th edition. New York: McGraw-Hill Education.
Pfleiderer, C. 1949. *Die Kreiselpumpen*. Berlin: Springer-Verlag.
Shepherd, D.G. 1956. *Principles of Turbomachinery*. New York: MacMillan.
Sultanian, B.K. 2019. *Logan's Turbomachinery: Flowpath Design and Performance Fundamentals*, 3rd edition. Boca Raton, FL: Taylor & Francis.
Swamee, P. and A. Jain. 1976. Explicit equations for pipe-flow problems. *Journal of the Hydraulics Division (ASCE)*. 102(5):657–664.

BIBLIOGRAPHY

Bleier, F.P. 1997. *Fan Handbook: Selection, Application, and Design*, 1st edition. New York: McGraw-Hill Education.
Gülich, J.F. 2014. *Centrifugal Pumps*. New York: Springer.
Sultanian, B.K. 2015. *Fluid Mechanics: An Intermediate Approach*. Boca Raton, FL: Taylor & Francis.

9 Centrifugal Compressors

REVIEW OF KEY CONCEPTS

Centrifugal compressors are widely used in ground-vehicle power plants, auxiliary power units, and other small units. Although slightly less efficient than axial-flow compressors, a single-stage centrifugal compressor can produce five times higher pressure ratio than a single-stage axial-flow compressor. We concisely present here some key concepts of centrifugal compressors. More details on each topic are given, for example, in Sultanian (2019).

ENERGY TRANSFER AND MECHANICAL EFFICIENCY

Unlike centrifugal pumps and fans, centrifugal compressors feature gas flows at high Mach numbers $(M > 0.3)$ that must be modeled using compressible flow equations. The total change in air total enthalpy in a centrifugal compressor changes due to aerodynamic energy transfer in the impeller and some additional rotational work transfer from frictional losses associated with bearing, seal, and disk friction. We assume that the mechanical energy lost through frictional processes reappears as enthalpy in the outflowing gas. Thus, we write the aerodynamic specific energy transfer in a centrifugal compressor as

$$E = \eta_m \left(h_{03} - h_{01} \right) = U_2 V_{\theta 2'} \tag{9.1}$$

where η_m denotes the mechanical efficiency.

For various thermodynamic calculations involved in the compressor analysis and design, it is imperative to understand the compressor h-s diagram shown in Figure 9.1. In this figure, points 1 and 2 indicate the gas state at the impeller inlet and outlet, respectively. The diffusion process occurs between points 2 and 3. As the flow kinetic energies are usually considerable, the corresponding stagnation properties are shown in the figure with subscripts 01, 02, and 03. The compressor efficiency

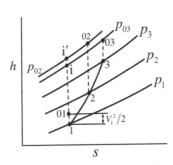

FIGURE 9.1 Enthalpy-entropy diagram.

represents the ratio of the ideal increase of fluid energy (work of an isentropic compression) divided by the actual energy input to the fluid needed to attain the actual final pressure p_{03}. We evaluate the work of the isentropic process from state 01 to state i shown in the figure from the equation

$$E_i = c_p(T_i - T_{01}) = c_p T_{01}\left[\left(\frac{p_{03}}{p_{01}}\right)^{(\gamma-1)/\gamma} - 1\right] \tag{9.2}$$

COMPRESSOR EFFICIENCY

We define the compressor aerodynamic efficiency, which is determined experimentally, as the ratio of ideal energy transfer to the actual energy transfer

$$\eta_c = \frac{\eta_m E_i}{E} = \frac{T_i - T_{01}}{T_{03} - T_{01}} \tag{9.3}$$

In Equations 9.1 and 9.3, we have assumed that the flow in the diffuser situated at the impeller outlet remains adiabatic, that is, $h_{02} = h_{03}$ and $T_{02} = T_{03}$.

IMPELLER EFFICIENCY

We define the impeller efficiency as

$$\eta_{im} = \frac{T_{i'} - T_{01}}{T_{02} - T_{01}} \tag{9.4}$$

where $T_{i'}$ is as shown in Figure 9.1.

COMPRESSOR TOTAL PRESSURE RATIO

Using Equations 9.1–9.3, we obtain the overall compressor total pressure ratio as

$$\frac{p_{03}}{p_{01}} = \left(1 + \frac{\eta_c E}{c_p T_{01} \eta_m}\right)^{\gamma/(\gamma-1)} = \left(1 + \frac{\eta_c U_2 V_{\theta 2'}}{c_p T_{01} \eta_m}\right)^{\gamma/(\gamma-1)} \tag{9.5}$$

Because of relative eddies between the blades, as in the case of centrifugal pumps, slip exists in the compressor impeller. Consequently, we calculate the actual tangential velocity component using the slip coefficient as

$$V_{\theta 2'} = \mu_s V_{\theta 2} \tag{9.6}$$

where we compute μ_s from the Stanitz equation

$$\mu_s = 1 - \frac{0.63\pi}{n_b}\left(\frac{1}{1 - \varphi_2 \cot\beta_2}\right) \tag{9.7}$$

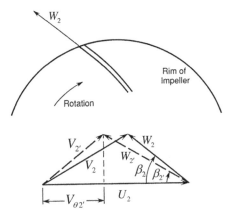

FIGURE 9.2 Velocity diagram at impeller exit.

Using Equations 9.6 and 9.7 in Equation 9.5 yields

$$\frac{p_{03}}{p_{01}} = \left(1 + \frac{\eta_c \mu_s U_2 V_{\theta 2}}{c_p T_{01} \eta_m}\right)^{\gamma/(\gamma-1)} \tag{9.8}$$

Thus, the total pressure ratio for a compressor stage can be determined from the knowledge of the ideal velocity triangle at the impeller exit shown in Figure 9.2, the number of blades, the inlet total temperature, the compressor stage efficiency, and the mechanical efficiency.

IMPELLER DESIGN

The impeller is usually designed with a number of unshrouded blades with axial flow at the inlet annulus $(V_1 = V_{m1})$ and a large exit tangential velocity component $V_{\theta 2'}$, which is less than the impeller tip speed U_2, but it has the same sense of direction. The blades are usually curved near the rim of the impeller, so that $\beta_2 < 90°$, but they are usually bent near the leading edge to conform to the direction of the relative velocity W_1 at the inlet.

From the velocity diagram at the impeller inlet, shown in Figure 9.3, we obtain $W_1 = \left(V_1^2 + U_1^2\right)^{1/2}$, which increases with radius. Therefore, at the inlet shroud of

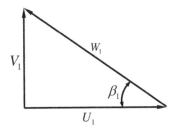

FIGURE 9.3 Velocity diagram at impeller inlet.

diameter D_{1s}, the relative velocity W_{1s} and the corresponding relative Mach number M_{R1s} are the highest. As U_1 increases with radius and V_1 remains constant, the angle β_1 decreases from hub to tip over the blade leading edge.

IMPELLER INLET DESIGN

Referring to Figure 9.3, choosing M_{R1s} allows us to carry out the impeller inlet design as follows:

Step 1. As the speed of sound remains the same in both absolute and relative reference frames, we compute the absolute Mach number M_1 from the relative Mach number M_{R1s} as

$$M_1 = M_{R1s}\sin 32° \tag{9.9}$$

giving

$$T_1 = \frac{T_{01}}{1+0.5(\gamma-1)M_1^2} \tag{9.10}$$

$$p_1 = \frac{p_{01}}{\left[1+0.5(\gamma-1)M_1^2\right]^{\gamma/(\gamma-1)}} \tag{9.11}$$

$C_1 = \sqrt{\gamma RT_1}$, $V_1 = M_1 C_1$, and $W_{1s} = M_{R1s} C_1$.

Step 2. Calculate U_{1s} from $U_{1s} = W_{1s}\cos 32°$ and the shroud diameter from $D_{1s} = 2U_{1s}/N$.

Step 3. Calculate the density from the equation of state of a perfect gas $\left(\rho_1 = p_1/RT_1\right)$ first and then calculate the hub diameter from the mass flow equation at the impeller inlet as

$$D_{1h} = \left(D_{1s}^2 - \frac{4m}{\pi\rho_1 V_1}\right)^{\frac{1}{2}} \tag{9.12}$$

Step 4. Referring to Figure 9.3, calculate the gas angle at the hub as

$$\beta_{1h} = \tan^{-1}\left(\frac{V_1}{U_{1h}}\right) \tag{9.13}$$

where the blade speed at the hub is given by $U_{1h} = ND_{1h}/2$.

IMPELLER OUTLET DESIGN

For the impeller outlet design, we follow the procedure as follows:

Step 1. Calculate the dimensional specific speed N_s as

$$N_s = \frac{NQ^{\frac{1}{2}}}{H^{\frac{3}{4}}}$$ (9.14)

where N is in rpm, Q in cubic feet per second, and H in feet.

Step 2. Using Table 9.1, determine the highest possible compressor efficiency and the corresponding dimensional specific diameter $D_s = D_2 H^{\frac{1}{4}} / Q^{\frac{1}{2}}$, where D_2 and H are in feet and Q in cubic feet per second.

Step 3. Calculate the impeller diameter D_2 from the specific diameter (obtained in Step 2) as

$$D_2 = \frac{D_s Q^{\frac{1}{2}}}{H^{\frac{1}{4}}}$$ (9.15)

Step 4. Calculate the impeller tip speed $U_2 = ND_2/2$, $E = \eta_m E_i / \eta_c$, and $V_{\theta 2'} = E/U_2$.

Step 5. Using $0.85 < \mu_s < 0.90$, calculate $V_{\theta 2} = V_{\theta 2'}/\mu_s$ and $W_{\theta 2} = U_2 - V_{\theta 2}$.

TABLE 9.1

Compressor Specific Diameter (D_s) in $\mathrm{ft}^{1.25}/(\mathrm{ft}^3/\mathrm{s})^{0.5}$ as a Function of Specific Speed (N_s) in $(\mathrm{rpm})(\mathrm{ft}^3/\mathrm{s})^{0.5}/\mathrm{ft}^{0.75}$ and Compressor Efficiency (η_c)

| N_s | Compressor Efficiency (η_c) | | | | |
	0.40	0.50	0.60	0.70	0.80
50	2.42	2.65	2.91	–	–
60	1.94	2.14	2.26	–	–
65	1.77	1.92	2.02	2.14	–
70	1.66	1.82	1.89	1.96	–
80	1.44	1.55	1.63	1.68	–
85	1.36	1.48	1.53	1.57	1.70
90	1.30	1.39	1.43	1.46	1.59
100	1.16	1.25	1.29	1.32	1.41
110	1.07	1.14	1.17	1.21	1.29
120	1.00	1.06	1.10	1.15	1.22
130	0.91	1.00	1.03	1.08	1.18
140	0.87	0.96	1.00	1.06	–
150	0.83	0.94	1.00	1.07	–
160	0.80	0.91	1.00	1.04	–
170	0.80	0.91	1.00	1.11	–
180	0.80	0.91	1.00	–	–
190	0.79	0.91	1.01	–	–
200	0.79	0.91	–	–	–

Source: Scheel, L. F. 1972. *Gas Machinery.* Gulf Publishing Co., Houston.

Step 6. Using $0.23 < \varphi_2 < 0.35$, calculate $W_{m2} = \varphi_2 U_2$. (Note $W_{m2} = W_{m2'} = V_{m2} = V_{m2'}$ in Figure 9.2.)

Step 7. Calculate the blade angle β_2 as

$$\beta_2 = \tan^{-1}\left(\frac{W_{m2}}{W_{\theta 2}}\right) \tag{9.16}$$

Step 8. Calculate the number of blades n_b as

$$n_b = \frac{0.63\pi}{\left(1 - \mu_s\right)\left(1 - 0.3\cot\beta_2\right)} \tag{9.17}$$

Step 9. Calculate impeller exit total temperature T_{02} as

$$T_{02} = T_{03} = T_{01} + \frac{E}{\eta_m c_p} \tag{9.18}$$

Step 10. Using $0.5 < \chi < 0.6$, where $\chi = (1 - \eta_{im})/(1 - \eta_c)$, which is the ratio of impeller losses to compressor losses, calculate the impeller efficiency $\eta_{im} = 1 - \chi(1 - \eta_c)$.

Step 11. Calculate the isentropic total temperature ratio for the impeller as

$$\frac{T_{i'}}{T_{01}} = 1 + \frac{\eta_{im}\left(T_{02} - T_{01}\right)}{T_{01}} \tag{9.19}$$

which is obtained from Equation 9.4. As shown in Figure 9.1, the total temperature $T_{i'}$ corresponds to the total pressure p_{02}.

Step 12. Calculate the impeller total pressure ratio as

$$\frac{p_{02}}{p_{01}} = \left(\frac{T_{i'}}{T_{01}}\right)^{\gamma/(\gamma-1)} \tag{9.20}$$

and hence calculate $p_{02} = p_{01}\left(p_{02}/p_{01}\right)$.

Step 13. Calculate $V_{2'} = \sqrt{V_{\theta 2'}^2 + V_{m2'}^2}$, $T_2 = T_{02} - V_{2'}^2/(2c_p)$, $C_V = V_{m2'}/V_{2'}$, giving

$$M_2 = \sqrt{\frac{2}{\gamma-1}\left(\frac{T_{02}}{T_2} - 1\right)} \tag{9.21}$$

and

$$\hat{F}_{f02} = \frac{M_2\sqrt{\gamma}}{\left(\dfrac{T_{02}}{T_2}\right)^{\gamma+1/2(\gamma-1)}} \qquad (9.22)$$

Step 14. Calculate the impeller exit flow area as

$$A_2 = \frac{m\sqrt{RT_{02}}}{p_{02}C_V\hat{F}_{f02}} \qquad (9.23)$$

and the impeller tip as

$$b_2 = \frac{A_2}{\pi D_2} \qquad (9.23)$$

IMPELLER DESIGN VALIDATION

For various impeller design parameters, Table 9.2 presents ranges that are considered optimal by Ferguson (1963) and Whitfield (1990). During or after the design process outlined in the foregoing, the designer should check the validity of their calculated results to ensure that these design parameters are within acceptable limits.

PROBLEM 9.1: EFFICIENCY OF A SINGLE-STAGE CENTRIFUGAL COMPRESSOR

The performance test data of a single-stage centrifugal air compressor, when running at 60,000 rpm, comprise $m = 1.0\,\text{kg/s}$, $p_{01} = 1.0135$ bar, $T_{01} = 289\,\text{K}$, and $p_{03} = 4.4$ bar. The geometric measurements of the impeller with 33 blades are $D_2 = 15.0368$ cm, $D_{1h} = 3.429$ cm, and $D_{1s} = 9.754$ cm. The blade inlet angle with respect to the tangent to the wheel is 90°. Calculate the efficiency of this compressor. Compare this calculated efficiency with that from Table 9.1. For air as a perfect gas, assume $\gamma = 1.4$ and $R = 287\,\text{J}/(\text{kg}\,\text{K})$.

TABLE 9.2
Design Parameters for Centrifugal Compressors

Parameter	Source	Recommended Range
Flow coefficient	Ferguson	$0.23 < \varphi_2 < 0.35$
Shroud-tip ratio	Whitfield	$0.5 < D_{1s}/D_2 < 0.7$
Absolute gas angle	Whitfield	$60° < \alpha_{2'} < 70°$
Diffusion ratio	Whitfield	$W_{1s}/W_{2'} < 1.9$

SOLUTION FOR PROBLEM 9.1

TIP SPEED

$$U_2 = \frac{ND_2}{2} = \frac{60,000 \times \pi \times 15.0368}{30 \times 2 \times 100} = 472.4\,\text{m/s}$$

As $\beta_2 = 90°$, we obtain $V_{\theta 2} = U_2 = 472.4\,\text{m/s}$.

SLIP COEFFICIENT

We compute the slip coefficient from Equation 9.7 as

$$\mu_s = 1 - \frac{0.63 \times \pi}{n_b}\left(\frac{1}{1 - \varphi\cot\beta_2}\right) = 1 - \frac{0.63 \times 3.1415}{33} = 0.94$$

giving

$$V_{\theta 2'} = \mu_s V_{\theta 2} = 0.94 \times 472.4 = 440.063 \text{ m/s}$$

Using Equation 5.1, we determine the aerodynamic specific work transfer in the compressor as

$$E = U_2 V_{\theta 2'} = 209773 \text{ J/kg}$$

Assuming $\eta_m = 0.96$ and using $c_p = R\gamma/(\gamma-1) = 287 \times 1.4/0.4 = 1004.5\,\text{J}/(\text{kg K})$, we calculate the actual total temperature rise in the compressor from Equation 9.1 as

$$T_{03} - T_{01} = \frac{E}{c_p\eta_m} = \frac{209,773}{1004.5 \times 0.96} = 217.535\,\text{K}$$

We determine the isentropic specific work transfer from Equation 9.2 as

$$E_i = c_p T_{01}\left[\left(\frac{p_{03}}{p_{01}}\right)^{(\gamma-1)/\gamma} - 1\right]$$

$$E_i = 1004.5 \times 289\left[\left(\frac{4.4}{1.0135}\right)^{(1.4-1)/1.4} - 1\right] = 151,061.514\,\text{J/kg}$$

giving the total temperature rise under isentropic compression as

$$T_i - T_{01} = \frac{E_i}{c_p} = \frac{151,061.514}{1004.5} = 150.4\,\text{K}$$

Finally, we calculate the compressor efficiency as

$$\eta_c = \frac{T_i - T_{01}}{T_{03} - T_{01}} = \frac{150.4}{217.5} = 0.691$$

FLOW AREA AT IMPELLER EYE

$$A_1 = \frac{\pi}{4}\left(D_{1s}^2 - D_{1h}^2\right) = \frac{3.1415}{4 \times 10,000}\left[(9.754)^2 - (3.429)^2\right] = 6.548 \times 10^{-3}\, m^2$$

INLET DENSITY AND VELOCITY

We determine the density and velocity at the inlet using the following iterative method:

Step 1. Assume $\rho_1 = 1.0\,kg/m^3$ $(\rho_1 = 1.130\,kg/m^3)$
Step 2. Calculate V_1 as

$$V_1 = \frac{m}{A_1\rho_1} = \frac{1.0}{6.548 \times 10^{-3} \times 1.0} = 152.393\,m/s \quad \left(V_1 = 134.861\,m/s\right)$$

Step 3. Calculate T_1

$$T_1 = T_{01} - \frac{V_1^2}{2c_p} = 289 - \frac{(153.393)^2}{2 \times 1004.5} = 277\,K \quad \left(T_1 = 279.5\,K\right)$$

Step 4. Calculate inlet static pressure

$$p_1 = p_{01}\left(\frac{T_1}{T_{01}}\right)^{\gamma/(\gamma-1)} = 1.0135 \times \left(\frac{277}{289}\right)^{3.5} = 0.878\,bar \quad \left(p_1 = 0.907\,bar\right)$$

Step 5. Calculate inlet density from the equation of state

$$\rho_1 = \frac{p_1}{RT_1} = \frac{0.878 \times 10^5}{287 \times 277} = 1.105\,kg/m^3 \quad \left(\rho_1 = 1.130\,kg/m^3\right)$$

Step 6. Repeat steps from 1 to 5 until ρ_1 calculated in step 5 is nearly equal to that assumed in step 1.

Note that the converged value is each step is shown within the parentheses.

VOLUMETRIC FLOW RATE

$$Q_1 = V_1\rho_1 = 134.861 \times 1.130 = 0.883\,m^3/s = 31.028\,ft^3/s$$

Output Head H

$$H = \frac{E_i}{g} = \frac{151{,}061.514}{9.81} = 15{,}398.727 \, \text{m} = 50{,}520.760 \, \text{ft}$$

Dimensional Specific Speed

$$N_s = \frac{NQ_1^{\frac{1}{2}}}{(H)^{\frac{3}{4}}} = \frac{60{,}000 \times (31.028)^{\frac{1}{2}}}{(50{,}520.760)^{\frac{3}{4}}} = 99.180 \, (\text{rpm}) \left(\text{ft}^3/\text{s} \right)^{0.5} / \text{ft}^{0.75}$$

Dimensional Specific Diameter

$$D_s = \frac{D_2 H^{\frac{1}{4}}}{Q^{\frac{1}{2}}} = \frac{(15.0368 \times 0.0328)(50{,}520.760)^{\frac{1}{4}}}{(31.028)^{\frac{1}{2}}} = 1.328 \; \text{ft}^{1.25} / \left(\text{ft}^3/\text{s} \right)^{0.5}$$

Compressor Efficiency from Table 9.1

For $N_s = 99.180 \, (\text{rpm}) \left(\text{ft}^3/\text{s} \right)^{0.5} / \text{ft}^{0.75}$ and $D_s = 1.328 \, \text{ft}^{1.25} / \left(\text{ft}^3/\text{s} \right)^{0.5}$, we obtain from Table 9.1 $\eta_c = 0.688$, which is slightly lower than $\eta_c = 0.691$ we computed in the foregoing.

PROBLEM 9.2: NONDIMENSIONAL POWER FUNCTION AT THE INLET OF A CENTRIFUGAL COMPRESSOR IMPELLER

For a centrifugal air compressor with axial flow at its impeller inlet, show that

$$\phi_{P1} = \frac{M_{R1s}^3 \cos^2\beta_{1s} \sin\beta_{1s}}{\left[1 + \dfrac{\gamma-1}{2} M_{R1s}^2 \sin^2\beta_{1s} \right]^{\frac{3\gamma-1}{2(\gamma-1)}}} = \frac{C_1 \dot{m} N^2}{\left(1-k^2\right) P_{01} \sqrt{T_{01}}} = \frac{\hat{C}_1 \dot{m}_{\text{corr}} N_{\text{corr}}^2}{\left(1-k^2\right)}$$

where $N_{\text{corr}} = N / \sqrt{\theta}$ is the corrected rpm, $\dot{m}_{\text{corr}} = \dot{m}\sqrt{\theta}/\delta$ is the corrected mass flow rate, $\theta = T_{01}/T_{0\text{ref}}$ is the dimensionless inlet total temperature, $\delta = p_{01}/p_{0\text{ref}}$ is the dimensionless inlet total pressure, $T_{0\text{ref}} = 288\,\text{K}$, $p_{0\text{ref}} = 101\,\text{kPa}$, and $k = D_{1h}/D_{1s}$, which is the hub-tip ratio at the inlet. Determine the constant C_1 and $\hat{C}_1$ in this equation.

SOLUTION FOR PROBLEM 9.2

Let us write

$$\phi_{P1} = \frac{M_{R1s}^3 \cos^2\beta_{1s} \sin\beta_{1s}}{\left[1 + \dfrac{\gamma - 1}{2} M_{R1s}^2 \sin^2\beta_{1s}\right]^{\frac{3\gamma - 1}{2(\gamma - 1)}}}$$

$$= \frac{V_1 U_1^2}{C^3 \left(\dfrac{T_{01}}{T_1}\right)^{\frac{\gamma}{(\gamma - 1)}} \left(\dfrac{T_{01}}{T_1}\right)^{\frac{1}{2}}}$$

$$= \frac{\left(\rho_1 A_1 V_1\right) N^2 D_{1s}^2}{4\rho_1 A_1 \left(\gamma R T_1\right)^{\frac{3}{2}} \left(\dfrac{p_{01}}{p_1}\right) \left(\dfrac{T_0}{T_1}\right)^{\frac{1}{2}}}$$

$$= \frac{\dot{m} N^2 D_{1s}^2}{\pi \left(\dfrac{p_1}{RT_1}\right) \left(D_{1s}^2 - D_{1h}^2\right) \left(\gamma R T_1\right)^{\frac{3}{2}} \left(\dfrac{p_{01}}{p_1}\right) \left(\dfrac{T_0}{T_1}\right)^{\frac{1}{2}}}$$

$$= \frac{\dot{m} N^2}{\left(\pi \gamma^{\frac{3}{2}} \sqrt{R}\right)\left(1 - k^2\right) p_{01} \sqrt{T_{01}}}$$

$$\phi_{P1} = \frac{M_{R1s}^3 \cos^2\beta_{1s} \sin\beta_{1s}}{\left[1 + \dfrac{\gamma - 1}{2} M_{R1s}^2 \sin^2\beta_{1s}\right]^{\frac{3\gamma - 1}{2(\gamma - 1)}}} = \frac{C_1 \dot{m} N^2}{\left(1 - k^2\right) p_{01} \sqrt{T_{01}}}$$

where

$$C_1 = \frac{1}{\pi \gamma^{\frac{3}{2}} \sqrt{R}}$$

Now we convert the equation for ϕ_{P1} in terms of $\dot{m}_{corr}$ and N_{corr} as

$$\phi_{P1} = \frac{\left(\dfrac{\dot{m}\sqrt{T_{01}}}{p_{01}}\right)\left(\dfrac{N}{\sqrt{T_{01}}}\right)^2}{\left(\pi \gamma^{\frac{3}{2}} \sqrt{R}\right)\left(1 - k^2\right)}$$

$$\phi_{P1} = \left(\frac{1}{\pi \gamma^{\frac{3}{2}} p_{0ref} \sqrt{RT_{0ref}}}\right) \frac{\dot{m}_{corr} N_{corr}^2}{\left(1 - k^2\right)} = \frac{\hat{C}_1 \dot{m}_{corr} N_{corr}^2}{\left(1 - k^2\right)}$$

where

$$\hat{C}_1 = \frac{1}{\pi \gamma^{\frac{3}{2}} p_{0ref} \sqrt{RT_{0ref}}}$$

For $\gamma = 1.4$, $p_{0ref} = 101 \times 10^3 \text{Pa}$, $T_{0ref} = 288 \text{K}$, we obtain $C_1 = 11.343 \times 10^{-3}$ $s\sqrt{K}/m$ and $\hat{C}_1 = 6.618 \times 10^{-9} \text{s}^3/\text{kg}$.

PROBLEM 9.3:   OPTIMUM ROTATIONAL SPEED OF A CENTRIFUGAL COMPRESSOR

Plot the dimensionless power function ϕ_{P1} as a function of β_{1s} with M_{R1s} as a parameter with values from 0.5 to 1.0. Show that the maximum value of ϕ_{P1} ranges from $\beta_{1s} = 30°$ for $M_{R1s} = 0.5$ to $\beta_{1s} = 32°$ for $M_{R1s} = 1.0$. For a centrifugal air compressor with axial flow at its impeller inlet, we are given with $M_{R1s} = 0.9$, $p_{01} = 101.325 \text{kPa}$, $T_{01} = 288 \text{K}$, $\dot{m} = 1.2 \text{kg/s}$, the hub-tip ratio $k = 0.4$. Using the ϕ_{P1} equation in Problem 9.2, find the maximum rpm corresponding to the optimum value of β_{1s}. For air as a perfect gas, assume $\gamma = 1.4$ and $R = 287 \text{J}/(\text{kg K})$.

SOLUTION FOR PROBLEM 9.3

The ϕ_{P1} equation is plotted in Figure 9.4, which shows a maximum value of ϕ_{P1} ranging from $\beta_{1s} = 30°$ for $M_{R1s} = 0.5$ to $\beta_{1s} = 32°$ for $M_{R1s} = 1.0$. For $M_{R1s} = 0.9$, we obtain $\beta_{1s} = 30°$. For the equation

$$\phi_{P1} = \frac{M_{R1s}^3 \cos^2\beta_{1s}\sin\beta_{1s}}{\left[1 + \frac{\gamma-1}{2}M_{R1s}^2\sin^2\beta_{1s}\right]^{\frac{3\gamma-1}{2(\gamma-1)}}} = \frac{\hat{C}_1 \dot{m}_{corr} N_{corr}^2}{\left(1 - k^2\right)}$$

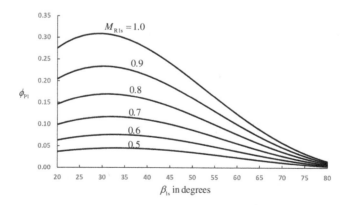

FIGURE 9.4   Plot of ϕ_{P1} versus β_{1s} with M_{R1s} as the parameter (Problem 9.3).

we compute

$$\phi_{P1} = \frac{M_{R1s}^3 \cos^2\beta_{1s}\sin\beta_{1s}}{\left[1+\dfrac{\gamma-1}{2}M_{R1s}^2\sin^2\beta_{1s}\right]^{\frac{3\gamma-1}{2(\gamma-1)}}} = \frac{(0.9)^3\times(\cos 30°)^2\times\sin 30°}{1+0.2\times(0.9\times\sin 30°)^2}$$

$$\phi_{P1} = \frac{0.2742}{1.1754} = 0.2333$$

$$\theta = \frac{T_{01}}{T_{0ref}} = \frac{288}{288} = 1.0$$

$$\delta = \frac{p_{01}}{p_{0ref}} = \frac{101.325}{101} = 1.003$$

$$\dot{m}_{corr} = \frac{\dot{m}\sqrt{\theta}}{\delta} = \frac{1.2\times\sqrt{1.0}}{1.003} = 1.196$$

Now, we compute rotational speed as follows:

$$N_{corr}^2 = \frac{\phi_{P1}\left(1-k^2\right)}{\hat{C}_1\dot{m}_{corr}} = \frac{0.2333\times(1-0.4\times0.4)}{6.618\times10^{-9}\times1.196} = 24,753,000$$

$$N_{corr} = \sqrt{25,612,479} = 4975\,\text{rad/s}$$

$$N = N_{corr}\times\sqrt{\theta} = 4975\times\sqrt{1.0} = 4975.239\,\text{rad/s}$$

$$N = \frac{4975\times30}{\pi} = 47,510\,\text{rpm}$$

PROBLEM 9.4: BASIC DIMENSIONS OF A CENTRIFUGAL AIR COMPRESSOR IMPELLER

For a single-sided centrifugal air compressor some of the design and operational data are as follows:

Mechanical efficiency (η_m)	0.96
Slip coefficient (μ_s)	0.90
Rotational speed (N)	17400 rpm
Inlet hub diameter (D_{1h})	0.15 m
Inlet shroud diameter (D_{1s})	0.30 m
Impeller diameter (D_2)	0.50 m
Air mass flow rate $(\dot{m})$	9.0 kg/s
Inlet total temperature (T_{01})	295 K
Inlet total pressure (p_{01})	1.1 bar
Isentropic efficiency (η_c)	0.78

a. Find the compressor pressure ratio and the power required to drive it. Assume that the air velocity at inlet is axial, and the relative velocity at impeller exit is radial. (Hint: $V_{\theta 2} = U_2$)

b. Assuming uniform air velocity at inlet, find the impeller blade angles at the hub and shroud radii of the eye.

c. Assuming $V_{r2} = V_1$ and the ratio of impeller losses to compressor losses (χ) = 0.5, find the axial width of the impeller channels at its periphery.

For air as a perfect gas, assume $\gamma = 1.4$ and $R = 287\,\mathrm{J}/(\mathrm{kg\,K})$.

SOLUTION FOR PROBLEM 9.4

(a) COMPRESSOR PRESSURE RATIO AND REQUIRED POWER

Rotational Speed

$$N = \frac{17{,}400 \times \pi}{30} = 1822.124\,\mathrm{rad/s}$$

$$U_2 = \frac{D_2 \times N}{2} = \frac{0.5 \times 1822.124}{2} = 455.531\,\mathrm{m/s}$$

Compressor Pressure Ratio

We compute the overall compressor pressure ratio as follows:

$$T_{03} - T_{01} = \frac{U_2 V_{\theta 2'}}{c_p \eta_m} = \frac{\mu_s U_2^2}{c_p \eta_m} = \frac{0.9 \times (455.531)^2}{1004.5 \times 0.96} = 193.7\,\mathrm{K}$$

$$T_{03} = 295 + 193.7 = 488.7\,\mathrm{K}$$

$$\frac{p_{03}}{p_{01}} = \left(1 + \frac{\eta_c (T_{03} - T_{01})}{T_{01}}\right)^{\gamma/(\gamma-1)} = \left(1 + \frac{0.78 \times 193.7}{295}\right)^{3.5} = 4.251$$

Required Power

$$P = \dot{m} c_p (T_{03} - T_{01}) = 9.0 \times 1004.5 \times 193.7 = 1750.852\,\mathrm{kW}$$

(b) IMPELLER BLADE ANGLES AT INLET

Impeller Inlet Area

$$A = \frac{\pi}{4}\left(D_{1s}^2 - D_{1h}^2\right) = \frac{3.1415}{4} \times \left[(0.3)^2 - (0.15)^2\right] = 0.0530\,\mathrm{m}^2$$

Inlet Absolute Mach Number and Velocity

From the mass flow rate equation at the impeller inlet

$$\dot{m} = \frac{A_1 \hat{F}_{f01} p_{01}}{\sqrt{RT_{01}}}$$

we obtain

$$\hat{F}_{f01} = \frac{\dot{m}\sqrt{RT_{01}}}{A_1 p_{01}} = \frac{9.0 \times \sqrt{287 \times 295}}{0.530 \times 110,000} = 0.4491$$

where

$$\hat{F}_{f01} = M_1 \sqrt{\frac{\gamma}{\left(1 + \dfrac{\gamma - 1}{2} M_1^2\right)^{\frac{\gamma+1}{\gamma-1}}}}$$

giving $M_1 = 0.4211$ through an iterative solution using, for example, the "Goal Seek" in MS Excel. Knowing $M_1 = 0.4211$, we obtain

$$\frac{T_{01}}{T_1} = 1 + \frac{\gamma - 1}{2} M_1^2 = 1 + 0.2 \times (0.4211)^2 = 1.0355$$

which yields

$$T_1 = \frac{295}{1.0355} = 284.9 \,\text{K}$$

$$C_1 = \sqrt{\gamma RT_1} = \sqrt{1.4 \times 287 \times 284.9} = 338.335 \,\text{m/s}$$

$$V_1 = M_1 C_1 = \sqrt{\gamma RT_1} = \sqrt{1.4 \times 287 \times 284.9} = 142.482 \,\text{m/s}$$

Relative Angle at Impeller Inlet Hub

$$U_{1h} = \frac{ND_{1h}}{2} = \frac{1822.124 \times 0.15}{2} = 136.659 \,\text{m/s}$$

$$\beta_{1h} = \tan^{-1}\left(V_1/U_{1h}\right) = \tan^{-1}\left(142.482/136.659\right) = 46.2°$$

Relative Angle at Impeller Inlet Shroud

$$U_{1s} = \frac{ND_{1s}}{2} = \frac{1822.124 \times 0.30}{2} = 273.319 \,\text{m/s}$$

$$\beta_{1h} = \tan^{-1}\left(V_1/U_{1h}\right) = \tan^{-1}\left(142.482/273.319\right) = 27.5°$$

(c) Axial Width of Impeller Channels at Their Periphery

With $V_{r2} = V_1 = 142.482\,\text{m/s}$, $U_2 = V_{\theta 2} = 455.531\,\text{m/s}$, and $T_{02} = T_{03} = 488.7\,\text{K}$, we obtain

$$V_{\theta 2'} = \mu_s V_{\theta 2} = 0.9 \times 455.531 = 409.978\,\text{m/s}$$

which yields

$$V_{2'} = \sqrt{V_{r2}^2 + V_{\theta 2'}^2} = \sqrt{(142.482)^2 + (409.978)^2} = 434.031\,\text{m/s}$$

From the definition of χ as

$$\chi = \frac{1 - \eta_{im}}{1 - \eta_c}$$

for $\chi = 0.5$, we obtain

$$\eta_{im} = 1 - \chi(1 - \eta_c) = 1 - 0.5 \times (1 - 0.78) = 0.89$$

giving

$$\frac{p_{02}}{p_{01}} = \left(1 + \frac{\eta_{im}(T_{02} - T_{01})}{T_{01}}\right)^{\gamma/(\gamma-1)} = \left(1 + \frac{0.89 \times 193.7}{295}\right)^{3.5} = 5.005$$

$$p_{02} = p_{01}\left(\frac{p_{02}}{p_{01}}\right) = 110{,}000 \times 5.005 = 550{,}565\,\text{Pa}$$

$$T_2 = T_{02} - \frac{V_{2'}^2}{2c_p} = 488.668 - \frac{(434.031)^2}{2 \times 1004.5} = 394.9\,\text{K}$$

$$\frac{T_{02}}{T_2} = \frac{488.7}{394.9} = 1.238$$

$$\frac{p_{02}}{p_2} = \left(\frac{T_{02}}{T_1}\right)^{\gamma/(\gamma-1)} = (1.238)^{3.5} = 2.108$$

$$p_2 = \frac{p_{02}}{p_{02}/p_2} = \frac{550{,}565}{2.108} = 261{,}192\,\text{Pa}$$

$$\rho_2 = \frac{p_2}{RT_2} = \frac{261{,}192}{287 \times 394.9} = 2.305\,\text{kg/m}^3$$

$$A_2 = \frac{\dot{m}}{\rho_2 V_r} = \frac{9.0}{2.305 \times 142.482} = 0.0274\,\text{m}^2$$

Thus, we calculate the width of the impeller channels at the periphery as

$$b_2 = \frac{A_2}{\pi D_2} = \frac{0.0274}{3.1415 \times 0.5} = 0.0174\,\text{m} = 1.74\text{ cm}$$

PROBLEM 9.5:   FINDING THE ROTATIONAL SPEED OF A CENTRIFUGAL AIR COMPRESSOR

The inducer eye of a centrifugal air compressor has the hub-tip ratio of 0.5. With a uniform axial flow at the inlet, the compressor delivers the maximum mass flow rate of 5.0 kg/s at the relative flow Mach number of 0.9 at its inlet shroud, a total pressure of 1.02 bar, and a total temperature of 290 K. Determine the rotational speed for the compressor, inlet axial velocity, and eye hub and shroud diameters. For air as a perfect gas, assume $\gamma = 1.4$ and $R = 287\,\text{J}/(\text{kg K})$.

SOLUTION FOR PROBLEM 9.5

OPTIMAL ROTATIONAL SPEED

For solving this problem, we use the following equation developed in the solution for Problem 9.2:

$$\phi_{P1} = \frac{M_{R1s}^3 \cos^2 \beta_{1s} \sin \beta_{1s}}{\left[1 + \dfrac{\gamma-1}{2} M_{R1s}^2 \sin^2 \beta_{1s}\right]^{\frac{3\gamma-1}{2(\gamma-1)}}} = \frac{C_1 \dot{m} N^2}{\left(1-k^2\right) p_{01} \sqrt{T_{01}}}$$

which we write as

$$\frac{C_1 \dot{m} N^2}{\left(1-k^2\right) p_{01} \sqrt{T_{01}}} = \frac{M_{R1s}^3 \cos^2 \beta_{1s} \sin \beta_{1s}}{\left[1 + \dfrac{\gamma-1}{2} M_{R1s}^2 \sin^2 \beta_{1s}\right]^{\frac{3\gamma-1}{2(\gamma-1)}}}$$

where, for $\gamma = 1.4$ and $R = 287\,\text{J}/(\text{kg K})$, we have

$$C_1 = \frac{1}{\pi \gamma^{\frac{3}{2}} \sqrt{R}} = 11.343 \times 10^{-3}\,\text{s}\sqrt{\text{K}}/\text{m}$$

For $M_{R1s} = 0.9$, the maximum value of the right-hand side of the equation corresponds to $\beta_{1s} = 30°$. The equation then yields

$$N^2 = \left[\frac{\left(1-k^2\right) p_{01} \sqrt{T_{01}}}{C_1 \dot{m}}\right] \frac{M_{R1s}^3 \cos^2 \beta_{1s} \sin \beta_{1s}}{\left[1 + \dfrac{\gamma-1}{2} M_{R1s}^2 \sin^2 \beta_{1s}\right]^{\frac{3\gamma-1}{2(\gamma-1)}}}$$

$$N^2 = \left[\frac{\left(1-(0.5)^2\right) \times 1.01 \times 10^5 \sqrt{290}}{11.343 \times 10^{-3} \times 5.0}\right] \frac{(0.9)^3 \times \cos^2 30° \times \sin 30°}{\left[1 + 0.2 \times (0.9)^2 \sin^2 30°\right]^4} = 5304995$$

$$N = \sqrt{5{,}304{,}995} = 2303.257\,\text{rad/s} = 21{,}994\,\text{rpm}$$

Uniform Inlet Axial Velocity

$$M_1 = M_{R1s}\sin\beta_{1s} = 0.9 \times \sin 30° = 0.45$$

$$\frac{T_{01}}{T_1} = 1 + \frac{\gamma-1}{2}M_1^2 = 1 + 0.2 \times (0.45)^2 = 1.041$$

$$T_1 = \frac{T_{01}}{1.041} = \frac{290}{1.041} = 278.7\,\text{K}$$

$$C_1 = \sqrt{\gamma R T_1} = \sqrt{1.4 \times 287 \times 278.7} = 334.644\,\text{m/s}$$

$$V_1 = M_1 C_1 = 0.45 \times 334.644 = 150.59\,\text{m/s}$$

Inlet Hub and Shroud Diameters

$$\hat{F}_{f01} = M_1\sqrt{\frac{\gamma}{\left(1+\dfrac{\gamma-1}{2}M_1^2\right)^{\frac{\gamma+1}{\gamma-1}}}} = 0.45 \times \sqrt{\frac{1.4}{\left(1+0.2\times(0.45)^2\right)^6}} = 0.4727$$

$$A_1 = \frac{m\sqrt{RT_{01}}}{p_{01}\hat{F}_{f01}} = \frac{5\times\sqrt{287\times290}}{101{,}000\times0.4727} = 0.03022\,\text{m}^2$$

$$D_{1s} = \left[\frac{4A_1}{\pi\left(1-k^2\right)}\right]^{\frac{1}{2}} = \frac{4\times0.03022}{3.1415\times0.75} = 0.2265\text{m} = 22.648\,\text{cm}$$

$$D_{1h} = 0.5\times22.65 = 11.324\,\text{cm}$$

PROBLEM 9.6:   MASS FLOW RATE THROUGH A CENTRIFUGAL AIR COMPRESSOR

A centrifugal air compressor with the impeller tip speed of 350 m/s has the exit flow area of 0.1 m². The flow enters the compressor at the total pressure of 1.013 bar and total temperature of 292 K and exits the impeller with the radial velocity component of 30 m/s and the tangential velocity component with the slip coefficient of 0.9. The impeller efficiency is 0.89. Determine the compressor mass flow rate. For air as a perfect gas, assume $\gamma = 1.4$ and $R = 287\,\text{J/}(\text{kgK})$.

SOLUTION FOR PROBLEM 9.6

Absolute Exit Mach Number

We compute the actual tangential velocity component of the exit absolute velocity as

$$V_{\theta 2'} = \mu_s V_{\theta 2} = \mu_s U_2 = 0.90 \times 350 = 315\,\text{m/s}$$

which gives the absolute exit velocity

$$V_{2'} = \sqrt{V_{r2}^2 + V_{\theta 2'}^2} = \sqrt{(30)^2 + (315)^2} = 316.425\,\text{m/s}$$

and the velocity coefficient

$$C_V = \frac{V_{r2}}{V_{2'}} = \frac{30}{316.425} = 0.0948$$

From the impeller work transfer, we obtain the exit air total temperature

$$T_{02} = T_{01} + \frac{\mu_s U_2^2}{c_p} = 292 + \frac{0.9 \times (350)^2}{1004.5} = 401.8\,\text{K}$$

and the corresponding static temperature

$$T_2 = T_{02} - \frac{V_{2''}^2}{2c_p} = 401.8 - \frac{(316.425)^2}{2 \times 1004.5} = 351.9\,\text{K}$$

giving

$$\frac{T_{02}}{T_2} = \frac{401.8}{351.9} = 1.142$$

and the impeller exit absolute Mach number

$$M_2 = \left[\left(\frac{2}{\gamma - 1}\right)\left(\frac{T_{02}}{T_2} - 1\right)\right]^{1/2} = \left[\left(\frac{2}{1.4 - 1}\right)(1.142 - 1)\right]^{1/2} = 0.841$$

MASS FLOW RATE

We compute the total-pressure mass flow function at the impeller exit as

$$\hat{F}_{f02} = M_2 \sqrt{\frac{\gamma}{\left(1 + \frac{\gamma - 1}{2} M_2^2\right)^{\frac{\gamma + 1}{\gamma - 1}}}} = 0.841 \times \sqrt{\frac{1.4}{\left(1 + 0.2 \times (0.841)^2\right)^6}} = 0.669$$

and the total pressure ratio

$$\frac{p_{02}}{p_{01}} = \left[1 + \frac{\eta_{im}(T_{02} - T_{01})}{T_{01}}\right]^{\frac{\gamma}{\gamma - 1}} = \left[1 + \frac{0.89 \times (401.8 - 292)}{292}\right]^{3.5} = 2.746$$

giving

$$p_{02} = p_{01} \times 2.746 = 101,300 \times 2.746 = 278,137\,\text{Pa}$$

and, finally, the mass flow rate

$$\dot{m} = \frac{A_1 C_V \hat{F}_{f02} p_{02}}{\sqrt{RT_{02}}} = \frac{0.1 \times 0.095 \times 0.669 \times 278{,}137}{\sqrt{287 \times 401.8}} = 5.2 \, \text{kg/s}$$

PROBLEM 9.7: OPERATIONAL PARAMETERS OF A CENTRIFUGAL COMPRESSOR WITH RADIAL BLADES

A centrifugal air compressor having an impeller with radial blades is designed for a rotational speed of 2500 rpm. The compressor requires 1.1 MW of power to discharge 10 kg/s of compressed air. The flow enters the compressor axially at the total pressure of 1.013 bar and the total temperature of 290 K exits the impeller radially. Assume a slip coefficient of 0.9 at the impeller exit and the specific speed of $N_s = \phi^{0.5}/\psi^{0.75} = 0.8$ where $\phi = V_{x1}/U_2^2$ and $\psi = E/U_2^2$. Find (1) the impeller tip speed, (2) inlet absolute velocity, (3) inlet absolute Mach number, and (d) inlet area. For air as a perfect gas, assume $\gamma = 1.4$ and $R = 287 \, \text{J}/(\text{kg K})$.

SOLUTION FOR PROBLEM 9.7

(a) IMPELLER TIP SPEED

Using the power required to drive the compressor $P = mE = 1.1 \times 10^6$ W and 100% mechanical efficiency, we obtain the aerodynamic specific energy transfer as

$$E = \frac{P}{m} = \frac{1.1 \times 10^6}{10} = 11 \times 10^4 \, \text{m}^2/\text{s}^2$$

which yields

$$U_2 = \sqrt{\frac{E}{\mu_s}} = \sqrt{\frac{1.1 \times 10^6}{0.9}} = 349.603 \, \text{m/s}$$

(b) INLET ABSOLUTE VELOCITY

We obtain

$$\psi = \frac{E}{U_2^2} = \mu_s = 0.9$$

From the given specific speed $N_s = \phi^{0.5}/\psi^{0.75} = 0.8$, we compute

$$\phi = \left(N_s \psi^{0.75}\right)^2 = \left[0.8 \times (0.9)^{0.75}\right]^2 = 0.5464$$

which yields

$$V_1 = \phi U_2 = 0.5464 \times 349.603 = 191.038 \, \text{m/s}$$

(c) INLET ABSOLUTE MACH NUMBER

We compute the static temperature at inlet as

$$T_1 = T_{01} - \frac{V_1^2}{2c_p} = 290 - \frac{(191.038)^2}{2 \times 1004.5} = 271.8\,\text{K}$$

which yields the absolute Mach number at the compressor inlet as

$$M_1 = \frac{V_1}{\sqrt{\gamma R T_1}} = \frac{191.038}{\sqrt{1.4 \times 1004.5 \times 271.8}} = 0.578$$

(d) INLET AREA

$$\frac{T_{01}}{T_1} = 1 + \frac{\gamma - 1}{2} M_1^2 = 1 + 0.2 \times (0.578)^2 = 1.0668$$

$$\hat{F}_{f01} = M_1 \sqrt{\frac{\gamma}{\left(1 + \frac{\gamma - 1}{2} M_1^2\right)^{\frac{\gamma+1}{\gamma-1}}}} = \frac{M_1 \sqrt{\gamma}}{\left(\frac{T_{01}}{T_1}\right)^{\frac{\gamma+1}{2(\gamma-1)}}} = \frac{0.578\sqrt{1.4}}{(1.0668)^3} = 0.563$$

Thus, we finally compute the compressor inlet area as

$$A_1 = \frac{\dot{m}\sqrt{R T_{01}}}{C_V p_{01} \hat{F}_{f01}} = \frac{10 \times \sqrt{287 \times 290}}{1.0 \times 101{,}300 \times 0.563} = 0.0506\,\text{m}^2$$

PROBLEM 9.8: DISCHARGE OF A CENTRIFUGAL AIR COMPRESSOR IMPELLER WITH RADIAL BLADES

A single-stage centrifugal air compressor, designed to operate at 17,000 rpm, has its impeller with tip diameter 0.42 m and passage width of 2.0 cm. The impeller has 18 blades, operating at the impeller efficiency of 93%. The flow enters the compressor axially at the total pressure of 1.013 bar and the total temperature of 291 K and exits the impeller radially at 25 m/s. Find (1) the absolute Mach number at impeller tip, (2) total pressure of air exiting the impeller, and (3) compressor mass flow rate. For air as a perfect gas, assume $\gamma = 1.4$ and $R = 287\,\text{J}/(\text{kg}\,\text{K})$.

SOLUTION FOR PROBLEM 9.8

(a) ABSOLUTE MACH NUMBER AT IMPELLER EXIT

For the given rotational speed $N = 17{,}000\,\text{rpm}$, we obtain

$$N = \frac{17{,}000 \times \pi}{30} = 1780.236\,\text{rad/s}$$

which yields

$$U_2 = \frac{1780.236 \times 0.42}{2} = 373.850 \, \text{m/s}$$

To find the slip coefficient, we use the Stanitz equation with $\beta_2 = 90°$

$$\mu_s = 1 - \frac{0.63\pi}{n_b} \left(\frac{1}{1 - \varphi_2 \cot \beta_2} \right) = 1 - \frac{0.63 \times \pi}{18} = 0.890$$

giving

$$V_{\theta 2'} = \mu_s V_{\theta 2} = \mu_s U_2 = 0.890 \times 373.850 = 332.743 \, \text{m/s}$$

$$V_{2'} = \sqrt{V_{r2}^2 + V_{\theta 2'}^2} = \sqrt{(25)^2 + (332.743)^2} = 330.680 \, \text{m/s}$$

From the impeller work transfer, at the impeller exit, we obtain the total temperature

$$T_{02} = T_{01} + \frac{\mu_s U_2^2}{c_p} = 292 + \frac{0.9 \times (373.850)^2}{1004.5} = 414.8 \, \text{K}$$

the static temperature

$$T_2 = T_{02} - \frac{V_{2'}^2}{2c_p} = 414.8 - \frac{(316.425)^2}{2 \times 1004.5} = 359.4 \, \text{K}$$

and the absolute Mach number

$$M_2 = \frac{V_{2'}}{\sqrt{\gamma R T_2}} = \frac{330.680}{\sqrt{1.4 \times 287 \times 359.4}} = 0.878$$

(b) Air Total Pressure at Impeller Exit

We compute the ratio of total pressures at impeller exit and inlet as

$$\frac{p_{02}}{p_{01}} = \left[1 + \frac{\eta_{im}(T_{02} - T_{01})}{T_{01}} \right]^{\gamma/\gamma - 1} = \left[1 + \frac{0.93 \times (414.8 - 291)}{291} \right]^{3.5} = 3.213$$

giving

$$p_{02} = p_{01} \left(\frac{p_{02}}{p_{01}} \right) = 101,300 \times 3.213 = 325,432 \, \text{Pa}$$

(c) COMPRESSOR MASS FLOW RATE

We first compute the impeller exit flow area as

$$A_2 = \pi D_2 b = 3.1415 \times 0.42 \times 0.02 = 0.0264 \, \text{m}^2$$

the total-to-static temperature ratio as

$$\frac{T_{02}}{T_2} = \frac{414.8}{359.4} = 1.154$$

the total-pressure mass flow function at impeller exit as

$$\hat{F}_{f02} = \frac{M_2 \sqrt{\gamma}}{\left(\dfrac{T_{02}}{T_2}\right)^{\gamma+1/2(\gamma-1)}} = 0.878 \times \frac{0.878\sqrt{1.4}}{(1.154)^3} = 0.676$$

and the velocity coefficient as

$$C_V = \frac{V_{r2}}{V_{2'}} = \frac{25}{330.680} = 0.0749$$

Thus, we obtain the mass flow rate from the equation

$$\dot{m} = \frac{A_2 C_V \hat{F}_{f02} p_{02}}{\sqrt{RT_{02}}} = \frac{0.0264 \times 0.0745 \times 0.676 \times 325,432}{\sqrt{287 \times 414.8}} = 1.260 \, \text{kg/s}$$

PROBLEM 9.9: BASIC DIMENSIONS AND EXIT FLOW PROPERTIES OF A CENTRIFUGAL COMPRESSOR IMPELLER

A single-stage centrifugal compressor receives air at a stagnation pressure of 1.013 bar and a stagnation temperature of 288 K. The rotor tip speed is 500 m/s, the velocity at the impeller eye is 100 m/s, and $V_{2'} = 456$ m/s. The energy transfer in the impeller is 220 kJ/kg. The slip coefficient is 0.9. The impeller and compressor efficiencies are 0.9 and 0.8, respectively. The mechanical efficiency is 0.98. Find (1) the overall compressor pressure ratio, (2) the blade angle β_2 at the impeller tip, (3) T_{02}, (4) T_2, (5) p_{02}, and (6) p_2.

SOLUTION FOR PROBLEM 9.9

(a) OVERALL COMPRESSOR PRESSURE RATIO

$$\frac{p_{03}}{p_{01}} = \left(1 + \frac{\eta_c E}{\eta_m c_p T_{01}}\right)^{\gamma/\gamma-1} = \left(1 + \frac{0.8 \times 220 \times 1000}{0.98 \times 1004.5 \times 288}\right)^{3.5} = 5.421$$

(b) The Blade Angle at Impeller Tip (β_2)

$$V_{\theta 2'} = \frac{E}{U} = \frac{220 \times 1000}{500} = 440 \, \text{m/s}$$

$$V_{m2'} = \sqrt{V_2^2 - V_{\theta 2'}^2} = \sqrt{(456)^2 - (440)^2} = 119.733 \, \text{m/s}$$

$$W_{m2} = V_{m2'} = 119.733 \, \text{m/s}$$

$$V_{\theta 2} = \frac{V_{\theta 2'}}{\mu_s} = \frac{440}{0.90} = 488.889 \, \text{m/s}$$

$$W_{\theta 2} = U_2 - V_{\theta 2} = 500 - 488.889 = 11.111 \, \text{m/s}$$

$$\beta_2 = \tan^{-1} \frac{W_{m2}}{W_{\theta 2}} = \tan^{-1} \frac{119.733}{11.111} = 84.7°$$

(c) Total Temperature at Impeller Exit (T_{02})

$$T_{03} - T_{01} = \frac{E}{\eta_m c_p}$$

$$T_{03} = T_{01} + \frac{E}{\eta_m c_p} = 288 + \frac{220 \times 1000}{0.98 \times 1004.5} = 511.5 \, \text{K}$$

(d) Static Temperature at Impeller Exit (T_2)

$$T_2 = T_{02} - \frac{V_{2'}^2}{2c_p} = 511.5 - \frac{(456)^2}{2 \times 1004.5} = 408 \, \text{K}$$

(e) Total Pressure at Impeller Exit (p_{02})

$$p_{02} = p_{01} \left[1 + \frac{\eta_{im}(T_{02} - T_{01})}{T_{01}} \right]^{\gamma/\gamma-1} = 1.013 \times 10^5 \left[1 + \frac{0.91 \times (511.8 - 288)}{288} \right]^{3.5} = 657,155 \, \text{Pa}$$

(f) Static Pressure at Impeller Exit (p_2)

$$\frac{p_{02}}{p_2} = \left(\frac{T_{02}}{T_2} \right)^{\gamma/\gamma-1} = \left(\frac{511.5}{408} \right)^{3.5} = 2.206$$

$$p_2 = \frac{657,155}{2.206} = 297,850 \, \text{Pa}$$

PROBLEM 9.10: PRELIMINARY IMPELLER DESIGN FOR A CENTRIFUGAL AIR COMPRESSOR

A single-stage centrifugal compressor draws in 3.3 kg of air per second at a total pressure of 100,000 Pa and a total temperature of 306 K. It discharges the air at a total pressure of 400,000 Pa. The compressor runs at 40,000 rpm. Find the basic dimensions of the impeller if its tip velocity is 547 m/s. Assume a mechanical efficiency of 0.96. Make reasonable assumptions as needed.

SOLUTION FOR PROBLEM 9.10

IMPELLER INLET

Rotational Speed

$$N = \frac{40,000 \times 3.1416}{30} = 4188.8 \, \text{rad/s}$$

which is constant for the impeller.

Inlet Volumetric Flow Rate

Assuming $M_{R1s} = 1.1$ and $\beta_{1s} = 30^\circ$, we calculate the inlet volumetric flow rate as follows:

$$M_1 = M_{R1s} \sin \beta_{1s} = 1.1 \sin 30^\circ = 0.550$$

$$\frac{T_{01}}{T_1} = 1 + \frac{\gamma - 1}{2} M_1^2 = 1 + 0.2 \times (0.550)^2 = 1.0605$$

$$T_1 = \frac{306}{1.0605} = 288.5 \, \text{K}$$

$$C_1 = \sqrt{\gamma R T_1} = \sqrt{1.4 \times 287 \times 288.5} = 340.495 \, \text{m/s}$$

$$V_1 = M_1 C_1 = 0.550 \times 340.495 = 187.272 \, \text{m/s}$$

$$W_{1s} = M_{R1s} C_1 = 1.1 \times 340.495 = 374.544 \, \text{m/s}$$

$$U_{1s} = W_{1s} \cos \beta_{1s} = 374.544 \times \cos 30^\circ = 324.365 \, \text{m/s}$$

$$p_1 = \frac{p_{01}}{(T_{01}/T_1)^{\gamma/\gamma-1}} = \frac{100,000}{(1.0605)^{3.5}} = 81,417 \, \text{Pa}$$

$$\rho_1 = \frac{p_1}{R T_1} = \frac{81,417}{287 \times 288.5} = 0.983 \, \text{kg/m}^3$$

$$Q_1 = \frac{\dot{m}}{\rho_1} = \frac{3.3}{0.983} = 3.357 \, \text{m}^3/\text{s} = 117.933 \, \text{cfs}$$

Impeller Hub and Shroud Diameters

$$D_{1s} = \frac{2U_{1s}}{N} = \frac{2 \times 324.365}{4188.790} = 0.155\,\text{m} = 155\,\text{mm}$$

$$D_{1h} = \left[D_{1s}^2 - \frac{4\dot{m}}{\pi \rho_1 V_1} \right]^{1/2} = \left[(0.155)^2 - \frac{4 \times 3.3}{3.1415 \times 0.983 \times 187.272} \right]^{1/2} = 0.0341\,\text{m} = 34\,\text{mm}$$

Ideal Output Head (H)

$$E_i = c_p T_{01} \left[\left(\frac{p_{03}}{p_{01}} \right)^{(\gamma-1)/\gamma} - 1 \right]$$

$$E_i = 1004.5 \times 306 \times \left[\left(\frac{400,000}{100,000} \right)^{(1/3.5)} - 1 \right] = 149,383 \text{ m}^2/\text{s}^2$$

$$H = \frac{E_i}{g} = \frac{149,383}{9.81} = 15,228\,\text{m} = 49,959\,\text{ft}$$

Dimensional Specific Speed (N_s)

$$N_s = \frac{N Q_1^{1/2}}{H^{3/4}} = \frac{40,000 \times (117.933)^{1/2}}{(49,959)^{3/4}} = 130.0$$

Impeller Outer Diameter (D_2) and Specific Diameter (D_s)

$$D_2 = \frac{2U_2}{N} = \frac{2 \times 547}{4188.790} = 0.261\,\text{m} = 0.857\,\text{ft}$$

$$D_s = \frac{D_2 H^{1/4}}{Q_1^{1/2}} = \frac{0.857 \times (49,959)^{1/4}}{(117.933)^{1/2}} = 1.180$$

Compressor Efficiency (η_c)

From Table 9.1, for $N_s = 130$ and $D_s = 1.180$, we obtain $\eta_c = 0.8$.

Actual Aerodynamic Specific Energy Transfer (E)

Assuming $\eta_m = 0.96$, we compute the actual aerodynamic energy transfer in the impeller as

$$E = \frac{\eta_m E_i}{\eta_c} = \frac{0.96 \times 149,383}{0.8} = 179,260\,\text{m}^2/\text{s}^2$$

Actual Absolute Tangential Velocity at Impeller $(V_{\theta 2'})$

$$V_{\theta 2'} = \frac{E}{U_2} = \frac{179{,}260}{547} = 327.715\,\text{m/s}$$

Ideal Absolute Tangential Velocity at Impeller $(V_{\theta 2})$

Assuming $\mu_s = 0.9$, we obtain

$$V_{\theta 2} = \frac{V_{\theta 2'}}{\mu_s} = \frac{327.715}{0.9} = 364.128\,\text{m/s}$$

giving

$$W_{\theta 2} = U_2 - V_{\theta 2} = 547 - 364.128 = 182.872\,\text{m/s}$$

Assuming $\varphi_2 = 0.3$, we obtain

$$W_{m2} = \varphi_2 U_2 = 0.3 \times 547 = 164.1\,\text{m/s}$$

Blade Angle at Impeller Exit (β_2)

$$\beta_2 = \tan^{-1}\frac{W_{m2}}{W_{\theta 2}} = \tan^{-1}\left(\frac{164.1}{182.872}\right) = 41.9°$$

Absolute Flow Angle $(\beta_{2'})$

$$\beta_{2'} = \tan^{-1}\left(\frac{V_{\theta 2'}}{V_{m2'}}\right) = \tan^{-1}\left(\frac{327.715}{182.872}\right) = 63.4°$$

Number of Blades

We obtain the number of blades n_b from Equation 9.17 as

$$n_b = \frac{0.63\pi}{(1 - \mu_s)(1 - 0.3\cot\beta_2)} = \frac{0.63\pi}{(1 - 0.9)(1 - 0.3 \times \cot 41.9°)} = 29.7$$

which yields

$$n_b = 30$$

Impeller Efficiency (η_{im})

Using $\chi = 0.55$, we obtain

$$\eta_{im} = 1 - \chi(1 - \eta_c) = 1 - 0.55 \times (1 - 0.8) = 0.890$$

Impeller Exit Total Temperature (T_{02})

Noting that $T_{02} = T_{03}$, we obtain from Equation 9.1

$$T_{02} = T_{03} = T_{01} + \frac{E}{\eta_m c_p} = 306 + \frac{179260}{1004.5 \times 0.96} = 492\,\text{K}$$

Impeller Tip Width (b_2)

We compute the total pressure ratio over the impeller as

$$\frac{p_{02}}{p_{01}} = \left[1 + \frac{\eta_{im}(T_{02} - T_{01})}{T_{01}}\right]^{\gamma/\gamma-1} = \left[1 + \frac{0.8 \times (492 - 306)}{306}\right]^{3.5} = 4.539$$

giving

$$p_{02} = 4.539\,p_{01} = 4.539 \times 100,000 = 453,900\,\text{Pa}$$

Now, we compute the Mach number at impeller exit as follows:

$$V_{2'} = \sqrt{V_{\theta 2'}^2 + V_{m2'}^2} = \sqrt{(327.715)^2 + (164.1)^2} = 366.505\,\text{m/s}$$

$$T_2 = T_{02} - \frac{V_{2'}^2}{2c_p} = 492 - \frac{(366.505)^2}{2 \times 1004.5} = 425\,\text{K}$$

$$\frac{T_{02}}{T_2} = \frac{492}{425} = 1.157$$

$$M_2 = \sqrt{\frac{2}{\gamma - 1}\left(\frac{T_{02}}{T_2} - 1\right)} = \sqrt{\frac{2}{1.4 - 1} \times (1.157 - 1)} = 0.887$$

giving

$$\hat{F}_{f02} = \frac{M_2\sqrt{\gamma}}{\left(\dfrac{T_{02}}{T_2}\right)^{\gamma+1/2(\gamma-1)}} = 0.887 \times \frac{0.887\sqrt{1.4}}{(1.157)^3} = 0.677$$

and the velocity coefficient

$$C_V = \frac{V_{m2'}}{V_{2'}} = \frac{164.1}{366.505} = 0.448$$

Now, we compute the impeller exit flow area as

$$A_2 = \frac{\dot{m}\sqrt{RT_{02}}}{p_{02}C_V \hat{F}_{f02}} = \frac{3.3 \times \sqrt{287 \times 492}}{453{,}923 \times 0.448 \times 0.677} = 0.00901\,\text{m}^2$$

giving the impeller tip width

$$b_2 = \frac{A_2}{\pi D_2} = \frac{0.00901}{3.1416 \times 0.261} = 0.011\,\text{m} = 11\,\text{mm}$$

Diffusion Ratio

$$W_{1s} = C_1 M_{R1s} = 340.495 \times 1.1 = 374.544 \ \text{m/s}$$

$$W_{\theta 2'} = U_2 - V_{\theta 2'} = 547 - 327.715 = 219.285 \ \text{m/s}$$

$$W_{2'} = \left(W_{\theta 2'}^2 + W_{m2}^2\right)^{\frac{1}{2}} = \left[(219.285)^2 + (164.1)^2\right]^{\frac{1}{2}} = 273.888\,\text{m/s}$$

$$\frac{W_{1s}}{W_{2'}} = \frac{374.544}{273.888} = 1.368$$

which is within the recommended range given in Table 9.2.

NOMENCLATURE

b_2 Blade width at $r = r_2$ in compressor impeller
c_p Specific heat at constant pressure
C Local speed of sound in gas
C_1 Speed of sound at impeller inlet
C_2 Speed of sound at impeller outlet
D_2 Impeller tip diameter
D_s Specific diameter
D_{1h} Hub diameter at impeller inlet
D_{1s} Shroud diameter at impeller inlet
E Energy transfer from impeller to gas
E_i Ideal energy transfer for isentropic compression from p_{01} to p_{03} ($=gH$)
g Gravitational acceleration
H Output head ($=E_i/g$)
h_{01} Total enthalpy of gas entering impeller
h_{02} Total enthalpy of gas leaving impeller
h_{03} Total enthalpy of gas leaving diffuser
k Hub-tip ratio at impeller inlet ($=D_{1h}/D_{1s}$)
m Mass flow rate of gas discharged from compressor
M Mach number
M_1 Absolute Mach number at impeller inlet ($=V_1/C_1$)
M_R Relative Mach number ($=W/C$)

M_{R1s} Relative Mach number at impeller inlet shroud $(=W_{1s}/C_1)$

N Rotor speed

N_s Specific speed

n_b Number of blades in the impeller

P Power to impeller shaft

p_1 Static pressure at impeller inlet

p_{01} Total pressure at stage inlet

p_{02} Total pressure at impeller outlet

p_{03} Total pressure at diffuser outlet

Q_1 Volumetric flow rate at impeller inlet

r Radial position measured from the axis of rotation

r_2 Radial position at impeller tip $(=D_2/2)$

r_3 Radial position at exit from vaneless diffuser

r_{sh} Shaft radius

R Gas constant

T Gas static temperature

T_0 Gas total (stagnation) temperature

T_1 Static temperature at impeller inlet

T_{01} Total temperature of gas entering impeller

T_{02} Total temperature of gas leaving impeller

T_{03} Total temperature of gas leaving diffuser

T_i Total temperature of gas at the end of an isentropic compression from p_{01} to p_{03}

$T_{i'}$ Total temperature of gas at the end of an isentropic compression from p_{01} to p_{02}

U Blade speed at any r

U_1 Impeller speed at the blade leading edge

U_{1s} U_1 at shroud diameter

U_{1h} U_1 at hub diameter

U_2 Impeller speed at blade tip

V Absolute velocity of gas

V_1 Absolute velocity at the blade leading edge

V_2 Absolute velocity of gas leaving the impeller with an infinite number of blades

V_m Meridional component of V at any r $(=V_r)$

$V_{2'}$ Absolute velocity of gas leaving the impeller with a finite number of blades

$V_{\theta 1}$ Tangential component of V_1

V_{m2} Meridional component of V_2 or $V_{2'}$ $(=W_{m2})$

V_r Radial component of V at any r $(=V_m)$

V_θ Tangential component of V at any r

$V_{\theta 2}$ Tangential component of V_2

$V_{\theta 2'}$ Tangential component of $V_{2'}$

W Velocity relative to impeller blade

W_1 Relative velocity at impeller blade leading edge

W_2 Relative velocity of gas leaving the impeller with an infinite number of blades

$W_{2'}$ Relative velocity of gas leaving the impeller with a finite number of blades

W_{m2} Meridional component of W_2, $W_{2'}$, V_2, or $V_{2'}$

W_{1s} Relative velocity of gas entering impeller at shroud

GREEK SYMBOLS

α Absolute gas angle (=the angle between V and V_r)

$\alpha_{2'}$ Absolute gas angle at $r = r_2$; $\tan^{-1}\left(V_{\theta 2'}/W_{m2}\right)$

α_3 Absolute gas angle at $r = r_3$

β_1 Angle between W_1 and U_1

β_{1s} Angle between W_{1s} and U_{1s}

β_2 Angle between W_2 and U_2; also the blade angle at its trailing edge

$\beta_{2'}$ Angle between $W_{2'}$ and U_2; actual fluid angle

χ Ratio of impeller loss to compressor losses

γ Ratio of specific heats

η_c Compressor efficiency

η_m Mechanical efficiency

η_{im} Impeller efficiency

μ_s Slip coefficient

ρ Gas density

ρ_1 Gas density at impeller inlet

ρ_2 Gas density at impeller exit

ϕ_P Power function

φ_2 Flow coefficient at impeller exit ($=W_{m2}/U_2$)

Ω Rotor angular velocity

REFERENCES

Ferguson, T.B. 1963. *The Centrifugal Compressor Stage*. London: Butterworths.

Whitfield, A. 1990. *Journal of Power and Energy*. London, UK: Institution of Mechanical Engineers.

Sultanian, B.K. 2019. *Logan's Turbomachinery: Flowpath Design and Performance Fundamentals*, 3rd edition. Boca Raton, FL: Taylor & Francis.

BIBLIOGRAPHY

Aungier, R.H. 2000. *Centrifugal Compressors: A Strategy for Aerodynamic Design and Analysis*. New York: ASME Press.

Boyce, M.P. 2002. *Centrifugal Compressors: A Basic Guide*. Tulsa: PennWell Corporation.

Braembussche, R.V. 2018. *Design and Analysis of Centrifugal Compressors* (Wiley-ASME Series), 1st edition. New York: Wiley.

Dixon, S.L. and C.A. Hall. 2013. *Fluid Mechanics and Thermodynamics of Turbomachinery*, 7th edition. London, UK: Elsevier.

Japiske, D. 1996. *Centrifugal Compressor Design and Performance*. White River Junction: Concepts Eti.

Shepherd, D.G. 1956. *Principles of Turbomachinery*. New York: MacMillan.

Sultanian, B.K. 2015. *Fluid Mechanics: An Intermediate Approach*. Boca Raton: Taylor & Francis.

Sultanian, B.K. 2018. *Gas Turbines: Internal Flow Systems Modeling* (Cambridge Aerospace Series). Cambridge: Cambridge University Press.

10 Axial-Flow Pumps, Fans, and Compressors

REVIEW OF KEY CONCEPTS

We concisely present here some key concepts of axial-flow compressors, fans, and pumps. More details on each topic are given, for example, in Sultanian (2019). Axial-flow compressors are ubiquitous in today's high-performance gas turbines used for aircraft propulsion and power generation. In these machines, compared to turbines, compressors feature many more stages to prevent flow separation on airfoils due to inevitable adverse pressure gradient in the flow.

Figure 10.1 shows that a compressor blade deflects the fluid through only a fraction of the angle that a turbine blade does. The figure also shows that the concave side of the compressor blade moves ahead of the convex side; the reverse is true of the turbine blade.

The approach to compressor stage design is the same as that used for axial-flow pumps and fans, except that the compressibility (density change) of the gas must be considered in the overall process of multistage machines. Axial-flow pumps and fans handle liquids and gases with nearly constant density. Like propellers, their blades have small curvature and cause little deflection of the flow velocity relative to the rotor. Unlike propellers, however, they are enclosed by a casing.

BLADE PROFILE AND AERODYNAMICS

In Figure 10.2, the blade motion is to the right. The lift force L is primarily responsible for the transfer of energy, and the drag force D, which is directed parallel to W_m, is primarily associated with blade losses. We obtain the tangential component $F_{b\theta}$ of the blade force as

$$F_{b\theta} = L\cos\beta_\mathrm{m} + D\sin\beta_\mathrm{m} \tag{10.1}$$

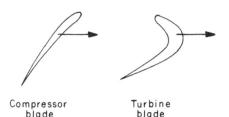

Compressor Turbine
blade   blade

FIGURE 10.1 Comparison of axial-flow compressor and turbine blades.

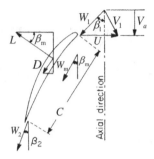

FIGURE 10.2 Blade motion and flow aerodynamics

giving the rate of energy transfer $\dot{E}$ as

$$\dot{E} = U\left(L\cos\beta_m + D\sin\beta_m\right) \tag{10.2}$$

Note that $F_{b\theta}$ in Equations 10.1 and $\dot{E}$ in Equation 10.2 are per blade per its unit length.

Referring to Figure 10.3, we obtain the energy transfer per unit mass as

$$E = U\left(V_{\theta 2} - V_{\theta 1}\right) \tag{10.3}$$

in which the blade velocity U is the same at the inlet and exit planes. The figure also shows that

$$\Delta V_\theta = V_{\theta 2} - V_{\theta 1} = W_{\theta 1} - W_{\theta 2} = \Delta W_\theta \tag{10.4}$$

By multiplying Equation 10.3 by the mass flow rate per blade per its unit length $\left(\dot{m} = \rho V_a S\right)$, we obtain an alternative expression for $\dot{E}$

$$\dot{E} = U\left(V_{\theta 2} - V_{\theta 1}\right)\left(\rho V_a S\right) \tag{10.5}$$

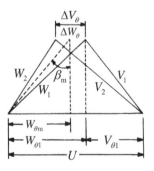

FIGURE 10.3 Velocity diagrams.

where V_a is the axial component of absolute velocity and S is the spacing between two adjacent blades in a row.

Equating Equation 10.2 with Equation 10.5 and nondimensionalizing, we obtain

$$C_L = 2\frac{\cos^2\beta_1}{\cos\beta_m}\frac{S}{C}\left(\tan\beta_1 - \tan\beta_2\right) \tag{10.6}$$

where we have neglected D because $D \ll L$, and the lift coefficient C_L is

$$C_L = \frac{L}{\tfrac{1}{2}\rho W_1^2 C} \tag{10.7}$$

where C is the chord, as shown in Figure 10.2. Equation 10.6 is for an ideal cascade. In the presence of boundary layers and nonuniform flow deflection in actual cascades, we use an empirical correlation equation.

STAGE PRESSURE RISE

The first stage of an axial-flow compressor, fan or pump may include inlet guide vanes, which deflect the fluid from an axial path by the angle α_1 and increase its velocity from V_a to V_1 (Figure 10.2), so that the fluid enters the rotor blades at the proper angle. Figure 10.3 shows that $V_2 > V_1$, which means that the rotor blades increase the flow kinetic energy. In a single-stage machine, outlet guide vanes may turn the flow back to the axial direction. In multistage machines, the vanes redirect the flow to its original direction, so that the flow leaves the stage at absolute the flow angle $\alpha_3 = \alpha_1$ with the absolute velocity $V_3 = V_1$. A complete stage accelerates the flow in the rotor and decelerates it in the stator, accompanied with a pressure rise in both.

Let us consider a control volume, surrounding a single moving blade, of width S and of unit height along the blade, as shown in Figure 10.4. Assuming no change in

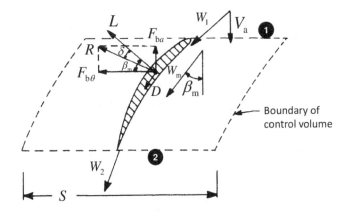

FIGURE 10.4   Control volume for a cascade blade.

the flow axial velocity from inlet to outlet and constant mass flow rate, we can write the force equilibrium equation as

$$(p_2 - p_1)S - F_{ba} = 0 \tag{10.8}$$

Expressing F_{ba} in terms of lift and drag, we obtain for the pressure rise across the blade row

$$(p_2 - p_1)_{rotor} = \frac{L}{S}\sin\beta_m - \frac{D}{L}\cos\beta_m \tag{10.9}$$

In nondimensional form, Equation 10.9 becomes

$$\frac{(p_2 - p_1)_{rotor}}{\tfrac{1}{2}\rho W_1^2} = \frac{C}{S}\left(C_L\sin\beta_m - C_D\cos\beta_m\right) \tag{10.10}$$

As derived in Sultanian (2019), an alternate expression for pressure rise across the blade is given by

$$(p_2 - p_1)_{rotor} = \varphi U^2 \varphi \Lambda_b \frac{R - \varphi\delta}{\varphi + \delta R} \tag{10.11}$$

where Λ_b is the blade loading coefficient $\Delta V_\theta / U$ and $\delta = C_D/C_L$, and that for the stage by

$$(\Delta p)_{st} = \rho U^2 \varphi \Lambda_b \left[\frac{R - \varphi\delta}{\varphi + \delta R} + \frac{1 - R - \varphi\delta}{\varphi + \delta(1 - R)}\right] \tag{10.12}$$

Defining the stage efficiency as the ratio of pressure rise with drag to that without drag $(\delta = 0)$, we obtain from Equation 10.12

$$\eta_{st} = \frac{(\Delta p)_{st}}{(\Delta p)_{st\_ideal}} = \varphi\left[\frac{R - \varphi\delta}{\varphi + \delta R} + \frac{1 - R - \varphi\delta}{\varphi + \delta(1 - R)}\right] \tag{10.13}$$

The drag-lift ratio δ used in this equation must be modified to account for several additional losses and written as

$$\delta = \frac{C_D + C_{D'} + C_{D''} + C_{D'''}}{C_L} \tag{10.14}$$

where, as presented in Sultanian (2019), we express profile drag coefficient C_D as

$$C_D = \frac{\zeta_p \cos^3 \beta_m}{\sigma} \tag{10.15}$$

where ζ_p is the profile loss coefficient and $\sigma = C/S$ is the solidity at mean radius; the blade surface frictional drag coefficient $C_{D'}$ as

$$C_{D'} = 0.02\frac{S}{h} \tag{10.16}$$

the drag coefficient $C_{D''}$ due to secondary flows as

$$C_{D''} = 0.018 C_L^2 \tag{10.17}$$

and, finally, the drag coefficient $C_{D'''}$ associated with balde tip clearance and leakage as

$$C_{D'''} = 0.018 \frac{c_{tip}}{h} C_L^{3/2} \tag{10.18}$$

Note that the evaluation of the drag-lift ratio δ from Equation 10.14 requires the use of Equation 10.6 in the form

$$C_L = \frac{2\cos\beta_m \left(\tan\beta_1 - \tan\beta_2\right)}{\sigma} \tag{10.19}$$

For additional details on the preliminary design of axial-flow pumps, fans, and compressors, readers are encouraged to review the material in Sultanian (2019).

WORK-DONE FACTOR

In a multistage axial-flow compressor, the annular walls create a boundary layer that causes peaking of the axial velocity V_a near the mean radius. This requires that we must multiply the aerodynamic energy transfer equation by a factor λ, called the work-done factor. The resulting equation becomes

$$E = \lambda U V_a \left(\tan\beta_1 - \tan\beta_2\right) \tag{10.20}$$

which we can use for each stage with a constant value of λ, approximated from

$$\lambda = 0.85 + 0.15\exp\left[\frac{-\left(n_{st}-1\right)}{2.73}\right] \tag{10.21}$$

where n_{st} is the number of stages in the compressor.

ISENTROPIC EFFICIENCY

The overall compressor pressure ratio can be determined from the product of individual stage pressure ratios. Similarly, the overall total temperature rise is the sum of stage total temperature rises. We express the isentropic efficiency of a compressor stage as

$$\eta_{st} = \left(\frac{T_{01}}{T_{03} - T_{01}} \right) \left(\Pi_{st}^{\frac{\gamma-1}{\gamma}} - 1 \right) \tag{10.22}$$

and for a multistage compressor as a whole

$$\eta_c = \left(\frac{T_{01}}{\Delta T_0} \right) \left(\Pi_c^{\frac{\gamma-1}{\gamma}} - 1 \right) \tag{10.23}$$

where Π_c denotes the total pressure ratio and ΔT_0 is the total temperature rise for the whole machine.

POLYTROPIC EFFICIENCY

The isentropic efficiency of a compressor depends upon its operating conditions. For example, if two identical compressors operate in series with equal pressure ratio across each, their isentropic efficiency will be equal. This isentropic efficiency, however, will be higher than its value when the same compressor operates at a higher-pressure ratio. If we divide the compression process across a compressor into a large number of very small, consecutive compression, then the isentropic efficiency across each is called the polytropic efficiency, or small-stage efficiency, which may be assumed constant for the machine, reflecting its state-of-the-art design engineering.

Referring to Figure 10.5, we assume that the index of polytropic compression along 01–03 is n, which yields

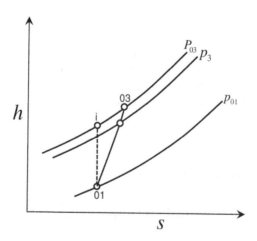

FIGURE 10.5 Enthalpy-entropy diagram.

$$\frac{T_{03}}{T_{01}} = \left(\frac{p_{03}}{p_{01}}\right)^{\frac{n-1}{n}} = \Pi_c^{\frac{n-1}{n}} \tag{10.24}$$

For $n = \gamma$, this equation yields the isentropic compression along 01-i.
  We define the compressor polytropic efficiency as

$$\eta_p = \frac{(dT_0)_{\text{isentropic}}}{(dT_0)_{\text{actual}}} = \frac{(dT_0/T_0)_{\text{isentropic}}}{(dT_0/T_0)_{\text{actual}}} \tag{10.25}$$

Integrating this equation between the compressor inlet and exit yields

$$\eta_p \int_{01}^{03} (dT_0/T_0)_{\text{actual}} = \int_{01}^{i} (dT_0/T_0)_{\text{isentropic}}$$

$$\left(\frac{T_{03}}{T_{01}}\right)^{\eta_p} = \frac{T_{03}}{T_{01}}$$

$$\Pi_c^{\frac{(n-1)\eta_p}{n}} = \Pi_c^{\frac{\gamma-1}{\gamma}}$$

giving

$$\frac{n-1}{n} = \frac{\gamma-1}{\eta_p \gamma} \tag{10.26}$$

To establish a relation between the compressor isentropic efficiency and its polytropic efficiency, we write

$$\eta_c = \frac{T_i - T_{01}}{T_{03} - T_{01}} = \frac{T_i/T_{01} - 1}{T_{03}/T_{01} - 1}$$

$$\eta_c = \frac{\Pi_c^{\frac{\gamma-1}{\gamma}} - 1}{\Pi_c^{\frac{n-1}{n}} - 1}$$

which with the substitution from Equation 10.26 yields

$$\eta_c = \frac{\Pi_c^{\frac{\gamma-1}{\gamma}} - 1}{\Pi_c^{\frac{\gamma-1}{\eta_p \gamma}} - 1} \tag{10.27}$$

which expresses compressor isentropic efficiency in terms of its polytropic efficiency and pressure ratio. Alternatively, to express compressor polytropic efficiency in terms of its isentropic efficiency and pressure ratio, we can rewrite Equation 10.27 as

$$\eta_p = \frac{\ln\left(\Pi_c^{\frac{\gamma-1}{\gamma}}\right)}{\ln\left(1 + \dfrac{\Pi_c^{\frac{\gamma-1}{\gamma}} - 1}{\eta_c}\right)} \qquad (10.28)$$

PROBLEM 10.1: ANALYSIS OF A HIGH-PERFORMANCE AXIAL-FLOW AIR COMPRESSOR

An axial-flow, high performance, single-stage, experimental air compressor has no inlet guide vanes and runs at 12,000 rpm with the stage efficiency of 0.87. The blade speed at the tip is 403 m/s, the hub-tip ratio is 0.70, and the solidity is 1.375. Find (1) the mean blade speed, (2) the absolute and relative Mach numbers of air entering the rotor if the relative flow angle at entry is 60° and the total temperature is 300 K, and (3) energy transfer and stage total pressure ratio if the relative flow angle at the rotor exit is 35°.

SOLUTION FOR PROBLEM 10.1

(a) MEAN BLADE SPEED

Rotor Angular Velocity

$$\Omega = \frac{\pi N}{30} = \frac{3.1416 \times 12,000}{30} = 1256.637\,\text{rad/s}$$

Tip and Hub Radii

$$r_{tip} = \frac{U_{tip}}{\Omega} = \frac{403}{1256.637} = 0.321\,\text{m}$$

$$r_{hub} = r_{tip}\left(\frac{r_{hub}}{r_{tip}}\right) = 0.321 \times 0.7 = 0.225\,\text{m}$$

Mean Radius and Blade Speed

The circle of mean radius divides the compressor annular area into two equal areas. Thus, we calculate the mean radius as

$$r_m = \left(\frac{r_{hub}^2 + r_{tip}^2}{2}\right)^{\frac{1}{2}} = \left\{\frac{(0.225)^2 + (0.321)^2}{2}\right\}^{\frac{1}{2}} = 0.277\,\text{m}$$

giving

$$U_m = r_m\Omega = 0.277 \times 1256.637 = 347.843\,\text{m/s}$$

(b) Inlet Absolute and Relative Mach Numbers

As V_1 is axially directed, we obtain

$$V_1 = U_m\cot\beta_1 = 347.843 \times \cot 60° = 200.827\,\text{m/s}$$

and

$$W_1 = \frac{U_m}{\sin\beta_1} = \frac{347.847}{\sin 60°} = 401.654\,\text{m/s}$$

Static Temperature

$$T_1 = T_{01} - \frac{V_1^2}{2c_p} = 300 - \frac{(200.827)^2}{2 \times 1004.5} = 280\,\text{K}$$

Speed of Sound

$$C_1 = \sqrt{\gamma R T_1} = \sqrt{1.4 \times 287 \times 280} = 335.371\,\text{m/s}$$

Absolute Mach Number

$$M_1 = \frac{V_1}{C_1} = \frac{200.827}{335.371} = 0.599$$

Relative Mach Number

$$M_{R1} = \frac{W_1}{C_1} = \frac{401.654}{335.371} = 1.198$$

(c) Energy Transfer and Stage Total Pressure Ratio

Energy Transfer
As the inlet flow is axial, we obtain

$$W_{\theta 1} = U_m = 347.843\,\text{m/s}$$

With constant axial velocity (flow velocity) between stage inlet and exit, which equals $V_1 = 200.827\,\text{m/s}$, we obtain

$$W_{\theta 2} = V_1 \tan\beta_2 = 200.827 \times \tan 35° = 140.621\,\text{m/s}$$

giving

$$E = U_m(W_{\theta 1} - W_{\theta 2}) = 347.843 \times (347.843 - 140.621) = 72,081\,\text{m}^2/\text{s}^2$$

and

$$\Pi_{st} = \frac{p_{03}}{p_{01}} = \left[1 + \frac{\eta_{st}E}{c_p T_{01}}\right]^{\gamma/\gamma-1} = \left[1 + \frac{0.87 \times 72,081}{1004.5 \times 300}\right]^{3.5} = 1.938$$

PROBLEM 10.2:   OVERALL COMPRESSOR EFFICIENCY FOR A FOUR-STAGE COMPRESSOR

Find the overall compressor efficiency for a four-stage compressor in which each stage has the same stage efficiency of 0.87 with $V_1 = V_2 = V_a = 200\,\text{m/s}$, $W_{\theta 1} = U = 350\,\text{m/s}$, and $\beta_2 = 35°$. The air total temperature at compressor inlet is 300 K.

SOLUTION FOR PROBLEM 10.2

WORK-DONE FACTOR

For the four-stage compressor $(n_{st} = 4)$, we compute the work-done factor from

$$\lambda = 0.85 + 0.15\exp\left[-(n_{st} - 1)/2.73\right] = 0.90$$

RELATIVE TANGENTIAL VELOCITY AT BLADE EXIT IN EACH STAGE

$$W_{\theta 2} = V_a \tan\beta_2 = 200 \times \tan 35° = 140.042\,\text{m/s}$$

SPECIFIC AERODYNAMIC ENERGY TRANSFER IN EACH STAGE

$$E = \lambda U(W_{\theta 1} - W_{\theta 2}) = 0.90 \times 350 \times (350 - 140.042) = 66,135.864\,\text{m}^2/\text{s}^2$$

AIR TOTAL TEMPERATURE RISE IN EACH STAGE

$$\Delta T_{0st} = \frac{E}{c_p} = \frac{66,135.864}{1004.5} = 65.8\,\text{K}$$

Stage 1: Total Pressure Ratio

$$\Pi_{st1} = \left[1 + \frac{\eta_{st}\Delta T_{0st}}{T_{01}}\right]^{\gamma/\gamma-1} = \left[1 + \frac{0.87 \times 65.8}{300}\right]^{3.5} = 1.843$$

Stage 2: Total Pressure Ratio

$$T_{02} = T_{01} + \Delta T_{0st} = 300 + 65.8 = 365.8\,\text{K}$$

$$\Pi_{st2} = \left[1 + \frac{\eta_{st}\Delta T_{0st}}{T_{02}}\right]^{\gamma/\gamma-1} = \left[1 + \frac{0.87 \times 65.8}{365.8}\right]^{3.5} = 1.664$$

Stage 3: Total Pressure Ratio

$$T_{03} = T_{02} + \Delta T_{0st} = 365.8 + 65.8 = 431.6\,\text{K}$$

$$\Pi_{st3} = \left[1 + \frac{\eta_{st}\Delta T_{0\_st}}{T_{03}}\right]^{\gamma/\gamma-1} = \left[1 + \frac{0.87 \times 65.8}{431.7}\right]^{3.5} = 1.547$$

Stage 4: Total Pressure Ratio

$$T_{04} = T_{03} + \Delta T_{0st} = 431.6 + 65.8 = 497.4\,\text{K}$$

$$\Pi_{st4} = \left[1 + \frac{\eta_{s}\Delta T_{0st}}{T_{04}}\right]^{\gamma/\gamma-1} = \left[1 + \frac{0.87 \times 65.8}{497.5}\right]^{3.5} = 1.464$$

COMPRESSOR OVERALL TOTAL PRESSURE RATIO

$$\Pi_c = \Pi_{st1}\Pi_{st2}\Pi_{st3}\Pi_{st4} = 1.843 \times 1.664 \times 1.547 \times 1.464 = 6.946$$

COMPRESSOR OVERALL ISENTROPIC EFFICIENCY (η_c)

$$\eta_c = \frac{T_{01}}{4\Delta T_0}\left(\Pi_c^{\gamma-1/\gamma} - 1\right) = \frac{300}{4 \times 65.8}(6.946)^{0.2857} = 0.843$$

COMPRESSOR OVERALL POLYTROPIC EFFICIENCY (η_p)

We first compute the compressor overall total temperature ratio as

$$\frac{T_{0out}}{T_{01}} = \frac{T_{01} + 4\Delta T_{0st}}{T_{01}} = \frac{300 + 4 \times 65.8}{300} = 1.878$$

and then calculate the compressor overall polytropic efficiency as

$$\eta_p = \frac{(\gamma-1)\ln \Pi_c}{\gamma \ln\left(T_{0out}/T_{01}\right)} = \frac{\ln(6.946)}{3.5 \times \ln(1.878)} = 0.879$$

Note that the compressor polytropic and stage efficiencies are nearly equal.

PROBLEM 10.3: COMPARING EFFICIENCIES OF TWO AXIAL COMPRESSORS FOR BIDDING

Two compressors are being compared for bidding. Compressor A is submitted with a total pressure ratio of 4.5 and isentropic efficiency of 85%. Compressor B is submitted with a discharge static pressure of 620 kPa, static temperature of 523 K, and velocity of 150 m/s as tested at the ambient conditions of 101 kPa and 291 K. Which compressor is more efficient? For air as a perfect gas, assume $\gamma = 1.4$ and $c_p = 1004.5 \text{ J}/(\text{kg K})$.

SOLUTION FOR PROBLEM 10.3

COMPRESSOR A

For compressor A, we are given:

Pressure ratio $(\Pi_{cA}) = 4.5$
Isentropic efficiency $(\eta_{cA}) = 85\%$

Calculations for Compressor A
Polytropic efficiency (η_{pA})

$$\eta_{pA} = \frac{\ln\left(\Pi_{cA}^{\frac{\gamma-1}{\gamma}}\right)}{\ln\left(1 + \frac{\Pi_{cA}^{\frac{\gamma-1}{\gamma}} - 1}{\eta_{cA}}\right)} = \frac{\ln\left(4.5^{0.2857}\right)}{\ln\left(1 + \frac{4.5^{0.2857} - 1}{0.85}\right)} = 87.8\%$$

COMPRESSOR B

For compressor B, we are given:

Inlet total pressure $(p_{01}) = 101 \text{ kPa}$
Inlet total temperature $(T_{01}) = 291 \text{ K}$
Exit static pressure $(p_3) = 620 \text{ kPa}$
Exit static temperature $(T_3) = 523 \text{ K}$
Exit velocity $(V_3) = 150 \text{ m/s}$

Calculations for Compressor B

Exit total temperature

$$T_{03} = T_3 + \frac{V_3^2}{2c_p} = 523 + \frac{150 \times 150}{2 \times 1004.5} = 534.2 \text{ K}$$

Exit total pressure

$$\frac{p_{03}}{p_2} = \left(\frac{T_{03}}{T_2}\right)^{\gamma/\gamma-1} = \left(\frac{534.2}{523}\right)^{3.5} = 1.077$$

$$P_{03} = 1.077 \times 620 = 667.75 \text{ kPa}$$

Polytropic efficiency (η_{pB})

$$\frac{T_{03}}{T_{01}} = \left(\frac{p_{03}}{p_{01}}\right)^{n-1/n}$$

where n is the exponent of polytropic compression. We obtain

$$\frac{n-1}{n} = \frac{\ln\left(\dfrac{T_{03}}{T_{01}}\right)}{\ln\left(\dfrac{p_{03}}{p_{01}}\right)} = \frac{\ln\left(\dfrac{534.2}{291}\right)}{\ln\left(\dfrac{667.75}{101}\right)} = \frac{0.6075}{1.8888} = 0.3216$$

For a compressor, the exponents of polytropic and isentropic compressions are related by the equation

$$\frac{n-1}{n} = \frac{\gamma-1}{\gamma \eta_{\text{pB}}}$$

giving

$$\eta_{\text{pB}} = \frac{\left(\dfrac{\gamma-1}{\gamma}\right)}{\left(\dfrac{n-1}{n}\right)} = \frac{0.2857}{0.3216} = 88.8\%$$

As the polytropic efficiency η_{pB} for compressor B is higher than the polytropic efficiency η_{pA} for compressor A, compressor B is more efficient than compressor A.

PROBLEM 10.4:   TWO AXIAL-FLOW COMPRESSORS OPERATING IN SERIES

At the total pressure ratio of 4, compressor A operates at the overall isentropic efficiency of 84.2%. Compressor B, operating at pressure ratio of 8, has the overall isentropic efficiency of 82.7%. Show that both compressors have an equal polytropic efficiency of ~87%. If we use these two compressors in series, calculate the overall isentropic efficiency of the combined unit. For air as a perfect gas, assume $\gamma = 1.4$ and $c_p = 1004.5 \ \text{J}/(\text{kg K})$.

SOLUTION FOR PROBLEM 10.4

We calculate the polytropic efficiency of compressor A with the total pressure ratio of 4 as

$$\eta_{pA} = \frac{\ln\left(\Pi_{cA}^{\frac{\gamma-1}{\gamma}}\right)}{\ln\left(1 + \frac{\Pi_{cA}^{\frac{\gamma-1}{\gamma}} - 1}{\eta_{cA}}\right)} = \frac{\ln\left(4^{0.2857}\right)}{\ln\left(1 + \frac{4^{0.2857} - 1}{0.842}\right)} = 86.928\% = \sim 87\%$$

Similarly, we obtain the polytropic efficiency of the compressor B with the total pressure ratio of 8 as

$$\eta_{pB} = \frac{\ln\left(\Pi_{cB}^{\frac{\gamma-1}{\gamma}}\right)}{\ln\left(1 + \frac{\Pi_{cB}^{\frac{\gamma-1}{\gamma}} - 1}{\eta_{cB}}\right)} = \frac{\ln\left(8^{0.2857}\right)}{\ln\left(1 + \frac{8^{0.2857} - 1}{0.827}\right)} = 86.928\% = \sim 87\%$$

Thus, we calculate the overall isentropic efficiency of the combined compressor unit operating at the total pressure ratio of 32 with $\eta_p = \eta_{pA} = \eta_{pB} = 0.87$ as

$$\eta_c = \frac{\Pi^{\frac{\gamma-1}{\gamma}} - 1}{\Pi^{\frac{\gamma-1}{(\eta_p \gamma)}} - 1} = \frac{(32)^{\frac{1.4-1}{1.4}} - 1}{(32)^{\frac{1.4-1}{0.87\times1.4}} - 1} = 79.8\%$$

PROBLEM 10.5:   COMPRESSOR PRESSURE RATIO FOR MAXIMUM NET WORK OUTPUT IN A GAS TURBINE ENGINE

An ideal, basic, air standard gas turbine engine has a compressor inlet temperature of 288 K and a turbine inlet temperature of 1400 K. If air enters the compressor at the rate of 1 kg/s, calculate (1) the pressure ratio that gives the maximum net work output, (2) the compressor work, (3) turbine work, (4) heat added, (5) thermal efficiency for the pressure ratio determined in (1), and (6) the power developed by the turbine in kilowatts. For air as a perfect gas, assume $\gamma = 1.4$ and $c_p = 1005 \text{ J}/(\text{kg K})$.

SOLUTION FOR PROBLEM 10.5

(a) TOTAL PRESSURE RATIO FOR MAXIMUM NET WORK OUTPUT

For maximum work output in the Brayton cycle with the fixed compressor inlet temperature (T_{01}) and the fixed turbine inlet temperature (T_{03}), we have

$$T_{02} = \sqrt{T_{01}T_{03}}$$

giving

$$T_{02} = \sqrt{288 \times 1400} = 635 \text{ K}$$

Using the isentropic relation between pressure ratio and temperature ratio, we obtain

$$\frac{p_{02}}{p_{01}} = \left(\frac{T_{02}}{T_{01}}\right)^{\frac{\gamma}{\gamma-1}} = \left(\frac{635}{288}\right)^{\frac{1.4}{1.4-1}} = (2.205)^{3.5} = 15.914$$

(b) SPECIFIC COMPRESSOR WORK

$$w_c = c_p(T_{02} - T_{01}) = 1005 \times (635 - 288) = 348.715 \text{ kJ/kg}$$

(c) SPECIFIC TURBINE WORK

$$\frac{T_{03}}{T_{04}} = \frac{T_{02}}{T_{01}} = 2.205$$

giving

$$T_{04} = \frac{T_{03}}{2.205} = \frac{1400}{2.205} = 635 \text{ K}$$

$$w_t = c_p(T_{04} - T_{03}) = 1005 \times (1400 - 635) = 768.845 \text{ kJ/kg}$$

(d) HEAT ADDED

$$q = c_p(T_3 - T_2) = 1005 \times (1400 - 635) = 768.847 \text{ kJ/kg}$$

(e) THERMAL EFFICIENCY

$$\eta_{th} = \frac{w_t - w_c}{q} = \frac{768.847 - 348.715}{768.847} = 54.6\%$$

(f) TURBINE POWER

$$P_t = \dot{m}\, w_t = 1 \times 768.845 = 768.845 \text{ kW}$$

PROBLEM 10.6: COMPRESSOR PRESSURE RATIO FOR MAXIMUM WORK OUTPUT IN A GAS TURBINE WITH REHEAT

In an ideal gas turbine cycle with reheat, air at state (p_{01}, T_{01}) is compressed to pressure $\Pi_c p_{01}$ and heated to T_{03}. The air is then expanded in turbine in two stages, each turbine having the same pressure ratio, with reheat to T_{03} between the stages. Assuming that the working fluid is a perfect gas with constant specific heats and that the compression and expansion are isentropic, show that the specific work output will be a maximum when Π_c is given by

$$\Pi_c = \left(\frac{T_{03}}{T_{01}} \right)^{2\gamma / 3(\gamma - 1)}$$

SOLUTION FOR PROBLEM 10.6

As shown in Figure 10.6, with equal pressure ratio across each turbine, we can write

$$\frac{T_{03}}{T_{04}} = \frac{T_{05}}{T_{06}}$$

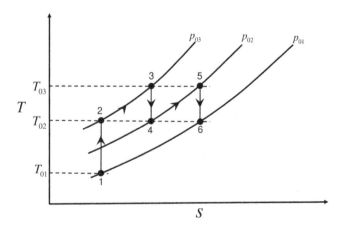

FIGURE 10.6   Gas turbine cycle with reheat.

Further, as $T_{03} = T_{05}$, we also obtain $T_{04} = T_{06}$.

Thus, we obtain the specific work output of both turbines as

$$w_t = c_p \left[(T_{03} - T_{04}) + (T_{05} - T_{06}) \right] = 2c_p (T_{03} - T_{04})$$

Compressor specific work input

$$w_c = c_p (T_{02} - T_{01})$$

Net specific work output of the cycle

$$w_{net} = w_t - w_c = 2c_p (T_{03} - T_{04}) - c_p (T_{02} - T_{01})$$

As the pressure ratio across each turbine stage is equal and the overall pressure ratio across both turbines equals that across the compressor, we can write

$$\frac{T_{03}}{T_{04}} = \left(\frac{T_{02}}{T_{01}} \right)^{\frac{1}{2}}$$

giving

$$T_{04} = T_{03} \left(\frac{T_{01}}{T_{02}} \right)^{\frac{1}{2}}$$

Substituting for T_{04}, the expression for w_{net} becomes

$$w_{net} = 2c_p T_{03} - 2c_p T_{03} \left(\frac{T_{01}}{T_{02}} \right)^{\frac{1}{2}} - c_p T_{02} + c_p T_{01}$$

where T_{01}, T_{03}, and c_p are constant.

For the maximum specific work output, we can write,

$$\frac{dw_{net}}{dT_{02}} = 0$$

which yields

$$\frac{d}{dT_{02}} \left(-2 \frac{T_{03}}{T_{01}} \left(\frac{T_{01}}{T_{02}} \right)^{\frac{1}{2}} - \frac{T_{02}}{T_{01}} \right) = 0$$

$$\frac{T_{03}}{T_{01}} = \left(\frac{T_{02}}{T_{01}} \right)^{\frac{3}{2}} = \Pi_c^{\frac{3(\gamma-1)}{2\gamma}}$$

$$\Pi_c = \left(\frac{T_{03}}{T_{01}} \right)^{\frac{2\gamma}{3(\gamma-1)}}$$

PROBLEM 10.7:   ROOT, MEAN, AND TIP VELOCITY DIAGRAMS FOR AN AXIAL-FLOW COMPRESSOR

An axial-flow compressor stage has blade root, mean and tip velocities of 150, 200, and 250 m/s, respectively. The stage is to be designed for stagnation temperature rise of 20 K and an axial velocity of 150 m/s, both being constant from root to tip. The work-done factor is 0.93. For a free vortex $(rV_\theta = \text{constant})$ design with 50% reaction at the mean radius, calculate (1) the degree of reaction at root and tip, (2) the stage air angles $(\alpha_1, \alpha_2, \beta_1 \text{ and } \beta_2)$ at root, mean and tip, and (3) sketch dimensionless velocity diagrams at root, mean, and tip. Assume $c_p = 1005 \text{ J}/(\text{kg K})$.

SOLUTION FOR PROBLEM 10.7

(a) THE DEGREE OF REACTION AT ROOT AND TIP

Reaction at Root

For a free vortex design, we calculate the reaction at the root from

$$\frac{1 - R_{\text{root}}}{1 - R_{\text{m}}} = \left(\frac{r_{\text{m}}}{r_{\text{root}}}\right)^2 = \left(\frac{U_{\text{m}}}{U_{\text{root}}}\right)^2$$

giving

$$R_{\text{root}} = 1 - (1 - R_{\text{m}})\left(\frac{U_{\text{m}}}{U_{\text{root}}}\right)^2 = 1 - (1 - 0.5)\left(\frac{200}{150}\right)^2 = 0.111$$

Reaction at Tip

For a free vortex design, we calculate the reaction at the tip from

$$R_{\text{tip}} = 1 - (1 - R_{\text{m}})\left(\frac{U_{\text{m}}}{U_{\text{tip}}}\right)^2 = 1 - (1 - 0.5)\left(\frac{200}{250}\right)^2 = 0.680$$

(b) STAGE AIR ANGLES $(\alpha_1, \alpha_2, \beta_1 \text{ and } \beta_2)$ AT ROOT, MEAN, AND TIP

We use the following equations, developed in Appendix B, to compute stage air angles:

$$\tan\alpha_1 = \frac{0.5\psi + (1 - R)}{\varphi}$$

$$\tan\alpha_2 = -\frac{0.5\psi - (1 - R)}{\varphi}$$

$$\tan\beta_1 = \frac{0.5\psi - R}{\varphi}$$

$$\tan\beta_2 = -\frac{0.5\psi + R}{\varphi}$$

These equations involve three design parameters, namely, $\varphi = V_a/U$, $\psi = \Delta T_0 c_p/(\lambda U^2)$, and R, which we have computed at root and tip radii in (a). All the computed values are summarized in Table 10.1.

(c) DIMENSIONLESS VELOCITY DIAGRAMS AT ROOT, MEAN, AND TIP

Figure 10.7 shows the dimensionless velocity diagrams at blade root, mean, and tip radii. At each radius, flow velocities are normalized by the local blade velocity. For each velocity triangle, the nondimensional blade velocity equals unity.

PROBLEM 10.8: MEANLINE ANALYSIS OF AN AXIAL-FLOW COMPRESSOR

Air at 101.3 kPa and 288 K enters and leaves an axial-flow compressor stage with a velocity of 150 m/s without swirl. The stage has a tip radius of 56.0 cm, a hub (root) radius of 35.0 cm and rotates at 8000 rpm. The air is turned through 12.0 degrees as it passes through the rotor with negligible incidence and deviation from the blades. Assume $\gamma = 1.4$ and $R = 287 \, \text{J}/(\text{kg K})$.

a. Determine the blade angles (β_1, β_2) and absolute flow angles (α_1, α_2) at the leading and trailing edges of the rotor blades at the mean radii.
b. Calculate the mass flow rate through the compressor.
c. Calculate loading coefficient, flow coefficient, and the degree of reaction.

TABLE 10.1
Summary of Computed Parameters (Problem 10.7)

Quantity	Root	Mean	Tip
R	0.111	0.50	0.68
φ	1.000	0.75	0.60
ψ	−0.961	−0.54	−0.346
α_1	22.2°	17.0°	13.8°
α_2	53.9°	45.8°	39.4°
β_1	−30.6°	−45.8°	−54.9°
β_2	20.3°	−17.0°	−40.2°

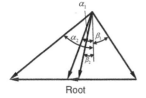
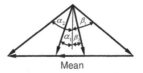
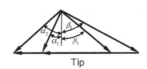

Root   Mean   Tip

FIGURE 10.7   Dimensionless velocity diagrams at blade root, mean, and tip radii (Problem 10.7).

d. Construct a dimensionless velocity diagram to scale showing velocities at rotor inlet and outlet with a common apex.
e. Assuming a polytropic efficiency of 0.9 and the work-done factor of 0.95, compute the total pressure ratio of this stage and the corresponding stage isentropic efficiency.

SOLUTION FOR PROBLEM 10.8

PRELIMINARY CALCULATIONS

For $\gamma = 1.4$, we obtain

$$\frac{\gamma - 1}{\gamma} = \frac{1.4 - 1}{1.4} = 0.2857$$

$$\frac{\gamma}{\gamma - 1} = \frac{1.4}{1.4 - 1} = 3.5$$

and

$$c_p = \frac{R\gamma}{\gamma - 1} = 287 \times 3.5 = 1004.5 \text{ J}/(\text{kg K})$$

Blade Angular Velocity

$$\Omega = \frac{\pi N}{30} = 837.758 \text{ rad}/\text{s}$$

(a) CALCULATION OF BLADE ANGLES AND ABSOLUTE FLOW ANGLES

Using the mean radius

$$r_m = 0.5 \times (0.35 + 0.56) = 0.455 \text{ m}$$

we compute the blade velocity at the mean radius as

$$U_m = r_m \Omega = 0.455 \times 837.758 = 381.180 \text{ m}/\text{s}$$

Blade Leading Edge

With no swirl at the blade inlet, we have $\alpha_1 = 0$ and $V_{\theta 1} = 0$, which yields

$$W_{\theta 1} = -U_m = -381.180 \text{ m}/\text{s}$$

and

$$\beta_1 = \tan^{-1}\left(\frac{W_{\theta 1}}{V_{x1}}\right) = \tan^{-1}\left(\frac{-381.180}{150}\right) = -68.52°$$

Blade Trailing Edge

With $\beta_2 = \beta_1 + 12 = -68.52 + 12 = -56.52°$, we obtain

$$W_{\theta 2} = V_{a2}\tan\beta_2 = 150 \times \tan(-56.52°) = -226.794 \text{ m/s}$$

$$V_{\theta 2} = W_{\theta 2} + U_m = -226.794 + 381.180 = 154.386 \text{ m/s}$$

and

$$\alpha_2 = \tan^{-1}\left(\frac{V_{\theta 2}}{V_{a2}}\right) = \tan^{-1}\left(\frac{154.386}{150}\right) = 45.826°$$

(b) COMPRESSOR MASS FLOW RATE

Annulus Area

$$A = \pi\left(r_{\text{tip}}^2 - r_{\text{hub}}^2\right) = \int \pi(0.56 \times 0.56 - 0.35 \times 0.35) = 0.600 \text{ m}^2$$

Inlet Density

$$\rho_1 = \frac{p_1}{RT_1} = \frac{101.3 \times 1000}{287 \times 288} = 1.226 \text{ kg/m}^3$$

Mass Flow Rate

$$\dot{m} = \rho_1 A V_{a1} = 1.226 \times 0.600 \times 150 = 110.366 \text{ kg/s}$$

(c) LOADING COEFFICIENT, FLOW COEFFICIENT, AND DEGREE OF REACTION

Loading Coefficient

$$\psi = \frac{V_{\theta 1} - V_{\theta 2}}{U_m} = \frac{0 - 154.386}{381.180} = -0.405$$

Flow Coefficient

$$\varphi = \frac{V_{a1}}{U_m} = \frac{150}{381.180} = 0.394$$

Degree of Reaction

$$R = -\frac{W_{\theta 1} + W_{\theta 2}}{2U_m} = -\frac{-381.180 - 226.794}{2 \times 381.180} = 0.797$$

(d) Dimensionless Velocity Diagram

$$V_1 = V_{a1} = V_{a2} = 150 \text{ m/s}$$

$$W_1 = \frac{V_{a1}}{\cos \beta_1} = \frac{150}{\cos(-68.52°)} = 409.632 \text{ m/s}$$

$$V_2 = \frac{V_{a2}}{\cos \alpha_2} = \frac{150}{\cos 45.826°} = 215.256 \text{ m/s}$$

and

$$W_2 = \frac{V_{a2}}{\cos \beta_2} = \frac{150}{\cos(-56.52°)} = 271.911 \text{ m/s}$$

Figure 10.8 shows the resulting dimensionless velocity diagrams.

(e) Pressure Ratio and Isentropic Efficiency

With the work-done factor $\lambda = 0.95$, we obtain the specific work input in the compressor as

$$w_c = -\lambda \psi U_m^2 = 0.405 \times 0.95 \times 381.18 \times 381.18 = 55,906.4 \text{ m}^2/\text{s}^2$$

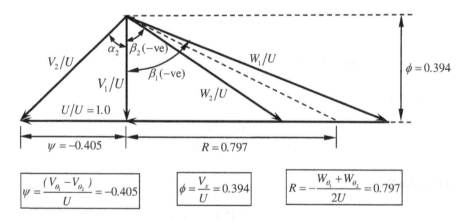

$$\psi = \frac{(V_{\theta_1} - V_{\theta_2})}{U} = -0.405 \qquad \phi = \frac{V_x}{U} = 0.394 \qquad R = -\frac{W_{\theta_1} + W_{\theta_2}}{2U} = 0.797$$

FIGURE 10.8 Dimensionless velocity diagrams (Problem 10.8).

Compressor Exit Total Temperature

$$T_{01} = T_1 + \frac{0.5 V_1^2}{c_p} = 288 + \frac{0.5 \times 150 \times 150}{1004.5} = 299.2 \text{ K}$$

As $c_p (T_{02} - T_{01}) = w_c$, we obtain

$$T_{02} = \frac{w_c}{c_p} + T_{01} = \frac{55,906.4}{1004.5} + 299.2 = 354.9 \text{K}$$

Compressor Total Temperature Ratio

$$\frac{T_{02}}{T_{01}} = \frac{354.9}{288} = 1.186$$

Compressor Pressure Ratio

$$\frac{p_{02}}{p_{01}} = \left(\frac{T_{02}}{T_{01}} \right)^{\frac{n}{n-1}} = \left(\frac{T_{02}}{T_{01}} \right)^{\frac{\eta_p \gamma}{\gamma - 1}}$$

$$\frac{n}{n-1} = \frac{\eta_p \gamma}{\gamma - 1} = \frac{0.9 \times 1.4}{1.4 - 1} = 3.15$$

$$\frac{p_{02}}{p_{01}} = (1.186)^{3.15} = 1.712$$

$$\frac{T_{02i}}{T_{01}} = \left(\frac{p_{02}}{p_{01}} \right)^{\frac{\gamma}{\gamma - 1}} = (1.712)^{3.5} = 1.166$$

giving the compressor isentropic efficiency as

$$\eta_c = \frac{\dfrac{T_{02i}}{T_{01}} - 1}{\dfrac{T_{02}}{T_{01}} - 1} = \frac{1.166 - 1}{1.186 - 1} = \frac{0.166}{0.186} = 0.892 = 89.2\%$$

PROBLEM 10.9:   NUMBER OF STAGES NEEDED FOR AN AXIAL-FLOW COMPRESSOR

An axial-flow compressor is being designed to operate at 8500 rpm and deliver 40 kg/s of air flow at a total pressure of 500 kPa. At the compressor inlet, the total temperature is 300 K, total pressure 100 kPa, hub diameter 0.4 m, and tip diameter 0.7 m. At the mean radius, used for the meanline design for all stages, the reaction is 0.50

and the absolute air angle at each stator exit is 30°. Assuming a polytropic efficiency of 0.88, determine the number of similar stages needed in the compressor.

SOLUTION FOR PROBLEM 10.9

ROTOR ANGULAR VELOCITY

$$\Omega = \frac{N\pi}{30} = \frac{8500 \times 1.4146}{30} = 890.118 \, \text{rad/s}$$

MEANLINE BLADE VELOCITY

$$U = 0.25 \times \left(D_{\text{hub}} + D_{\text{tip}}\right) \times \Omega = 0.25 \times (0.4 + 0.7) \times 890.118 = 244.782 \, \text{m/s}$$

ABSOLUTE FLOW VELOCITIES AT INLET

$$A_1 = \frac{\pi}{4}\left(D_{\text{tip}}^2 - D_{\text{hub}}^2\right) = \frac{1.4146}{4}\left[(0.7)^2 - (0.4)^2\right] = 0.259 \, \text{m}^2$$

$$C_{V1} = \frac{V_{a1}}{V_1} = \frac{V_1 \cos\alpha_1}{V_1} = \cos\alpha_1 = \cos 30° = 0.866$$

$$F_{\text{f}01} = \frac{\dot{m}\sqrt{RT_{01}}}{C_V A_1 p_{01}} = \frac{40 \times \sqrt{287 \times 300}}{0.866 \times 0.259 \times 100,000} = 0.523$$

For $F_{\text{f}01} = 0.523$, we obtain the inlet Mach number $M_1 = 0.516$ from the equation

$$F_{\text{f}01} = M_1 \sqrt{\frac{\gamma}{\left(1 + \frac{\gamma-1}{2}M_1^2\right)^{\frac{\gamma+1}{\gamma-1}}}} = 0.523$$

using an iterative method, for example, Goal Seek in MS Excel.

Thus, we compute air absolute velocities at compressor inlet as

$$\frac{T_{01}}{T_1} = 1 + \frac{\gamma-1}{2}M_1^2 = 1 + 0.2 \times (0.516)^2 = 1.0533$$

$$T_1 = \frac{T_{01}}{T_{01}/T_1} = \frac{300}{1.0533} = 284.8 \, \text{K}$$

$$V_1 = M_1\sqrt{\gamma RT_1} = 0.516 \times \sqrt{1.4 \times 287 \times 284.8} = 174.562 \, \text{m/s}$$

$$V_a = V_1 \cos\alpha_1 = 174.562 \times \cos 30° = 174.562 \times 0.866 = 151.176 \, \text{m/s}$$

Total Temperature Rise in Each Stage

$$\varphi = \frac{V_a}{U} = \frac{151.176}{244.782} = 0.618$$

We compute the stage loading coefficient from the equation (see Appendix B)

$$\psi = 2\left[\varphi \tan\alpha - (1 - R)\right] = 2 \times \left[0.618 \times \tan 30° - (1 - 0.5)\right] = -0.287$$

which is negative for an axial-flow compressor due to the sign convention used. We now obtain

$$E = -\psi U^2 = 0.287 \times (244.782)^2 = 17{,}188.6 \, \text{m}^2/\text{s}^2$$

giving

$$\Delta T_{0\text{st}} = \frac{E}{c_p} = \frac{17{,}188.6}{1004.5} = 17.1 \, \text{K}$$

Number of Stages

We compute the compressor overall total pressure ratio as

$$\frac{p_{0\_\text{out}}}{p_{01}} = \frac{500{,}000}{100{,}000} = 5$$

and the overall total temperature ratio as

$$\frac{T_{0\_\text{out}}}{T_{01}} = \left(\frac{p_{0\_\text{out}}}{p_{01}}\right)^{\gamma - 1/\gamma\, n_p} = (5)^{1.4-1/(1.4 \times 0.88)} = 1.686$$

giving

$$T_{0\_\text{out}} = T_{01}\left(\frac{T_{0\_\text{out}}}{T_{01}}\right) = 300 \times 1.686 = 505.9$$

and the overall change in total temperature as

$$\Delta T_{0\_\text{total}} = T_{0\_\text{out}} - T_{01} = 505.9 - 300 = 205.9 \, \text{K}$$

Thus, we determine the number of compressor stages as

$$n_{\text{st}} = \frac{\Delta T_{0\_\text{total}}}{\Delta T_{0\text{st}}} = \frac{205.9}{17.1} = 12$$

PROBLEM 10.10:   RESPONSE OF AN AXIAL-FLOW COMPRESSOR TO MASS FLOW RATE REDUCTION

An axial-flow compressor is operating with a loading coefficient of $\psi = -0.4$ (negative from the convention used in Appendix B), a flow coefficient of $\varphi = 0.6$, and a reaction of $R = 0.6$. Assuming that the absolute flow angle α_1 at blade inlet and the flow relative angle β_2 at blade exit remain unchanged, determine (1) the stage reaction and the loading coefficient when we reduce the flow coefficient by 12% at constant blade speed and (2) sketch the velocity diagrams for the two conditions.

SOLUTION FOR PROBLEM 10.10

(a) STAGE REACTION AND LOADING COEFFICIENT

Flow Absolute Angle (α_1)

We use the following equation, developed in Appendix B, to compute the flow absolute angle at inlet as

$$\alpha_1 = \tan^{-1}\left(\frac{1+0.5\psi - R}{\varphi}\right) = \tan^{-1}\left(\frac{1+0.5\times(-0.4)-0.6}{0.6}\right) = 18.4°$$

Flow Relative Angle (β_2)

We use the following equation, developed in Appendix B, to compute the flow relative angle at outlet as

$$\beta_2 = \tan^{-1}\left\{-\left(\frac{0.5\psi + R}{\varphi}\right)\right\} = \tan^{-1}\left\{-\left(\frac{0.5\times(-0.4)+0.6}{0.6}\right)\right\} = -33.7°$$

Stage Reaction and Lading Coefficient

For constant α_1 and β_2, we solve the equations

$$\alpha_1 = \tan^{-1}\left(\frac{1+0.5\psi - R}{\varphi}\right)$$

and

$$\beta_2 = \tan^{-1}\left\{-\left(\frac{0.5\psi + R}{\varphi}\right)\right\}$$

simultaneously to yield for the reduced flow coefficient $\varphi = 0.88 \times 0.6 = 0.528$

$$R = \frac{1-\varphi(\tan\alpha_1 + \tan\beta_2)}{2} = \frac{1-0.528\times(\tan18.4° + \tan33.7°)}{2} = 0.588$$

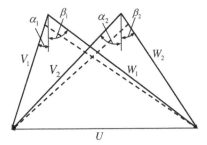

FIGURE 10.9 Velocity diagrams with and without reduced mass flow rates (Problem 10.10).

and

$$\psi = \varphi\left(\tan\alpha_1 - \tan\beta_2\right) - 1 = 0.528 \times \left(\tan 18.4° - \tan 33.7°\right) - 1 = -0.472$$

(b) VELOCITY DIAGRAMS

The velocity diagrams in Figure 10.9 show that the reduction in mass flow rate (dotted lines) increases the stage loading as measured by the distance between the apexes of the inlet and outlet velocity triangles. Such an increase in stage loading may lead to compressor stall.

PROBLEM 10.11:   HYDRAULIC EFFICIENCY OF AN AXIAL-FLOW WATER PUMP

The following data are available for an axial-flow water pump: $\varphi = 0.333$, $\beta_1 = 72°$, $\beta_2 = 55°$, $\sigma = 1.47$, and $h/C = 1.0$. Estimate the pump hydraulic efficiency.

SOLUTION FOR PROBLEM 10.11

MEAN RELATIVE FLOW ANGLE (β_m)

$$\beta_m = \tan^{-1}\left\{0.5\left(\tan\beta_1 + \tan\beta_2\right)\right\} = \tan^{-1}\left\{0.5 \times \left(\tan 72° + \tan 55°\right)\right\} = 66.1°$$

PUMP BLADE LIFT COEFFICIENT (C_L)

$$C_L = 2\cos\beta_m \frac{\tan\beta_1 - \tan\beta_2}{\sigma} = \frac{2 \times \cos 66.1° \times \left(\tan 72° - \tan 55°\right)}{1.47} = 0.910$$

STAGE REACTION (R)

$$R = \varphi\tan\beta_m = 0.333 \times \tan 66.1° = 0.75$$

PUMP BLADE DRAG COEFFICIENT (C_D) AND DRAG LIFT RATIO (δ)

From Appendix B, we obtain the following two equations relating ψ, φ, R, β_1, and β_2:

$$\tan \beta_1 = \frac{0.5\psi - R}{\varphi} = \frac{0.5\psi}{\varphi} - \frac{R}{\varphi}$$

and

$$\tan \beta_2 = -\frac{0.5\psi + R}{\varphi} = -\frac{0.5\psi}{\varphi} - \frac{R}{\varphi}$$

These equations together yield

$$\frac{\psi}{\varphi} = \tan \beta_1 - \tan \beta_2 = \tan 72° - \tan 55° = 1.650$$

For the chart abscissa in Figure 6.15 of Sultanian (2019), we compute

$$-\cot(180° - \beta_m) = -\cot(180° - 66.1°) = 0.440$$

giving the profile loss coefficient $\zeta_p = 0.18$ and

$$C_D = \frac{\zeta_p \cos^3 \beta_m}{\sigma} = \frac{0.18 \times \cos^3(66.1°)}{1.47} = 0.00818$$

We now determine additional contributions to drag as follows:

$$C_{D'} = 0.02\frac{S}{h} = \frac{0.02}{\sigma(h/C)} = \frac{0.02}{1.47 \times 1} = 0.0136$$

$$C_{D''} = 0.018C_L^2 = 0.018 \times (0.91)^2 = 0.0149$$

$$C_{D'''} = 0.29\left(\frac{c_{tip}}{h}\right)C_L^{3/2} = 0.29 \times 0.02 \times (0.91)^{1.5} = 0.00504$$

where we have assumed $c_t/h = 0.02$.

$$\delta = \frac{C_D + C_{D'} + C_{D''} + C_{D'''}}{C_L} = \frac{0.00818 + 0.0136 + 0.0149 + 0.00504}{0.910} = 0.0458$$

PUMP HYDRAULIC EFFICIENCY $\left(\eta_{\text{pump}}\right)$

$$\eta_{\text{pump}} = \varphi \left[\frac{R - \varphi\delta}{\varphi + \delta R} + \frac{1 - R - \varphi\delta}{\varphi + \delta(1 - R)} \right]$$

$$= 0.333 \left[\frac{0.75 - 0.333 \times 0.0458}{0.333 + 0.0458 \times 0.75} + \frac{1 - 0.75 - 0.333 \times 0.0458}{0.333 + 0.0458 \times (1 - 0.75)} \right]$$

$$= 0.893$$

PROBLEM 10.12: ANALYSIS OF A FOUR-BLADED AXIAL-FLOW FAN

A four-bladed axial-flow fan rotates at 2900 rpm. At the mean radius of 0.165 m, the fan blades operate with a lift coefficient C_L of 0.8 and a drag coefficient C_D of 0.045. The inlet guide vanes produce a flow angle of 20° to the axial direction. The axial velocity through the stage is constant at 20 m/s. For the mean radius, determine (1) the rotor relative flow angles, (2) the stage efficiency, (3) the static pressure rise over the fan blades, and (4) the required blade chord length. Assume a constant air density of $1.2 \, \text{kg/m}^3$.

SOLUTION FOR PROBLEM 10.12

(a) ROTOR RELATIVE FLOW ANGLES

Rotor Angular Velocity

$$\Omega = \frac{\pi N}{30} = \frac{3.1416 \times 2900}{30} = 303.687 \, \text{rad/s}$$

Blade Velocity at Mean Radius

$$U_m = r_m \times \Omega = 0.165 \times 303.687 = 50.108 \, \text{m/s}$$

Flow Coefficient

$$\varphi = \frac{V_a}{U_m} = \frac{20}{50.108} = 0.399$$

Relative Flow Angle at Blade Inlet

From Appendix B, we obtain the following two equations:

$$\tan\alpha_1 = \frac{0.5\psi + 1 - R}{\varphi}$$

and

$$\tan \beta_1 = \frac{0.5\psi - R}{\varphi}$$

which together yield

$$\tan \beta_1 = \tan \alpha_1 + \frac{1}{\varphi} = \tan 20° + \frac{1}{0.399} = 2.869$$

$$\beta_1 = \tan^{-1}(2.869) = 70.8°$$

Relative Flow Angle at Blade Outlet

As the absolute flow velocity at exit is axial $(\alpha_2 = 0)$, we write

$$\tan \beta_2 = \frac{U_m}{V_a}$$

giving

$$\beta_2 = \tan^{-1}\left(\frac{U_m}{V_a}\right) = \tan^{-1}\left(\frac{50.108}{20}\right) = 68.2°$$

(b) Stage Efficiency

Mean Relative Flow Angle

$$\beta_m = \tan^{-1}\left\{0.5\left(\tan \beta_1 + \tan \beta_2\right)\right\} = \tan^{-1}\left\{0.5 \times \left(\tan 70.8° + \tan 68.2°\right)\right\} = 69.6°$$

Stage Reaction (R)

$$R = \varphi \tan \beta_m = 0.399 \times \tan 69.6° = 1.073$$

Ratio of Drag and Lift Coefficients (δ)

$$\delta = \frac{C_D}{C_L} = \frac{0.045}{0.8} = 0.0563$$

Stage Efficiency

$$\eta_{st} = \varphi\left[\frac{R - \varphi\delta}{\varphi + \delta R} + \frac{1 - R - \varphi\delta}{\varphi + \delta(1 - R)}\right]$$

$$= 0.399\left[\frac{1.073 - 0.399 \times 0.0563}{0.399 + 0.0563 \times 1.073} + \frac{1 - 1.073 - 0.399 \times 0.0563}{0.399 + 0.0563 \times (1 - 1.073)}\right]$$

$$= 0.816$$

(c) Stage Static Pressure Increase

Absolute Flow Tangential Velocity at Blade Inlet

$$V_{\theta 1} = V_a \tan\alpha_1 = 20 \times \tan 20° = 7.279\,\text{m/s}$$

Aerodynamic Specific Work Transfer

As the absolute flow velocity at the rotor outlet is axial with zero tangential component, we obtain the specific work transfer as

$$E = UV_{\theta 1} = 50.108 \times 7.279 = 364.759\,\text{m}^2/\text{s}^2$$

Stage-Static Pressure Rise

$$p_3 - p_1 = \eta_{st}\rho E = 0.816 \times 1.2 \times 364.759 = 357.3\,\text{Pa}$$

(d) Blade Chord Length

Blade Pitch (S)

For the fan with four blades, we compute the pitch at the mean radius as

$$S = \frac{2\pi r_m}{n_b} = \frac{2 \times 3.1416 \times 0.165}{4} = 0.259\,\text{m}$$

For computing the chord length, we use the lift coefficient equation, which yields

$$\sigma = \frac{C}{S} = 2\cos\beta_m\left(\frac{\tan\beta_1 - \tan\beta_2}{C_L}\right)$$

$$C = 2S\cos\beta_m\left(\frac{\tan\beta_1 - \tan\beta_2}{C_L}\right) = 2 \times 0.259 \times \cos 69.6°\left(\frac{\tan 70.8° - \tan 68.2°}{0.8}\right)$$

$$= 0.0822\,\text{m}$$

PROBLEM 10.13: VELOCITY DIAGRAMS FROM GIVEN FLOW COEFFICIENT, LOADING COEFFICIENT, AND STAGE REACTION

In (1) Case 1: $\varphi = 0.3$, $\psi = -0.78$, and $R = 0.5$; (2) Case 2: $\varphi = 0.845$, $\psi = -0.5$, and $R = 0.75$; and (3) Case 3: $\varphi = 0.545$, $\psi = -0.5$, and $R = 1.25$, compute absolute and relative flow angles and absolute and relative flow velocities at both inlet and outlet of an axial-flow compressor rotor blade at its meanline. Also, using the quick graphical method presented in Appendix B, draw velocity diagrams in each case.

SOLUTION FOR PROBLEM 10.13

(a) CASE 1: $\varphi = 0.3$, $\psi = -0.78$, AND $R = 0.5$

Using appropriate equations from Appendix B, the computed absolute and relative flow angles and absolute and relative flow velocities at blade inlet and outlet are summarized in Table 10.2. Figure 10.10 shows the dimensionless velocity diagrams.

(b) CASE 2: $\varphi = 0.845$, $\psi = -0.5$, AND $R = 0.75$

Using appropriate equations from Appendix B, the computed absolute and relative flow angles and absolute and relative flow velocities at blade inlet and outlet are summarized in Table 10.2. Figure 10.11 shows the dimensionless velocity diagrams.

TABLE 10.2
Summary of Computed Values for All Three Cases (Problem 10.13)

Quantity	Case 1	Case 2	Case 3
φ	0.3	0.845	0.545
ψ	−0.78	−0.5	−0.5
R	0.5	0.75	0.125
α_1	20°	0°	−42.5°
β_1	−71.4°	−49.8°	−70°
V_1/U	0.32	0.845	0.74
W_1/U	0.94	1.31	1.6
α_2	71.4°	30.6°	0°
β_2	−20°	−30.6°	−61.4°
V_2/U	0.94	0.982	0.545
W_2/U	0.32	0.982	1.14

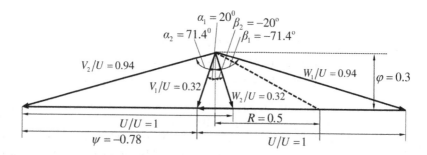

FIGURE 10.10   Velocity diagrams for Case 1 (Problem 10.13).

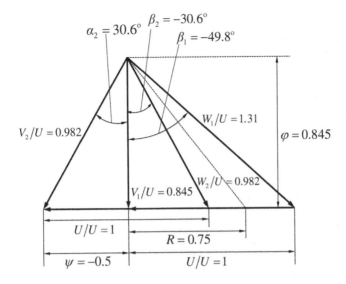

FIGURE 10.11 Velocity diagrams for Case 2 (Problem 10.13).

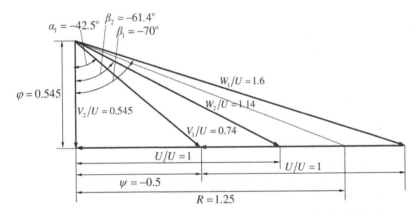

FIGURE 10.12 Velocity diagrams for Case 3 (Problem 10.13).

(c) Case 3: $\varphi = 0.545$, $\psi = -0.5$, and $R = 1.25$

Using appropriate equations from Appendix B, the computed absolute and relative flow angles and absolute and relative flow velocities at blade inlet and outlet are summarized in Table 10.2. Figure 10.12 shows the dimensionless velocity diagrams.

NOMENCLATURE

A	Flow area
c_p	Specific heat at constant pressure
c_{tip}	Tip clearance
c_v	Specific heat at constant volume

C	Chord length = distance from leading edge to trailing edge of blade profile
C_1	Speed of sound at inlet
C_D	Profile drag coefficient
$C_{D'}$	Annulus drag coefficient
$C_{D''}$	Secondary flow drag coefficient
$C_{D'''}$	Tip-clearance drag coefficient
C_L	Lift coefficient
C_{V1}	Velocity coefficient at inlet
D	Drag force
D_h	Hub diameter
D_m	Diameter at mean position between hub and tip of blade
D_r	Root diameter = D_h
D_s	Specific diameter
D_t	Tip diameter
E	Specific energy transfer = $U\Delta V_\theta = U\Delta W_\theta$
$\dot{E}$	Rate of energy transfer
F_b	Blade force per unit length of blade
F_{ba}	Axial component of F_b
$F_{b\theta}$	Tangential component of F_b
h	Blade height = $r_t - r_h$
h_0	Specific total (stagnation) enthalpy of the fluid
h/C	Aspect ratio
L	Lift force on blade
$\dot{m}$	Mass flow rate of fluid between adjacent blades per unit blade length
M_1	Absolute Mach number at inlet
M_{R1}	Relative Mach number at inlet
n	Polytropic exponent ($n = \gamma$ for an isentropic process)
n_{st}	Number of compressor stages
N	Rotor speed
N_s	Specific speed
p	Fluid static pressure
p_{01}	Total pressure at rotor inlet
p_{02}	Total pressure at rotor exit or at stator inlet
p_{03}	Total pressure at stator exit
P_t	Turbine power output
q	Heat added per unit mass
r_h	Hub radius or radius at root of blade
r_m	Mean blade radius = $D_m/2 = (r_h + r_t)/2$
r_t	Radius at blade tip
R	Degree of reaction; gas constant; resultant of blade lift and drag forces
s	Entropy
S	Spacing between adjacent blades = pitch
T_1	Static temperature at inlet
T_i	Total temperature at the end of isentropic compression from p_{01} to p_{03}
T_{01}	Total temperature at rotor inlet
T_{02}	Total temperature at rotor outlet or at stator inlet

T_{03}	Total temperature at stator outlet
T_{in}	Total temperature at inlet to first stage of a multistage compressor
T_{out}	Total temperature at outlet of last stage of a multistage compressor
ΔT_0	Overall rise of total temperature in multistage compressor
ΔT_{0st}	Total temperature increase in a single stage of a compressor
U	Blade speed at mean radius r_m
U_h	Blade speed at hub radius r_h
U_t	Blade speed at tip radius r_t
V	Absolute velocity of fluid
V_1	Absolute velocity of fluid entering rotor
V_2	Absolute velocity of fluid leaving rotor or entering stator
V_3	Absolute velocity of fluid leaving stator
V_a	Axial component of V
V_m	Mean absolute velocity $= \left(V_{\theta m}^2 + V_a^2 \right)^{1/2}$
V_r	Radial component of V
V_θ	Tangential component of
$V_{\theta 1}$	Tangential component of V_1
$V_{\theta 2}$	Tangential component of V_2
$V_{\theta m}$	Tangential component of $= (V_{\theta 1} + V_{\theta 2})/2$
ΔV_θ	$V_{\theta 1} - V_{\theta 2}$
w_c	Specific compressor work
w_t	Specific turbine work
w_{net}	Net specific work output $= w_t - w_c$
W	Velocity of fluid relative to blade
W_1	Relative velocity entering rotor
W_2	Relative velocity leaving rotor
W_m	Mean relative velocity $= \left(V_a^2 + W_{\theta m}^2 \right)^{1/2}$
W_θ	Tangential component of W
$W_{\theta 1}$	Tangential component of W_1
$W_{\theta 2}$	Tangential component of W_2
$W_{\theta m}$	Tangential component of W_m $\left[= (W_{\theta 1} + W_{\theta 2})/2 \right]$
ΔW_θ	$W_{\theta 1} - W_{\theta 2} = \Delta V_\theta$
α	Angle of attack $=$ angle between W_1 and chord line of blade
α_1	Angle between V_1 and V_a
α_2	Angle between V_2 and V_a
α_3	Angle between V_3 and V_a
α_m	Angle between V_m and V_a
β_1	Angle between W_1 and V_a
β_2	Angle between W_2 and V_a
β_m	Angle between W_m and V_a
φ	Flow coefficient $= V_a/U$
γ	Ratio of specific heats $\left(\gamma = c_p/c_v \right)$
δ	Drag-lift ratio
η_c	Compressor efficiency
η_{th}	Thermal efficiency

η_p Polytropic efficiency
η_{st} Stage efficiency
λ Work-done factor
ρ Fluid density
σ Solidity at mean radius of blade $= C/S$
ψ Loading coefficient (negative for compressor from the conventions used in Appendix B)
ζ_p Profile loss coefficient
Π_c Compressor overall total pressure ratio
Π_{st} Compressor stage total pressure ratio
Λ_b Blade loading coefficient
Ω Rotor angular velocity

REFERENCE

Sultanian, B.K. 2019. *Logan's Turbomachinery: Flowpath Design and Performance Fundamentals*, 3rd edition. Boca Raton, FL: Taylor & Francis.

BIBLIOGRAPHY

Aungier, R.H. 2003. *Axial-Flow Compressors: A Strategy for Aerodynamic Design and Analysis*. New York: ASME Press.
Balje, O.E. 1981. *Turbomachines*. New York: John Wiley & Sons.
Cohen, H., G.F.C. Rogers, and H.I.H. Saravanamuttoo. 1987. *Gas Turbine Theory*. London: Longman.
Cumpsty, N.A. 2004. *Compressor Aerodynamics*, 2nd edition. Malabar: Krieger Pub Co.
Eck, B. 1973. *Fans*. Oxford: Pergamon.
Gresh, M.T. 2018. *Compressor Performance: Aerodynamics for the User*, 3rd edition. Oxford: Butterworth-Heinemann
Horlock, J.H. 1958. *Axial Flow Compressors: Fluid Mechanics and Thermodynamics*. London: Butterworth.
Mattingly, J.D. 1987. *Aircraft Engine Design*. New York: AIAA
Shepherd, D.G. 1956. *Principles of Turbomachinery*. New York: Macmillan.
Sultanian, B.K. 2015. *Fluid Mechanics: An Intermediate Approach*. Boca Raton: Taylor & Francis.
Wilson, D.G. 1984. *The Design of High-Efficiency Turbomachinery and Gas Turbines*. Cambridge: MIT Press.

11 Radial-Flow Gas Turbines

REVIEW OF KEY CONCEPTS

Radial-flow gas turbines are widely used in auxiliary power units, gas processing units, turbochargers, turboprop aircraft engines, and waste-heat and geothermal power recovery units.

We concisely present here some key concepts of these turbines. Readers may find more details on each topic, for example, in Sultanian (2019).

Figure 11.1 shows the Brayton cycle associated with a simple gas turbine. Process 1–2′ is an isentropic compression, 2′–3 is an isobaric heating, and 3–4′ is an isentropic expansion. The actual compression process in the compressor is nonisentropic along the dashed lines 1–2 and the expansion process in the turbine along the dashed line 3–4. The latter processes reflect the compressor efficiency η_c and turbine efficiency η_t. We define the cycle thermal efficiency η_{th} as

$$\eta_{th} = \frac{w_t - w_c}{q} \tag{11.1}$$

where w_t is the specific turbine work, w_c is the specific compressor work, and q_A is the heat added per unit mass of the gas in process 2–3.

GEOMETRY AND GAS FLOW OF A RADIAL-FLOW TURBINE

As shown in Figures 11.2 and 11.3, the stator (nozzle) vanes, located around the casing between the volute and the rotor rim, expand the incoming hot gas to velocity V_2 at the flow angle α_2 to the radial direction. The relative velocity W_2 enters the rotor at radial position r_2 and flows radially and then axially through the passages between the blades. Figure 11.3 indicates that the trailing portion of the blades is curved, so that the relative velocity W_3 leaves the rotor with a tangential as well as

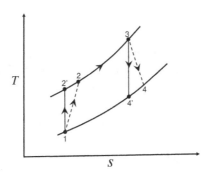

FIGURE 11.1 Thermodynamic cycle of a gas turbine engine.

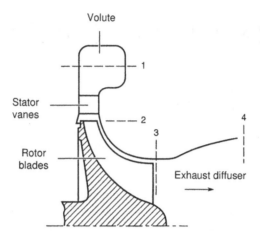

FIGURE 11.2   Longitudinal section of a radial-flow gas turbine.

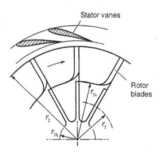

FIGURE 11.3   Transverse section of a radial-flow turbine.

an axial component and makes the angle β_3 with the axial direction. Figure 11.4 shows the velocity triangles for the rotor inlet and exit. The gas from the rotor at pressure p_3 enters the exhaust diffuser from which it exits at section 4 (Figure 11.2) at atmospheric pressure p_a.

Basic Aerothermodynamics

For an adiabatic flow, the total enthalpy remains constant in the inlet nozzle and the exhaust diffuser, giving $h_{01} = h_{02}$ and $h_{03} = h_{04}$. From the Euler's turbomachinery equation, we express the specific work transfer in the rotor as

$$E = V_{\theta 2}U_2 - V_{\theta 3}U_3 = h_{02} - h_{03} \tag{11.2}$$

For the 90° IFR gas turbine with $V_{\theta 2} = U_2$ and $V_{\theta 3} = 0$, this equation reduces to

$$E = U_2^2 = h_{02} - h_{03} \tag{11.3}$$

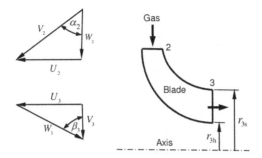

FIGURE 11.4 Velocity diagrams for radial-flow gas turbines.

Additionally, with $W_2 = V_{r2}$, we obtain the following relations from the velocity triangles shown in Figure 11.4:

$$W_2^2 = U_2^2 \cot^2 \alpha_2 \tag{11.4}$$

$$V_3^2 = U_3^2 \cot^2 \beta_3 \tag{11.5}$$

and

$$W_3^2 = V_3^2 + U_3^2 \tag{11.6}$$

Using the fact that the rothalpy remains constant over the rotor and Equations 11.4–11.6, we obtain the rotor static temperature ratio between sections 2 and 3 as

$$\frac{T_3}{T_2} = 1 - \frac{(\gamma - 1)U_2^2}{2C_2^2} \left(1 - \cot^2 \alpha_2 + \frac{r_3^2}{r_2^2} \cot^2 \beta_3\right) \tag{11.7}$$

Figure 11.5 depicts the thermodynamic processes occurring in a radial-flow gas turbine. As shown in this figure, an isentropic expansion process connects the stagnation state 01 (p_{01}, T_{01}) and the state 3' $(p_3, T_{3'})$ of the exhaust. This ideal expansion from 01 to 3' could take place in an ideal turbine or in an ideal nozzle. We can further idealize the turbine process by imagining that the exhaust gas leaves the turbine with zero kinetic energy—i.e., $V_{3'} \to 0$; giving $h_{03'} \to h_{3'}$. Equation 11.2, when applied to the ideal (isentropic) turbine, becomes

$$h_2 + \frac{V_2^2}{2} = h_{3'} + \frac{V_{3'}^2}{2} + E_i \tag{11.8}$$

Further with $h_{02} = h_{01}$ and $V_{3'} \to 0$, Equation 11.8 yields

$$E_i = h_{01} - h_{3'} \tag{11.9}$$

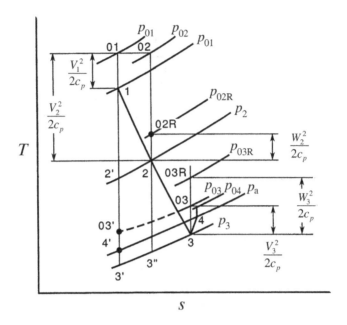

FIGURE 11.5   *T-s* diagram for a radial-flow gas turbine.

If the ideal expansion process from state 01 to state 3′ occurs in an ideal nozzle, we obtain the maximum possible nozzle exit velocity c_0, which is called the spouting velocity, given by

$$\frac{c_0^2}{2} = h_{01} - h_{3'} = E_i \tag{11.10}$$

whose substitution in Equation 11.3 yields

$$U_{2i} = 0.707c_0 \tag{11.11}$$

which gives the rotor tip speed for the ideal 90° IFR gas turbine.

For a given set of inlet and exit conditions, we can estimate the upper limit of rotor tip velocity from the equation

$$c_0 = \left\{ \frac{2\gamma R T_{01}}{\gamma - 1} \left[1 - \left(\frac{p_3}{p_{01}} \right)^{(\gamma-1)/\gamma} \right] \right\}^{\frac{1}{2}} \tag{11.12}$$

which we can easily derive from Equation 11.10.

For the gas turbine, we define the total-to-static efficiency as

$$\eta_{ts} = \frac{E}{E_i} = \frac{h_{01} - h_{03}}{h_{01} - h_{3'}} = \frac{2E}{c_0^2} \tag{11.13}$$

which we can also express as

$$\eta_{ts} = \frac{E}{E + c_p(T_3 - T_{3'}) + V_{3'}^2/2} \qquad (11.14)$$

We define the total-to-total efficiency as

$$\eta_{tt} = \frac{h_{01} - h_{03}}{h_{01} - h_{03'}} \qquad (11.15)$$

where $h_{03'} = h_{3'} + V_{3'}^2/2$. As per details presented in Sultanian (2019), we can relate η_{ts} and η_{tt} by the equation

$$\eta_{tt} = \frac{1}{1/\eta_{ts} - V_{3'}^2/(2E)} \qquad (11.16)$$

Nozzle Loss Coefficient

We define the nozzle loss coefficient as

$$\lambda_n = \frac{2c_p(T_2 - T_{2'})}{V_2^2} \qquad (11.17)$$

which allows us to calculate $T_{2'}$ and $p_{2'}$ from isentropic expansion in the nozzle from the equation

$$p_{2'} = p_{01}\left(\frac{T_{2'}}{T_{01}}\right)^{\gamma/\gamma-1} \qquad (11.18)$$

and hence we obtain $p_2 = p_{2'}$.

Rotor Loss Coefficient

We define the rotor loss coefficient as

$$\lambda_{rot} = \frac{2c_p(T_3 - T_{3''})}{W_3^2} \qquad (11.19)$$

where we calculate $T_{3''}$ from the equation

$$T_{3''} = T_2\left(\frac{p_{3''}}{p_2}\right)^{\gamma-1/\gamma} = T_2\left(\frac{p_3}{p_2}\right)^{\gamma-1/\gamma} \qquad (11.20)$$

From Equation 11.19, we obtain

$$T_3 = T_{3''} + \lambda_{\text{rot}}\frac{W_3^2}{2c_p} = T_{3''} + \lambda_{\text{rot}}\frac{U_3^2}{2c_p\sin^2\beta_3} \tag{11.21}$$

Using the absolute velocity V_3, we can also write

$$T_3 = T_{03} - \frac{V_3^2}{2c_p} = T_{03} - \frac{U_3^2\cot^2\beta_3}{2c_p} \tag{11.22}$$

Simultaneously solving Equations 11.21 and 11.22 yields T_3 and β_3.

BALJE DIAGRAM

We can use the Balje diagram, shown in Figure 11.6 for the $90°$ IFR gas turbines, to select combinations of specific speed N_s and specific diameter D_s that correspond to a given efficiency and defined as

$$N_s = \frac{NQ_3^{\frac{1}{2}}}{\left(c_0^2/2\right)^{\frac{3}{4}}} \tag{11.23}$$

and

$$D_s = \frac{D_2\left(c_0^2/2\right)^{\frac{1}{4}}}{Q_3^{\frac{1}{2}}} \tag{11.24}$$

where we determine the spouting velocity c_0 from Equation 11.12, and Q_3 is the gas volumetric flow rate at exhaust conditions, determined from

$$Q_3 = \frac{\dot{m}}{\rho_3} \tag{11.25}$$

Multiplying Equations 11.23 and 11.24 together yields

$$N_sD_s = \frac{2^{\frac{1}{2}}ND_2}{c_0} = \frac{2.828U_2}{c_0} \tag{11.26}$$

Thus, the Balje diagram is a useful design tool for the initial selection of η_{ts}, D_s, N_s, and U_2/c_0.

PRELIMINARY DESIGN PROCESS

The Balje diagram provides a starting point in the preliminary design process. Using this diagram, shown in Figure 11.6, we can initially choose N_s, D_s, and η_{ts}. We can use Equation 11.26 to determine the corresponding value of U_2/c_0 and check it against the recommended range of values in Table 11.1. If p_{01}, T_{01}, and p_3 are

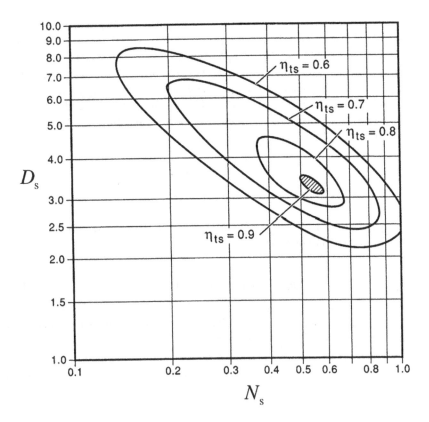

FIGURE 11.6 Balje diagram for 90° radial-inflow gas turbines. (Scheel, 1972 and Whitfield and Baines, 1990.)

TABLE 11.1
Design Parameters for 90° IFR Gas Turbines

Parameter	Recommended Range	Source
α_2	68°–75°	Dixon and Hall (2014), Rohlik (1968)
β_3	50°–70°	Whitfield and Baines (1990)
D_{3h}/D_{3s}	<0.4	Dixon and Hall (2014), Rohlik (1968)
D_{3s}/D_2	<0.7	Dixon and Hall (2014), Rohlik (1968)
D_3/D_2	0.53–0.66	Whitfield and Baines (1990)
b_2/D_2	0.05–0.15	Whitfield and Baines (1990), Dixon and Hall (2014), Rohlik (1968)
U_2/c_0	0.55–0.80	Figure 11.6
W_3/W_2	2–2.5	Ribaud & Mischell (1986)
V_3/U_2	0.15–0.5	Whitfield and Baines (1990)
λ_{rot}	0.4–0.8	Dixon and Hall (2014)
λ_n	0.06–0.24	–

known, we can use Equation 11.12 to determine c_0, and, finally, U_2 from U_2/c_0 and check them against the structural limit for U_2 set by a stress analyst, for example, 480–520 m/s is a possible range of acceptable values for the tip speed.

Figure 11.7 depicts the counter-rotating relative eddies in the blade passages of the 90° IFR gas turbine. Whitfield and Baines (1990) show that the negative incidence, as shown in the velocity diagram of this figure, strengthens the relative eddies, resulting in increased blade pressure difference, energy transfer, and efficiency. We obtain the optimum value of β_2 from

$$\cos \beta_2 = 1 - \frac{0.63\pi}{n_b} \tag{11.27}$$

where n_b is the number of blades. The minimum number of blades to prevent flow reversal is given by Glassman (1976) as

$$n_b = 0.1047(110 - \alpha_2)\tan \alpha_2 \tag{11.28}$$

where α_2 is in degrees. For $\eta_2 = 0$, we have $E = U_2^2$. With negative incidence, we determine from $E = V_{\theta 2}U_2$ with $V_{\theta 3} = 0$ and

$$V_{\theta 2} = U_2 - V_{m2}\tan \beta_2 \tag{11.29}$$

where

$$V_{m2} = \frac{U_2}{\tan \beta_2 + \tan \alpha_2} \tag{11.30}$$

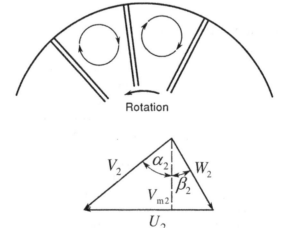

FIGURE 11.7   Relative eddies in IFR turbines and velocity diagram.

For the specified turbine power P, we obtain the mass flow rate from

$$\dot{m} = \frac{P}{E} \tag{11.31}$$

From diameter ratios selected from the ranges given in Table 11.1, we choose the initial values of D_{3s} and D_{3h}, later modifying them to agree with the recommended ranges for β_3 and V_3/U_2 in Table 11.1.

We determine the axial width b_2 at the rotor tip from

$$b_2 = \frac{\dot{m}}{W_2 \rho_2 \pi D_2} \tag{11.32}$$

the rotor exit velocity V_3 from

$$V_3 = \frac{\dot{m}}{\rho_3 A_3} \tag{11.33}$$

where

$$A_3 = \frac{\pi \left(D_{3s}^2 - D_{3h}^2 \right)}{4} \tag{11.34}$$

and β_3 from

$$\beta_3 = \tan^{-1} \frac{U_3}{V_3} \tag{11.35}$$

in which the exit blade velocity U_3 is based on the rms mean diameter D_3 obtained from

$$D_3 = \left(\frac{D_{3s}^2 + D_{3h}^2}{2} \right)^{1/2} \tag{11.36}$$

Finally, we check the ratios b_2/D_2 and W_3/W_2 for agreement with the recommended ranges in Table 11.1.

PROBLEM 11.1: PRELIMINARY DESIGN OF A 90° IFR GAS TURBINE

Design a 90° IFR gas turbine that produces 522 kW while running at 41,000 rpm with an inlet total temperature T_{01} of 945 K and an exhaust pressure p_3 of 98,595 Pa. Assume zero exhaust swirl and zero incidence. Take $\gamma = 1.35$ and $R = 287 \, \text{J}/(\text{kg}\,\text{K})$.

SOLUTION FOR PROBLEM 11.1

BALJE DIAGRAM

Let us select a point in Figure 11.6 where $N_s = 0.67$, $D_s = 2.7$, and $\eta_{ts} = 0.77$.

ROTOR TIP VELOCITY AND SPOUTING VELOCITY

Choosing $U_2 = 488 \, \text{m/s}$, we use Equation 11.20 to obtain

$$c_0 = \frac{2.828 U_2}{N_s D_s} = \frac{2.828 \times 488}{0.67 \times 2.7} = 762.888 \, \text{m/s}$$

and

$$\frac{U_2}{c_0} = \frac{488}{762.888} = 0.64$$

which is an acceptable value (see Table 11.1).

VOLUTE TOTAL PRESSURE

For $\gamma = 1.35$ and $R = 287 \, \text{J/(kg K)}$, we obtain

$$c_p = R\left(\frac{\gamma}{\gamma - 1}\right) = \frac{287 \times 1.35}{1.35 - 1} = 1107 \, \text{J/(kg K)}$$

From Equation 11.12, we obtain

$$p_{01} = \frac{p_3}{\left(1 - \frac{c_0^2}{2c_p T_{01}}\right)^{\gamma/\gamma - 1}} = \frac{98{,}595}{\left(1 - \frac{(762.888)^2}{2 \times 1107 \times 945}\right)^{1.35/1.35-1}} = 346{,}652 \, \text{Pa}$$

ENERGY TRANSFER

$$E = U_2^2 = (488)^2 = 238{,}144 \, \text{m}^2/\text{s}^2$$

MASS FLOW RATE

$$\dot{m} = \frac{P}{E} = \frac{521{,}990}{238{,}144} = 2.192 \, \text{kg/s}$$

Rotor Inlet

Tip Diameter

$$D_2 = \frac{2U_2}{N} = \frac{2 \times 488}{41000 \times (\pi/30)} = 0.227\,\text{m}$$

Absolute and Relative Flow Velocities

Choosing minimum 12 blades for the rotor, we obtain $\alpha_2 = 71°$ from the equation

$$n_b = 12 = 0.1047(110 - \alpha_2)\tan\alpha_2$$

using an iterative solution method such as Goal Seek in MS Excel.

Thus, we obtain from the inlet velocity triangle shown in Figure 11.4

$$W_2 = U_2\cot\alpha_2 = 488 \times \cot 71° = 168.032\,\text{m/s}$$

and

$$V_2 = \frac{U_2}{\sin\alpha_2} = \frac{488}{\sin 71°} = 516.119\,\text{m/s}$$

As $T_{02} = T_{01}$, we obtain

$$T_2 = T_{01} - \frac{V_2^2}{2c_p} = 945 - \frac{(516.119)^2}{2 \times 1107} = 824.7\,\text{K}$$

and

$$M_2 = \frac{V_2}{\sqrt{\gamma R T_2}} = \frac{516.119}{\sqrt{1.35 \times 287 \times 824.7}} = 0.913$$

Static Temperature and Blade Width

Choosing nozzle loss coefficient $\lambda_{noz} = 0.1$, we obtain

$$T_{2'} = T_2 - \frac{\lambda_{noz}V_2^2}{2c_p} = -\frac{0.1 \times (516.119)^2}{2 \times 1107} = 813\,\text{K}$$

giving

$$p_2 = p_{01}\left(\frac{T_{2'}}{T_{01}}\right)^{\gamma/\gamma-1} = 346,652 \times \left(\frac{813}{945}\right)^{1.35/1.35-1} = 193,708\,\text{Pa}$$

$$\rho_2 = \frac{p_2}{RT_2} = \frac{195,800}{287 \times 824.7} = 0.818\,\text{kg/m}^3$$

$$b_2 = \frac{\dot{m}}{W_2\rho_2\pi D_2} = \frac{2.192}{168.032 \times 0.818 \times 3.1416 \times 0.227} = 0.0223\,\text{m}$$

and

$$\frac{b_2}{D_2} = \frac{0.0223}{0.227} = 0.098$$

which is an acceptable value (see Table 11.1).

Rotor Outlet

Hub and Shroud (Tip) Diameters

Based on the recommended ranges of various parameters in Table 11.1, let us choose $D_{3s}/D_2 = 0.75$, which a little higher than the rule of Table 7.1 but appears to be necessary to reduce the exhaust velocity V_3, and $D_{3h}/D_{3s} = 0.35$, giving

$$D_{3s} = 0.699 \times D_2 = 0.75 \times 0.227 = 0.170\,\text{m}$$

and

$$D_{3h} = 0.35 \times D_{3s} = 0.35 \times 0.170 = 0.060\,\text{m}$$

Mean Diameter

We calculate the mean diameter from the equation

$$D_3 = \left[\left(D_{3h}^2 + D_{3s}^2\right)/2\right]^{1/2} = \left[\left\{(0.060)^2 + (0.170)^2\right\}/2\right]^{1/2} = 0.128\,\text{m}$$

and rotor outlet-inlet diameter ratio of

$$\frac{D_3}{D_2} = \frac{0.128}{0.227} = 0.562$$

which is an acceptable value (see Table 11.1).

Flow Area

$$A_3 = \frac{\pi}{4}\left(D_{3s}^2 - D_{3h}^2\right) = \frac{3.1416}{4} \times \left[(0.170)^2 - (0.060)^2\right] = 0.02\,\text{m}^2$$

Blade Velocity at Mean Diameter

$$U_3 = U_2\left(\frac{D_3}{D_2}\right) = 488 \times \left(\frac{0.128}{0.227}\right) = 274.195\,\text{m/s}$$

Absolute and Relative Flow Velocities

First, we calculate the total temperature T_{03} from the equation

$$T_{03} = T_{01} - \frac{E}{c_p} = 945 - \frac{238,144}{1107} = 730\,\text{K}$$

Combining the following three equations

$$\rho_3 = \frac{p_3}{RT_3}$$

$$V_3 = \frac{\dot{m}}{\rho_3 A_3}$$

$$T_3 = T_{03} - \frac{V_3^2}{2c_p}$$

results in the following quadratic equation for V_3:

$$V_3^2 + bV_3 + c = 0$$

where

$$b = \frac{2A_3 c_p p_3}{\dot{m}R} = \frac{2 \times 0.02 \times 1107 \times 98,595}{2.192 \times 287} = 6951.240$$

and

$$c = -2c_p T_{03} = -2 \times 1107 \times 730 = -1,615,942$$

with the solution

$$V_3 = \frac{-b + \sqrt{b^2 - 4c}}{2} = \frac{-6951.240 + \sqrt{(6951.240)^2 + 4 \times 1,615,942}}{2} = 225.174\,\text{m/s}$$

From the rotor exit velocity triangle in Figure 11.4, we obtain

$$\beta_3 = \tan^{-1} \frac{274.195}{225.174} = 50.6°$$

which is in the acceptable range,

$$W_3 = \left(U_3^2 + V_3^2\right)^{1/2} = \left\{(274.195)^2 + (225.174)^2\right\}^{1/2} = 354.804\,\text{m/s}$$

and

$$\frac{W_3}{W_2} = \frac{354.804}{168.032} = 2.112$$

which, according to Table 11.1, is also acceptable.

PROBLEM 11.2:   RADIAL-INFLOW TURBINE DRIVEN BY COMPRESSED AIR

The diameter of a radial-inflow turbine rotor is 0.25 m, and that of its nozzle exit is 0.255 m. The height of the nozzle vane at exit is 0.025 m. The nozzle flow velocity at exit makes an angle of 72.5° from the radial direction. The constant stagnation pressure and stagnation temperature in the nozzle are 2.05×10^5 Pa and 425 K, respectively. The rotor tip velocity is 95% of the tangential component of the nozzle exit velocity. The air mass flow rate through the turbine is 1.65 kg/s. Find (1) nozzle exit Mach number, static temperature, and static pressure and (2) rotor rpm. Assume air to be a perfect gas with $\gamma = 1.4$ and $R = 287 \, \text{J}/(\text{kg} \, \text{K})$.

SOLUTION FOR PROBLEM 11.2

(a) NOZZLE EXIT MACH NUMBER, STATIC TEMPERATURE, AND STATIC PRESSURE

Nozzle Exit Flow Area

$$A_{2noz} = \pi D_{2noz} b_2 = 3.1416 \times 0.255 \times 0.025 = 0.02 \, \text{m}^2$$

Nozzle Exit Flow Velocity Coefficient (C_V)

$$C_V = \frac{V_{2r}}{V_2} = \cos \alpha_2 = \cos 72.5° = 0.301$$

Total-Pressure Mass Flow Function F_{f02}

$$F_{f02} = \frac{\dot{m}\sqrt{RT_{02}}}{C_V A_{2noz} p_{02}} = \frac{1.65 \times \sqrt{287 \times 425}}{0.301 \times 0.02 \times 2.05 \times 10^5} = 0.467$$

Mach Number (M_2)

For $F_{f02} = 0.467$ and $\gamma = 1.4$, we solve the following equation iteratively (e.g., Goal Seek in MS Excel)

$$F_{f02} = M_2 \sqrt{\frac{\gamma}{\left(1 + \frac{\gamma - 1}{2} M_2^2\right)^{\frac{\gamma+1}{\gamma-1}}}}$$

to yield $M_2 = 0.443$.

Static Temperature and Pressure

$$\frac{T_{02}}{T_2} = 1 + 0.5(\gamma - 1)M_2^2 = 1 + 0.5 \times (1.4 - 1) \times (0.443)^2 = 1.0392$$

$$T_2 = \frac{T_{02}}{T_{02}/T_2} = \frac{425}{1.0392} = 409\,\text{K}$$

$$\frac{p_{02}}{p_2} = \left(\frac{T_{02}}{p_2}\right)^{\gamma/\gamma-1} = (1.0392)^{3.5} = 1.144$$

$$p_2 = \frac{p_{02}}{p_{02}/p_2} = \frac{2.05 \times 10^5}{1.144} = 1.792 \times 10^5\,\text{Pa}$$

(b) Rotor RPM

Nozzle Exit Velocity (V_2)

$$V_2 = M_2\sqrt{\gamma R T_2} = 0.443 \times \sqrt{1.4 \times 287 \times 409} = 179.464\,\text{m/s}$$

$$V_{\theta 2} = V_2 \sin\alpha_2 = 179.464 \times \sin 72.5° = 171.158\,\text{m/s}$$

$$U_2 = 0.95 \times V_{\theta 2} = 0.95 \times 171.158 = 162.6\,\text{m/s}$$

$$N = \frac{60U_2}{\pi D_2} = \frac{60 \times 162.6}{3.1416 \times 0.250} = 12{,}422\,\text{rpm}$$

PROBLEM 11.3: OPTIMUM-EFFICIENCY DESIGN OF AN IFR TURBINE ROTOR

An IFR turbine rotor with 13 blades needs to produce 400 kW from a supply of gas heated to a total temperature of 1100 K at a mass flow rate of 1.2 kg/s at the optimum total-to-static efficiency of $\eta_{ts} = 0.85$. Using the method of Whitfield and Baines (1990), find (1) the overall total-to-static pressure ratio and (2) the rotor tip speed and inlet Mach number. Assume $\gamma = 1.33$ and $c_p = 1187\,\text{J/(kg K)}$.

SOLUTION FOR PROBLEM 11.3

(a) Overall Total-to-Static Pressure Ratio

Specific Energy Transfer

$$E = \frac{P}{\dot{m}} = \frac{400 \times 1000}{1.2} = 333333.333\,\text{m}^2/\text{s}^2$$

Overall Total-to-Static Pressure Ratio

$$\frac{p_{01}}{p_3} = \left(1 - \frac{E}{\eta_{ts}c_pT_{01}}\right)^{-\left(\frac{\gamma}{\gamma-1}\right)} = \left(1 - \frac{333,333.333}{0.85 \times 1187 \times 1100}\right)^{-\left(\frac{1.33}{1.33-1}\right)} = 4.218$$

(b) ROTOR TIP SPEED AND INLET MACH NUMBER

Absolute Flow Angle (α_2)

Iteratively solving the following equation:

$$n_b = 13 = 0.1047(110 - \alpha_2)\tan\alpha_2$$

we obtain $\alpha_2 = 73.7°$.

Relative Flow Angle (β_2)

$$\beta_2 = \cos^{-1}\left(1 - \frac{0.63\pi}{n_b}\right) = \cos^{-1}\left(1 - \frac{0.63 \times 3.1416}{13}\right) = 32°$$

Rotor Tip Speed (U_2)

$$U_2 = \left(\frac{E}{\cos\beta_2}\right)^{\frac{1}{2}} = \left(\frac{333,333.333}{\cos 32°}\right)^{\frac{1}{2}} = 627.053\,\text{m/s}$$

Rotor Inlet Mach Number (M_2)

$$V_{\theta2} = U_2\cos\beta_2 = 627.053 \times \cos 32° = 531.587\,\text{m/s}$$

$$V_2 = \frac{V_{\theta2}}{\sin\alpha_2} = \frac{531.587}{\sin 73.7°} = 553.849\,\text{m/s}$$

$$T_2 = T_{02} - \frac{V_2^2}{2c_p} = 1100 - \frac{(553.849)^2}{2 \times 1187} = 970.8\,\text{K}$$

$$M_2 = \frac{V_2}{\sqrt{\gamma RT_2}} = \frac{V_2}{\sqrt{(\gamma-1)c_pT_2}} = \frac{V_2}{\sqrt{(1.33-1) \times 1187 \times 970.8}} = 0.898$$

PROBLEM 11.4: PERFORMANCE OF A RADIAL-FLOW GAS TURBINE

A radial-inflow gas turbine operates at 61,000 rpm with the overall total-to-static pressure ratio of 2.0. Its inlet diameter is 0.125 m. The hot gases at a total temperature of 1100 K enters the rotor radially and exits it axially at a mass flow rate of 0.35 kg/s. Find (1) the ratio of rotor tip speed to spouting velocity, (2) the total-to-static pressure ratio, and (3) turbine power. Assume $\gamma = 1.35$ and $R = 287\,\text{J}/(\text{kg K})$.

SOLUTION FOR PROBLEM 11.4

(a) Ratio of Rotor Tip Speed to Spouting Velocity

Rotor Tip Speed (U_2)

$$U_2 = \frac{ND_2}{2} = \frac{61,000 \times 3.1416 \times 0.125}{2 \times 30} = 399.244 \,\text{m/s}$$

Spouting Velocity (c_0)

$$\frac{T_{01}}{T_{3'}} = \left(\frac{p_{01}}{p_3} \right)^{\gamma-\frac{1}{\gamma}} = (2)^{1.35-\frac{1}{1.35}} = 1.197$$

$$T_{3'} = \frac{1000}{1.197} = 835.5\,\text{K}$$

$$E_i = c_p (T_{01} - T_{3'}) = \frac{R\gamma}{\gamma-1}(T_{01} - T_{3'}) = \left(\frac{287 \times 1.35}{1.35-1} \right) \times (1000 - 835.5) = 182,083 \,\text{m}^2/\text{s}^2$$

$$c_0 = \sqrt{2E_i} = \sqrt{2 \times 182,083} = 603.462 \,\text{m/s}$$

$$\frac{U_2}{c_0} = \frac{399.244}{603.462} = 0.662$$

(b) Total-to-Static Efficiency (η_{ts})

$$E = U^2 = (399.244)^2 = 159,396 \,\text{m}^2/\text{s}^2$$

$$\eta_{ts} = \frac{E}{E_i} = \frac{159,396}{182,083} = 0.875$$

(c) Turbine Power (P_t)

$$P_t = \frac{\dot{m}E}{1000} = \frac{0.35 \times 159,396}{1000} = 55.8 \,\text{kW}$$

PROBLEM 11.5: STANDARD 90° IFR GAS TURBINE WITH CHOKED NOZZLE

A standard 90° IFR gas turbine runs at 410,000 rpm. The rotor tip diameter is 0.218 m, and the tip blade height is 0.0222 m. Gas leaves the nozzle at an angle of 68° and a Mach number of 1.0 (choked). The total-to-static efficiency is 0.80, and the nozzle velocity coefficient $(\varphi = V_2/V_{2'})$ is 0.95. The exhaust pressure is 1 bar. Assuming $\gamma = 1.35$ and $R = 287 \,\text{J}/(\text{kgK})$, which yields $c_p = 1107 \,\text{J}/(\text{kgK})$ for the gas, find (1) the relative velocity entering the rotor, (2) the static temperature at the rotor inlet, (3) the total pressure at the nozzle inlet, (4) the static pressure at the rotor inlet, and (5) the turbine power in kW.

SOLUTION FOR PROBLEM 11.5

(a) Relative Velocity Entering the Rotor

Rotor Tip Speed

$$U_2 = \frac{\pi D_2 N}{60} = \frac{3.1416 \times 0.218 \times 41,000}{60} = 468 \, \text{m/s}$$

Relative Velocity at Rotor Inlet

$$W_2 = U_2 \cot \alpha_2 = 468 \times \cot 68° = 189.081 \, \text{m/s}$$

(b) Static Temperature at the Rotor Inlet

Absolute Velocity at Rotor Inlet

$$V_2 = \frac{U_2}{\sin \alpha_2} = \frac{468}{\sin 68°} = 504.746 \, \text{m/s}$$

Static Temperature at Rotor Inlet
For $M_2 = 1$, we obtain

$$T_2 = \frac{V_2^2}{\gamma R} = \frac{(504.746)^2}{1.35 \times 287} = 657.6 \, \text{K}$$

(c) Total Pressure at Nozzle Inlet

Total Temperature at Nozzle Inlet

$$T_{01} = T_{02} = \left(\frac{\gamma + 1}{2}\right) T_2 = \left(\frac{1.35 + 1}{2}\right) \times 657.6 = 772.6 \, \text{K}$$

Specific Energy Transfer (E)

$$E = U_2^2 = (468)^2 = 219,017 \, \text{m}^2/\text{s}^2$$

Overall Total-to-Static Pressure Ratio

$$\frac{p_{01}}{p_3} = \left(1 - \frac{E}{\eta_{ts} c_p T_{01}}\right)^{-\left(\frac{\gamma}{\gamma-1}\right)} = \left(1 - \frac{219,017}{0.80 \times 1107 \times 772.6}\right)^{-\left(\frac{1.35}{1.35-1}\right)} = 4.429$$

Total Pressure at Nozzle Inlet

$$p_{01} = p_3 \left(\frac{p_{01}}{P_3} \right) = 1.0 \times 10^5 \times 4.429 = 442,852 \, \text{Pa}$$

(d) STATIC PRESSURE AT ROTOR INLET

Ideal Absolute Velocity at Rotor Inlet

$$V_{2'} = \frac{V_2}{\varphi} = \frac{504.746}{0.95} = 531.312 \, \text{m/s}$$

Static Temperature at Nozzle Outlet from Isentropic Expansion $(T_{2'})$

$$T_{2'} = T_{01} - \frac{V_{2'}^2}{2c_p} = 772.6 - \frac{(531.312)^2}{2 \times 1107} = 645.1 \text{K}$$

Nozzle Total-to-Static Pressure Ratio

$$\frac{p_{01}}{p_{2'}} = \left(\frac{T_{01}}{T_{2'}} \right)^{\gamma/\gamma-1} = \left(\frac{772.6}{645.1} \right)^{1.35/(1.35-1)} = 2.005$$

Static Pressure at Rotor Inlet

$$p_2 = p_{2'} = \frac{p_{01}}{p_{01}/p_{2'}} = \frac{442,852}{2.005} = 220,871 \, \text{Pa}$$

(e) TURBINE POWER

Density at Rotor Inlet

$$\rho_2 = \frac{p_2}{RT_2} = \frac{220,871}{287 \times 657.6} = 1.170 \, \text{kg/m}^3$$

Mass Flow Rate

$$\dot{m} = \rho_2 W_2 \pi D_2 b_2 = 1.170 \times 189.081 \times 3.1416 \times 0.218 \times 0.0222 = 3.365 \, \text{kg/s}$$

Turbine Power

$$P_t = \dot{m}E = \frac{3.365 \times 219,017}{1000} = 737 \, \text{kW}$$

PROBLEM 11.6:   RADIAL-FLOW GAS TURBINE
WITH FLAT RADIAL BLADES

A radial-flow gas turbine with flat radial blades runs at 24,200 rpm. The gas enters the rotor at a radius of 0.152 m and exits at a mean radius of 0.076 m. Exhaust gases at a static pressure of 1 bar and a static temperature of 645 K leave axially at a relative Mach number of 0.75. Assuming $\gamma = 1.35$ and $R = 287\,\mathrm{J}/(\mathrm{kg\,K})$, which yields $c_p = 1107\,\mathrm{J}/(\mathrm{kg\,K})$ for the gas, calculate (1) the mass flow rate to produce 75 kW and (2) the blade height at the exit.

SOLUTION FOR PROBLEM 11.6

(a) TURBINE MASS FLOW RATE

Rotor Tip Speed

$$U_2 = \frac{\pi r_2 N}{30} = \frac{3.1416 \times 0.152 \times 24,200}{30} = 385.201\,\mathrm{m/s}$$

Specific Energy Transfer (E)

$$E = U_2^2 = (385.201)^2 = 148,380\,\mathrm{m^2/s^2}$$

Turbine Mass Flow Rate

$$\dot{m} = \frac{P_t}{E} = \frac{200}{148,380} = 0.505\,\mathrm{kg/s}$$

(b) BLADE HEIGHT AT ROTOR EXIT

Exit Flow Velocity

$$V_{a3} = M_{R3}\sqrt{\gamma R T_3} = 0.75 \times \sqrt{1.35 \times 287 \times 645} = 374.929\,\mathrm{m/s}$$

Exit Flow Density

$$\rho_3 = \frac{P_3}{R T_3} = \frac{10^5}{287 \times 645} = 0.540\,\mathrm{kg/m^3}$$

Rotor Exit Flow Area

$$A_3 = \frac{\dot{m}}{\rho_3 V_{a3}} = \frac{0.505}{0.540 \times 374.929} = 0.0025\,\mathrm{m^2}$$

Blade Height at Rotor Exit

$$b_3 = \frac{A_3}{2\pi r_{3m}} = \frac{0.0025}{2 \times 3.1416 \times 0.076} = 5.23 \times 10^{-3} \, \text{m} = 5.23 \, \text{mm}$$

PROBLEM 11.7:   SPECIFIC WORK OF A 90° IFR GAS TURBINE

Hot gas at a total pressure of 395 kPa and at a total temperature of 1120 K enters a 90° IFR gas turbine and exits axially at a static pressure of 101.3 kPa. The total-to-static efficiency for the turbine is 0.814. The nozzle flow is choked at the exit. Assuming $\gamma = 1.35$ and $R = 287 \, \text{J}/(\text{kg K})$, which yields $c_p = 1107 \, \text{J}/(\text{kg K})$ for the gas, find (1) the turbine specific energy transfer and (2) flow absolute angle at the rotor inlet.

SOLUTION FOR PROBLEM 11.7

(a) TURBINE SPECIFIC ENERGY TRANSFER

Turbine Overall Pressure Ratio

$$\frac{p_{01}}{p_3} = \frac{395 \times 10^3}{101.3 \times 10^3} = 3.899$$

Turbine Specific Energy Transfer

$$E = \eta_{ts} c_p T_{01} \left[1 - \left(\frac{p_3}{p_{01}} \right)^{\gamma-1/\gamma} \right] = 0.814 \times 1107 \times 1120 \times \left[1 - \left(\frac{1}{3.899} \right)^{1.35-1/1.35} \right]$$

$$= 300{,}026 \, \text{J/kg}$$

$$E = 300.026 \, \text{kJ/kg}$$

(b) ABSOLUTE FLOW ANGLE AT ROTOR INLET

Rotor Tip Speed

$$U_2 = \sqrt{E} = \sqrt{300{,}026} = 547.746 \, \text{m/s}$$

Static Temperature at Rotor Inlet

We calculate the static temperature from the equation

$$T_2 = \frac{T_{02}}{1 + \frac{\gamma-1}{2} M_2^2}$$

which for $T_{02} = T_{01}$ and $M_2 = 1$ yields

$$T_2 = \frac{2T_{01}}{\gamma + 1} = \frac{2 \times 1120}{1.35 + 1} = 953.2\,\text{K}$$

Flow Absolute Velocity at Rotor Inlet

$$V_2 = M_2\sqrt{\gamma RT_2} = \sqrt{\gamma RT_2} = \sqrt{1.35 \times 287 \times 953.2} = 607.712\ \text{m/s}$$

Flow Absolute Angle at Rotor Inlet

$$\alpha_2 = \sin^{-1}\left(\frac{U_2}{V_2}\right) = \sin^{-1}\left(\frac{547.746}{607.712}\right) = 64.3°$$

NOMENCLATURE

A_3	Annular flow area at rotor exit
b_2	Width of vane at $r = r_2$
c_0	Spouting velocity
c_p	Specific heat at constant pressure
C_V	Flow velocity coefficient
D_2	Rotor tip diameter
D_s	Specific diameter
D_{3h}	Hub diameter at rotor outlet
D_{3s}	Shroud diameter at rotor outlet
E	Energy transfer from fluid to rotor
E_i	Energy transfer for isentropic turbine
F_{f0}	Total-pressure mass flow function
h_1	Specific enthalpy of gas in volute
h_2	Specific enthalpy of gas entering rotor
h_3	Specific enthalpy of gas leaving rotor
h_4	Specific enthalpy of gas leaving diffuser
$h_{3'}$	Specific enthalpy of gas leaving ideal rotor
h_{01}	Specific total enthalpy of gas entering stator
h_{02}	Specific total enthalpy of gas leaving stator
h_{03}	Specific total enthalpy of gas leaving rotor
$h_{03'}$	Specific total enthalpy of gas leaving ideal rotor
h_{0R}	Specific relative total enthalpy
M_2	Absolute Mach number at rotor inlet
$\dot{m}$	Mass flow rate of gas
N	Rotor speed
N_s	Specific speed
n_b	Number of blades
P_t	Turbine power
p_1	Static pressure of gas in volute
p_2	Static pressure of gas leaving stator

$p_{2'}$	Static pressure of gas leaving ideal stator
p_3	Static pressure of gas leaving rotor
$p_{3'}$	Static pressure of gas leaving ideal rotor
p_{01}	Total pressure of gas in volute
p_{02}	Total pressure of gas at stator outlet
p_{03}	Total pressure of gas at rotor outlet
$p_{03'}$	Total pressure of gas leaving ideal rotor
p_{02R}	Relative total pressure entering rotor
p_{03R}	Relative total pressure leaving rotor
q	Heat added per unit mass in gas turbine cycle
Q_3	Volumetric flow rate based on gas density at rotor outlet
r	Radial position measured from axis of rotation
r_2	Rotor tip radius $= D_2/2$
r_3	RMS radius at rotor exit $= D_3/2$
R	Gas constant
T	Gas temperature
T_0	Total temperature of gas
T_1	Static temperature of gas in volute
T_2	Static temperature of gas leaving stator
$T_{2'}$	Static temperature of gas leaving ideal stator
T_3	Static temperature of gas leaving rotor
$T_{3'}$	Static temperature of gas leaving ideal turbine
$T_{3''}$	Static temperature of gas leaving ideal rotor with inlet temperature T_2
T_{01}	Total temperature of gas in volute
T_{02}	Total temperature of gas leaving stator
T_{03}	Total temperature of gas leaving rotor
T_{02R}	Relative total temperature of gas entering rotor
T_{03R}	Relative total temperature of gas leaving rotor
U_2	Rotor tip speed
U_{2i}	Tip speed of ideal rotor
U_{2h}	Rotor speed at exit hub diameter
U_{2s}	Rotor speed at exit shroud diameter
U_3	Rotor speed at exit rms mean diameter
V_1	Absolute velocity of gas in volute
V_2	Absolute velocity of gas at stator exit
V_3	Absolute velocity of gas leaving the rotor
V_4	Absolute velocity of gas at diffuser exit
$V_{3'}$	Absolute velocity of gas at exit of ideal turbine
V_{m2}	Meridional component of V_2; $(V_{m2} = V_{r2})$
V_{r2}	Radial component of V_{r2} at $r = r_2$
$V_{\theta2}$	Tangential component of V_2
V_{a3}	Axial component of V_3
V_{m3}	Meridional component of V_3; $(V_{m2} = V_{a3})$
$V_{\theta3}$	Tangential component of V_3
w_c	Specific compressor work
w_t	Specific turbine work

W Gas velocity relative to blade
W_2 Gas relative velocity at rotor tip
W_3 Relative velocity of gas leaving the rotor
W_{m2} Meridional component of W_2
$W_{\theta2}$ Tangential component of W_2
$W_{\theta3}$ Tangential component of W_3

GREEK SYMBOLS

α_2 Absolute gas angle at r_2, equals $\tan^{-1}\left(V_{\theta2}/V_{m2}\right)$
α_3 Absolute gas angle at r_3
β_2 Angle between W_2 and V_{m2}
β_3 Angle between W_3 and V_{m3}
γ Ratio of specific heats
η_{ts} Total-to-static turbine efficiency
η_{tt} Total-to-total turbine efficiency
η_{th} Thermal efficiency of gas turbine cycle
λ_n Nozzle loss coefficient
λ_{rot} Rotor loss coefficient
γ Nozzle velocity coefficient $V_2/V_{2'}$
ρ Gas density
ρ_2 Gas density leaving stator
ρ_3 Gas density leaving rotor

REFERENCES

Dixon, S.L. and C.A. Hall. 2014. *Fluid Mechanics and Thermodynamics of Turbomachinery*, 7th edition. Kidlington: Elsevier.
Ribaud, Y., and C. Mischell. 1986. *Study and Experiments of a Small Radial Turbine for Auxiliary Power Units. TP No. 1986–55.* ONERA, Chatillon.
Rohlik, H.E. 1968. *Analytical Determination of Radial Inflow Turbine Design Geometry for Maximum Efficiency.* NASA TN D-4384.
Scheel, L.F. 1972. *Gas Machinery.* Houston: Gulf Publishing Co.
Sultanian, B.K. 2019. *Logan's Turbomachinery: Flowpath Design and Performance Fundamentals*, 3rd edition. Boca Raton, FL: Taylor & Francis.
Whitfield, A., and N.C. Baines. 1990. *Design of Radial Turbomachines.* Essex: Longman.

BIBLIOGRAPHY

Aungier, R.H. 2006. *Turbine Aerodynamics: Axial-Flow and Radial-Flow Turbine Design and Analysis.* New York: ASME Press.
Glassman, A.J. 1976. *Computer Program for Design and Analysis of Radial Inflow Turbines.* NASA TN 8164.
Saravanamutto, H.I.H., G.F.C. Rogers, H. Cohen, P.V. Straznicky, and A.C. Nix. 20017. *Gas Turbine Theory*, 7th edition. Harlow: Pearson.
Shepherd, D.G. 1956. *Principles of Turbomachinery.* New York: Macmillan.
Sultanian, B.K. 2015. *Fluid Mechanics: An Intermediate Approach.* Boca Raton, FL: Taylor & Francis.

12 Axial-Flow Gas Turbines

REVIEW OF KEY CONCEPTS

We concisely present here some key concepts of axial-flow gas turbines. More details on each topic are given, for example, in Sultanian (2019). Problems and their solutions presented in this chapter primarily focus on the performance and preliminary design of an axial-flow reaction gas turbine stage, commonly found in multistage gas turbines used in stationary power plants as well as in gas-turbine engines to drive ships, trains, and aircraft.

AEROTHERMODYNAMICS

Figure 12.1 shows the sectional view an axial-flow turbine stage and the corresponding velocity diagrams. Euler's turbomachinery equation presented in Appendix A forms the basis for the aerothermodynamic analysis of these turbines. Appendix B presents a quick method to draw the composite dimensionless velocity diagram directly from the knowledge of the flow coefficient, loading coefficient, and degree of reaction.

ENTHALPY-ENTROPY DIAGRAM

Figure 12.2 shows the key thermodynamic states within a turbine stage. The actual end states in terms of static properties are 1 and 2 in the stator and 2 and 3 in the rotor. The corresponding total-property states—i.e., 01, 02, and 03—are found by constructing isentropic processes between the actual states and the corresponding

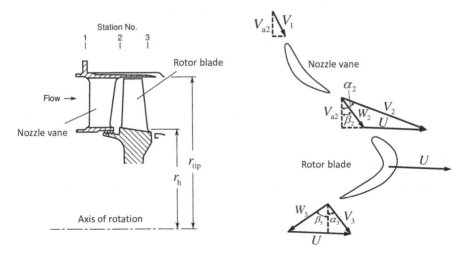

FIGURE 12.1 Sectional view of an axial-flow turbine stage and the corresponding velocity diagrams.

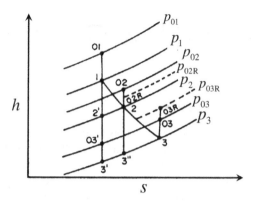

FIGURE 12.2 Enthalpy-entropy diagram.

total pressure lines (isobars). The states 2', 03', and 3' are those corresponding to an ideal (isentropic) expansion from the inlet to the exit pressure for the stage.

ISENTROPIC EFFICIENCIES

We define the total-to-total stage efficiency as

$$\eta_{tt} = \frac{h_{01} - h_{03}}{h_{01} - h_{03'}} \qquad (12.1)$$

which for constant c_p becomes

$$\eta_{tt} = \frac{T_{01} - T_{03}}{T_{01} - T_{03'}} \qquad (12.2)$$

We define the total-to-static efficiency for the stage by

$$\eta_{ts} = \frac{h_{01} - h_{03}}{h_{01} - h_{3'}} \qquad (12.3)$$

which for constant c_p becomes

$$\eta_{ts} = \frac{T_{01} - T_{03}}{T_{01} - T_{3'}} \qquad (12.4)$$

MEAN RADIUS

Referring to Figure 12.1, we define the mean radius, which divides the annulus area in half, by

$$r_m = \left(\frac{r_h^2 + r_{tip}^2}{2} \right)^{1/2} \qquad (12.5)$$

which we use for the stage meanline analysis and preliminary design.

POLYTROPIC EFFICIENCY

If we divide the expansion process across a turbine into a large number of very small, consecutive expansions, then the isentropic efficiency across each is called the polytropic efficiency, or small-stage efficiency, which may be assumed constant for a given gas turbine, reflecting its state-of-the-art design engineering. We earlier discussed the polytropic efficiency for a compressor in Chapter 10.

With reference to Figure 12.2, we assume that the index of polytropic expansion in the turbine along 1–3 is n, which yields

$$\frac{T_{01}}{T_{03}} = \left(\frac{p_{01}}{p_{03}}\right)^{\frac{n-1}{n}} = \Pi_t^{\frac{n-1}{n}} \tag{12.6}$$

Note that for $n = \gamma$, this equation yields the isentropic expansion along 1–3′.

We define the turbine polytropic efficiency as

$$\eta_{pt} = \frac{(dT_0)_{actual}}{(dT_0)_{isentropic}} = \frac{(dT_0/T_0)_{actual}}{(dT_0/T_0)_{isentropic}} \tag{12.7}$$

Integrating Equation 12.7 across the turbine stage yields

$$\eta_{pt} \int_1^{3'} (dT_0/T_0)_{isentropic} = \int_1^3 (dT_0/T_0)_{actual}$$

$$\left(\frac{T_{01}}{T_{03}}\right) = \left(\frac{T_{01}}{T_{03'}}\right)^{\eta_{pt}} \tag{12.8}$$

$$\Pi_t^{\frac{n-1}{n}} = \Pi_t^{\frac{\eta_{pt}(\gamma-1)}{\gamma}}$$

$$\frac{n-1}{n} = \frac{\eta_{pt}(\gamma-1)}{\gamma}$$

For relating the turbine total-to-total isentropic efficiency to its polytropic efficiency, we write

$$\eta_t = \eta_{tt} = \frac{T_{01} - T_{03}}{T_{01} - T_{03'}} = \frac{1 - T_{03}/T_{01}}{1 - T_{03'}/T_{01}}$$

$$\eta_t = \frac{1 - (1/\Pi_t)^{\frac{n-1}{n}}}{1 - (1/\Pi_t)^{\frac{\gamma-1}{\gamma}}}$$

which using Equation 12.8 yields

$$\eta_t = \frac{1-\left(1/\Pi_t\right)^{\frac{\eta_{pt}(\gamma-1)}{\gamma}}}{1-\left(1/\Pi_t\right)^{\frac{\gamma-1}{\gamma}}} \tag{12.9}$$

which expresses turbine isentropic efficiency in terms its polytropic efficiency and pressure ratio. Alternatively, to express turbine polytropic efficiency in terms of its isentropic efficiency and pressure ratio, we rewrite this equation as

$$\eta_{pt} = \frac{\ln\left(1-\eta_t\left\{1-\left(1/\Pi_t\right)^{\frac{\gamma-1}{\gamma}}\right\}\right)}{\left(\frac{\gamma-1}{\gamma}\right)\ln\left(1/\Pi_t\right)} \tag{12.10}$$

In Figure 12.3, we plot Equations 12.9 and 10.27, which is for the compressor, for $\eta_{pt} = \eta_{pc} = 0.9$. In this figure, we have assumed $\gamma = 1.4$ for the compressor and $\gamma = 1.33$ for the turbine. For the pressure ratio close to 1.0, both the compressor and the turbine have equal isentropic and polytropic efficiencies. The figure further shows that, for the compressor, the isentropic efficiency decreases with pressure ratio and, for the turbine, it increases with pressure ratio. The rate of decrease in η_c for the compressor, however, is higher than the corresponding rate of increase in η_t for the turbine.

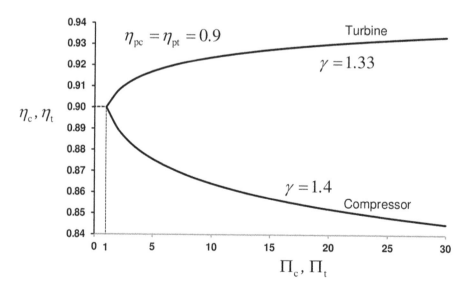

FIGURE 12.3 Variation of compressor and turbine isentropic efficiencies with pressure ratio for a constant value of their polytropic efficiency.

PROBLEM 12.1:   ANALYSIS OF AN AXIAL-FLOW GAS TURBINE STAGE

The absolute velocities entering and leaving a certain axial-flow gas turbine stage are in the axial direction with constant V_a. For the flow coefficient $\varphi = 0.5$ and the absolute angle of 67.3° at the rotor inlet, find (1) the stage-loading coefficient ψ, (2) the degree of reaction, and (3) the inlet and outlet flow angles relative to the rotor blade.

SOLUTION FOR PROBLEM 12.1

(a) Stage-Loading Coefficient

As in Appendix B, section 1 represents the blade inlet and section 2 represents the blade outlet. As the flow leaves the turbine stage axially, we have $\alpha_2 = 0$.

Equations B.17 yields

$$\tan 68.2° = \frac{0.5\psi + (1 - R)}{0.5}$$

$$0.5\psi - R = 0.5\tan 68.2° - 1 = 0.195$$

Similarly, Equation B.21 yields

$$0.5\psi + R = 1$$

Adding this equation to $0.5\psi - R = 0.195$ yields

$$\psi = 1.195$$

(b) Degree of Reaction

Subtracting $0.5\psi - R = 0.195$ from $0.5\psi + R = 1$ yields

$$2R = 1 - 0.195 = 0.805$$

$$R = \frac{0.805}{2} = 0.402$$

(c) Flow Angles Relative to the Rotor Blades

Relative Flow Angle at Rotor Inlet

From Equation B.25, we obtain

$$\beta_1 = \tan^{-1}\left(\frac{0.5\psi - R}{\varphi}\right) = \tan^{-1}\left(\frac{0.5 \times 1.195 - 0.402}{0.5}\right) = 21.3°$$

Relative Flow Angle at Rotor Outlet

From Equation B.29, we obtain

$$\beta_2 = \tan^{-1}\left(-\frac{0.5\psi + R}{\varphi}\right) = \tan^{-1}\left(-\frac{0.5 \times 1.195 + 0.402}{0.5}\right) = -63.4°$$

PROBLEM 12.2: FREE-VORTEX DESIGN OF AN AXIAL-FLOW GAS TURBINE STAGE

An axial flow turbine stage operating at 7500 rpm is to be designed for free-vortex conditions at exit from vane (nozzle) row and for a zero swirl at exit from the blade row (rotor). The gas entering the stage has a stagnation temperature of 1100 K and a mass flow rate of 25 kg/s, the root and tip diameters are 0.6 and 0.8 m, respectively. At the rotor tip the stage reaction is 55% and the axial velocity is constant at 150 m/s. The velocity of the gas entering the stage is equal to that leaving it. Determine (1) the power output of the turbine stage, (2) the stagnation and static temperatures at the stage exit, and (3) degree of reaction at the root section. Assume

$$c_p = 1147 \; \text{J}/(\text{kg K})$$

SOLUTION FOR PROBLEM 12.2

(a) POWER OUTPUT OF THE TURBINE STAGE

$$\Omega = \frac{\pi N}{30} = \frac{3.1416 \times 7500}{30} = 785.398 \; \text{rad/s}$$

$$r_{tip} = \frac{D_{tip}}{2} = \frac{0.8}{2} = 0.4 \; m$$

$$U_{tip} = r_{tip}\Omega = 0.40 \times 785.398 = 314.159 \; \text{m/s}$$

For $V_2 = V_a$, we have $\alpha_2 = 0$, and Equation B.21 reduces to

$$0.5\psi_{tip} + R_{tip} = 1$$

which for $R_{tip} = 0.55$ yields

$$\psi_{tip} = 2(1 - R_{tip}) = 2 \times (1 - 0.55) = 0.9$$

As the specific work output is uniform from root to tip for a free-vortex design, we obtain the turbine stage power output as

$$P_t = \dot{m}E = \dot{m}\psi_{tip}U_{tip}^2 = 25 \times 0.9 \times (314.159)^2 = 2,220,661\,\text{W} = 2220.661 \; \text{kW}$$

(b) STAGNATION AND STATIC TEMPERATURES AT STAGE EXIT

Stagnation Temperature at Stage Exit

$$T_{02} = T_{01} - \frac{P_t}{\dot{m}c_p} = 1100 - \frac{2,220,661}{25'1147} = 1022.6\,\text{K}$$

Static Temperature at Stage Exit

$$T_2 = T_{02} - \frac{V_2^2}{2c_p} = T_{02} - \frac{V_a^2}{2c_p} = 1022.6 - \frac{150'150}{2'1147} = 1012.7\,\text{K}$$

(c) ROOT SECTION REACTION

For a free vortex design, the reactions at blade tip and root sections are related by the equation

$$\frac{1 - R_{\text{root}}}{1 - R_{\text{tip}}} = \left(\frac{r_{\text{tip}}}{r_{\text{root}}}\right)^2$$

giving

$$R_{\text{root}} = 1 - \left(1 - R_{\text{tip}}\right)\left(\frac{r_{\text{tip}}}{r_{\text{root}}}\right)^2 = 1 - (1\text{-}0.55)\left(\frac{0.40}{0.30}\right)^2 = 0.2$$

PROBLEM 12.3: PERFORMANCE ANALYSIS OF AN AXIAL-FLOW GAS TURBINE STAGE

An axial-flow gas turbine stage develops 3.5 MW at a mass flow rate of 25 kg/s. At the nozzle inlet, the total temperature is 1100 K, and the total pressure is 800 kPa. The static pressure at the nozzle exit (rotor inlet) is 500 kPa. At the rotor inlet, the absolute flow angle measured from the axial direction is 50°. The axial velocity is constant across the stage, and the gas enters and leaves the stage without any absolute swirl velocity. For an isentropic flow in the nozzle, determine (1) the nozzle exit velocity, (2) the blade velocity, (3) the stage reaction, and (4) relative flow angles at rotor inlet and outlet. Assume $R = 287\,\text{J}/(\text{kg}\,\text{K})$ and $c_p = 1148\,\text{J}/(\text{kg}\,\text{K})$.

SOLUTION FOR PROBLEM 12.3

In this solution, we designate nozzle exit and rotor inlet by subscript 1 and rotor outlet by subscript 2.

(a) NOZZLE EXIT VELOCITY

Total-to-Static Temperature Ratio at Rotor Inlet

As the flow in the nozzle is isentropic, we have $p_{01} = 800,000\,\text{Pa}$. We obtain

$$\frac{T_{01}}{T_1} = \left(\frac{p_{01}}{p_1}\right)^{R/c_p} = \left(\frac{800,000}{500,000}\right)^{287/1148} = 1.125$$

Static Temperature at Rotor Inlet

$$T_1 = \frac{T_{01}}{\left(T_{01}/T_1\right)} = \frac{1100}{1.125} = 978.1\,\text{K}$$

Nozzle Exit Velocity (Absolute Velocity at Rotor Inlet)

$$V_1 = \sqrt{2c_p\left(T_{01} - T_1\right)} = \sqrt{2 \times 1148 \times (1100 - 978.1)} = 529.140\,\text{m/s}$$

(b) BLADE VELOCITY

Specific Energy Transfer

$$E = \frac{P_t}{\dot{m}} = \frac{3.5 \times 10^6}{25} = 140,000\,\text{J/kg}$$

Tangential Component of V_1

$$V_{\theta 1} = V_1 \sin \alpha_1 = 529.140 \times \sin 50° = 405.344\,\text{m/s}$$

Blade Velocity

For $\alpha_2 = 0$, we obtain

$$U = \frac{E}{V_{\theta 1}} = \frac{140,000}{405.344} = 345.385\,\text{m/s}$$

(c) STAGE REACTION

Loading Coefficient

$$\psi = \frac{E}{U^2} = \frac{140,000}{(345.385)^2} = 1.174$$

Stage Reaction

For $\alpha_2 = 0$, we obtain from Equation B.21

$$R = 1 - 0.5\psi = 1 - 0.5 \times 1.174 = 0.413$$

(d) RELATIVE FLOW ANGLES AT ROTOR INLET AND OUTLET

Flow Coefficient

$$\varphi = \frac{V_1 \cos \alpha_1}{U} = \frac{529.140 \times \cos 50°}{345.385} = 0.985$$

Relative Flow Angle at Rotor Inlet

From Equation B.25, we obtain

$$\beta_1 = \tan^{-1}\left(\frac{0.5\psi - R}{\varphi}\right) = \tan^{-1}\left(\frac{0.5 \times 1.174 - 0.413}{0.985}\right) = 10°$$

Relative Flow Angle at Rotor Outlet

From Equation B.29, we obtain

$$\beta_2 = \tan^{-1}\left(-\frac{0.5\psi + R}{\varphi}\right) = \tan^{-1}\left(-\frac{0.5 \times 1.174 + 0.413}{0.985}\right) = -45.4°$$

PROBLEM 12.4: NUMBER OF IMPULSE STAGES NEEDED IN A MULTISTAGE AXIAL-FLOW GAS TURBINE

Hot gases enter a multistage axial-flow gas turbine at a total pressure of 4.5 bar and a total temperature of 1150 K. The last stage turbine exit total pressure is 1.0 bar. The turbine operates at 9000 rpm. The blade velocity at the mean radius, which corresponds to half annulus area, is 275 m/s. The blade length is 0.10 m. Assuming zero interstage swirl and 100% total-to-total efficiency determine (1) the blade radii at tip and hub and (2) the number of stages required if all stages selected are impulse type. Assume $\gamma = 1.33$ and $R = 287\,\text{J}/(\text{kg K})$ for all stages.

SOLUTION FOR PROBLEM 12.4

(a) BLADE TIP AND HUB RADII

Rotor Angular Velocity

$$\Omega = \frac{\pi N}{30} = \frac{3.1416 \times 9000}{30} = 942.478 \text{ rad/s}$$

Mean Radius

$$r_m = \frac{U_m}{\Omega} = \frac{275}{942.478} = 0.292 \text{ m}$$

Tip Radius

The following two equations

$$r_m^2 = \frac{\left(r_{hub}^2 + r_{tip}^2\right)}{2}$$

$$r_{tip} - r_{hub} = b$$

for the blade geometry yield

$$r_{tip} = \frac{b + \sqrt{4r_m^2 - b^2}}{2} = \frac{0.1 + \sqrt{4 \times 0.292 \times 0.292 - 0.1 \times 0.1}}{2} = 0.337\,\text{m}$$

Hub Radius

$$r_{hub} = r_{tip} - b = 0.337 - 0.10 = 0.237\,\text{m}$$

(b) NUMBER OF TURBINE STAGES

Maximum Specific Work Output of Each Stage

For zero interstage swirl, we obtain from Equation B.21

$$\psi_m = 2(1 - R_m)$$

which for an impulse stage with $R = 0$ reduces to $\psi = 2$. Thus, we write

$$w_{stage} = \psi_m U_m^2 = 2 \times 275 \times 275 = 151{,}250 \text{ J/kg}$$

Total Temperature at Turbine Exit

$$\frac{\gamma - 1}{\gamma} = \frac{1.33 - 1}{1.33} = 0.248$$

$$c_p = \frac{R\gamma}{\gamma - 1} = \frac{287}{0.248} = 1156.7 \text{ J/(kg K)}$$

For an isentropic turbine, we obtain

$$T_{0e} = T_{01} \left(\frac{p_{0e}}{p_{01}}\right)^{\frac{\gamma - 1}{\gamma}} = 1150 \left(\frac{1}{4.5}\right)^{\frac{1.33 - 1}{1.33}} = 792 \text{ K}$$

Overall Turbine Specific Work Output

$$w_t = c_p(T_{01} - T_{0e}) = 1156.7 \times (1150 - 792) = 414{,}314 \text{ J/kg}$$

Number of Stages

$$N_{stage} = \frac{w_t}{w_{stage}} = \frac{414{,}314}{151{,}250} = 2.7 \approx 3$$

PROBLEM 12.5: TURBINE POLYTROPIC EFFICIENCY IN A GAS TURBINE ENGINE

In a gas turbine engine, the ideal (isentropic) and real specific work inputs in the compressor are 350 and 405 kJ/kg, respectively (ideal is lower than real). The corresponding specific work outputs in the turbine are 760 and 710 kJ/kg (ideal is higher than real). If the polytropic efficiency of the compressor is 90%, compute the polytropic efficiency of the turbine. Assume $\gamma = 1.4$ for the compressor and $\gamma = 1.333$ for the turbine.

SOLUTION FOR PROBLEM 12.5

COMPRESSOR ANALYSIS

Isentropic Efficiency

$$\eta_c = \frac{350}{405} = 0.864$$

Compressor Pressure Ratio

For $\eta_c = 0.864$ and the given polytropic efficiency $\eta_{pc} = 0.90$, we can compute the compressor pressure ratio Π_c from the equation

$$\eta_c = \frac{\Pi_c^{\gamma_c - 1/\gamma_c} - 1}{\Pi_c^{\gamma_c - 1/\eta_{pc}\gamma_c} - 1}$$

which becomes

$$0.864 = \frac{\Pi_c^{1.4 - 1/1.4} - 1}{\Pi_c^{1.4 - 1/(0.9 \times 1.4)} - 1}$$

For solving this equation, we use an iterative solution method—for example, Goal Seek in MS Excel—and obtain $\Pi_c = 10.0$.

TURBINE ANALYSIS

Isentropic Efficiency

$$\eta_t = \frac{710}{760} = 0.934$$

Polytropic Efficiency

We assume here that the pressure ratio across the turbine equals that across the compressor, that is, $\Pi_t = \Pi_c = 10.0$.

$$\eta_{pt} = \frac{\ln\left(\eta_t\left\{\left[\left(\dfrac{1}{\Pi_t}\right)^{\frac{\gamma_t-1}{\gamma_t}} - 1\right] + 1\right\}\right)}{\ln\left(\left(\dfrac{1}{\Pi_t}\right)^{\frac{\gamma_t-1}{\gamma_t}}\right)} = \frac{\ln\left(0.934 \times \left\{\left(\dfrac{1}{10}\right)^{\frac{1.333-1}{1.333}} - 1\right\} + 1\right)}{\ln\left(\left(\dfrac{1}{10}\right)^{\frac{1.333-1}{1.333}}\right)} = 0.913$$

Note that for the same pressure ratio, we have $\eta_{pc} > \eta_c$ and $\eta_{pt} < \eta_t$.

PROBLEM 12.6:   TOTAL-TO-STATIC EFFICIENCY OF AN AXIAL-FLOW GAS TURBINE

An axial-flow gas turbine operates at an overall total pressure ratio of 8 with a polytropic efficiency of 0.85. For this turbine, find (1) the total-to-total efficiency, (2) the total-to-static efficiency if the Mach number at stage exit is 0.3, and (3) inlet total temperature if the exit velocity is 160 m/s. Assume $R = 287\,\mathrm{J/(kg\,K)}$ and $c_p = 1175\ \mathrm{J/(kg\,K)}$.

SOLUTION FOR PROBLEM 12.6

In this solution, we designate turbine stage inlet by subscript 1 and outlet by subscript 3. For $R = 287\,\mathrm{J/(kg\,K)}$ and $c_p = 1175\ \mathrm{J/(kg\,K)}$, we obtain

$$\gamma = \frac{c_p}{c_p - R} = \frac{1175}{1175 - 287} = 1.323$$

(a) TOTAL-TO-TOTAL EFFICIENCY

With the turbine total pressure ratio $\Pi_t = 8$ and polytropic efficiency $\eta_{pt} = 0.85$, we compute the total-to-total efficiency as

$$\eta_{tt} = \frac{1-\left(\dfrac{1}{\Pi_t}\right)^{\frac{\eta_{pt}(\gamma-1)}{\gamma}}}{1-\left(\dfrac{1}{\Pi_t}\right)^{\frac{(\gamma-1)}{\gamma}}} = \frac{1-\left(\dfrac{1}{\Pi_t}\right)^{\frac{\eta_{pt}R}{c_p}}}{1-\left(\dfrac{1}{\Pi_t}\right)^{\frac{R}{c_p}}} = \frac{1-\left(\dfrac{1}{8}\right)^{\frac{0.85\times287}{1175}}}{1-(8)^{\frac{287}{1175}}} = 0.880$$

(b) Total-to-Static Efficiency

We express the relation between η_{tt} and η_{ts} by the following equation:

$$\frac{\eta_{ts}}{\eta_{tt}} = \frac{1-\left(\dfrac{p_{03}}{p_{01}}\right)^{\frac{\gamma-1}{\gamma}}}{1-\left(\dfrac{p_3}{p_{01}}\right)^{\frac{\gamma-1}{\gamma}}}$$

which yields

$$\eta_{ts} = \eta_{tt}\left[\frac{1-\left(\dfrac{p_{03}}{p_{01}}\right)^{R/c_p}}{1-\left(\dfrac{p_3}{p_{01}}\right)^{R/c_p}}\right] = \eta_{tt}\left[\frac{1-\left(\dfrac{p_{03}}{p_{01}}\right)^{R/c_p}}{1-\left(\dfrac{p_{03}}{p_{01}}\right)^{R/c_p}\left(\dfrac{p_3}{p_{03}}\right)^{R/c_p}}\right] = \eta_{tt}\left[\frac{\left(\Pi_t\right)^{R/c_p}-1}{\left(\Pi_t\right)^{R/c_p}-\dfrac{1}{1+\dfrac{\gamma-1}{2}M_3^2}}\right]$$

which further yields

$$\eta_{ts} = \eta_{tt}\left[\frac{\left(\Pi_t\right)^{R/c_p}-1}{\left(\Pi_t\right)^{R/c_p}-\dfrac{1}{1+\dfrac{\gamma-1}{2}M_3^2}}\right] = 0.880\times\left[\frac{(8)^{287/1175}-1}{(8)^{287/1175}-\dfrac{1}{1+\dfrac{(1.323-1)\times(0.3)^2}{2}}}\right]$$

$$= 0.862$$

(c) Inlet Total Temperature

Exit Static Temperature

$$T_3 = \frac{V_3^2}{\gamma R M_3^2} = \frac{(160)^2}{1.323\times287\times(0.3)^2} = 749\,\text{K}$$

Exit Total Temperature

$$T_{03} = T_{03} + \frac{V_3^2}{2c_p} = 749 + \frac{(160)^2}{2 \times 1175} = 760 \, \text{K}$$

Inlet Total Temperature

$$T_{01} = \frac{T_{03}}{1 - \eta_{tt}\left[1 - \left(\dfrac{1}{\Pi_t}\right)^{R/c_p}\right]} = \frac{760}{1 - 0.880 \times \left[1 - \left(\dfrac{1}{8}\right)^{287/1175}\right]} = 1170 \, \text{K}$$

PROBLEM 12.7: SPECIFIC WORK OUTPUT OF AN AXIAL-FLOW GAS TURBINE

In an axial-flow gas turbine stage with constant axial velocity $V_a = 300 \, \text{m/s}$, the absolute flow angle measured from the axial direction at the rotor inlet is $\alpha_1 = 52.5°$ and that the outlet is $\alpha_2 = -11.5°$. The rotor mean radius is $r_m = 0.15 \, \text{m}$, and it rotates at 21000 rpm. Find (a) the specific work output of the rotor and the drop in gas total temperature, (b) degree of reaction, and (c) the relative flow angles at rotor inlet and outlet. Assume $c_p = 1148 \, \text{J/(kg K)}$.

SOLUTION FOR PROBLEM 12.7

(a) SPECIFIC WORK OUTPUT AND DROP IN GAS TOTAL TEMPERATURE

Blade Velocity at the Mean Radius

$$U = \frac{\pi r_m N}{30} = \frac{3.1416 \times 0.15 \times 21{,}000}{30} = 329.867 \, \text{m/s}$$

Flow Coefficient

$$\varphi = \frac{V_a}{U} = \frac{300}{329.867} = 0.909$$

Loading Coefficient

From Equations B.17 and B.21, we obtain

$$\psi = \varphi\left(\tan \alpha_1 - \tan \alpha_2\right)$$

which yields

$$\psi = 0.909 \times \left[\tan 50.5° - \tan(-11.5°)\right] = 1.37$$

Specific Work Output

$$E = \psi U^2 = 1.37 \times (329.867)^2 = 149101 \text{ J/kg} = 149.101 \text{ kJ/kg}$$

Drop in Gas Total Temperature

$$T_{01} - T_{02} = \frac{E}{c_p} = \frac{149,101}{1148} = 130 \text{ K}$$

(b) DEGREE OF REACTION

From Equations B.17 and B.21, we obtain

$$R = 1 - \frac{\varphi\left(\tan\alpha_1 + \tan\alpha_2\right)}{2} = 1 - \frac{0.909 \times \left[\tan 50.5° + \tan(-11.5°)\right]}{2} = 0.5$$

(c) RELATIVE FLOW ANGLES AT ROTOR INLET AND OUTLET

Rotor Inlet
From Equation B.25, we obtain

$$\beta_1 = \tan^{-1}\left(\frac{0.5\psi - R}{\varphi}\right) = \tan^{-1}\left(\frac{0.5 \times 1.37 - 0.5}{0.909}\right) = 11.5°$$

Rotor Outlet
From Equation B.29, we obtain

$$\beta_2 = \tan^{-1}\left(-\frac{0.5\psi + R}{\varphi}\right) = \tan^{-1}\left(-\frac{0.5 \times 1.37 + 0.5}{0.909}\right) = -52.5°$$

Note that, for 50% reaction $(R = 0.5)$ computed here, we obtain in magnitude $\alpha_1 = |\beta_2| = 52.5°$ and $|\alpha_2| = \beta_1 = 11.5°$, giving symmetric velocity triangles at the rotor inlet and outlet at the mean radius.

PROBLEM 12.8:   SIMPLE-CYCLE PERFORMANCE OF AN AIR-COOLED GAS TURBINE

Figure 12.4 shows schematically a 200 MW modern gas turbine engine for power generation. The operating simple-cycle parameters for the baseline design are given as follows:

$$p_{01} = p_{04} = 101.325 \text{ kPa}$$

$$T_{01} = 288 \text{ K}$$

Compressor pressure ratio $\left(p_{02}/p_{01}\right) = 20$
Compressor pressure ratio at the point of turbine cooling flow extraction $\left(p_{02''}/p_{01}\right) = 10$
Turbine cooling flow $\left(\dot{m}_{\text{cool}}\right) = 20\%$ of the compressor inlet flow $\left(\dot{m}_{\text{engine}}\right)$
Combustor total pressure loss is 1% of compressor exit total pressure
Turbine inlet temperature $\left(T_{03}\right) = 2000 \text{ K}$
Compressor polytropic efficiency $\left(\eta_{\text{pc}}\right) = 92\%$
Turbine polytropic efficiency $\left(\eta_{\text{pt}}\right) = 92\%$
Ratio of specific heats across compressor $\left(\gamma_{\text{c}}\right) = 1.4$
Ratio of specific heats across turbine $\left(\gamma_{\text{t}}\right) = 1.33$

Assuming $R = 287 \text{ J}/\left(\text{kg K}\right)$, calculate compressor inlet flow rate $\left(\dot{m}_{\text{engine}}\right)$ and cycle thermodynamic efficiency $\left(\eta_{\text{th}}\right)$ for (1) the baseline design, (2) upgraded baseline design with $\eta_{\text{pc}} = \eta_{\text{pt}} = 93\%$, (3) upgraded baseline design with $p_{02''}/p_{01} = 9$, and (4) upgraded baseline design with $\dot{m}_{\text{cool}} = 0.18\,\dot{m}_{\text{engine}}$. Assume that the turbine cooling air flow does not contribute to turbine work. Show details of all your calculations for the baseline design only. Tabulate your calculation results for all four cases.

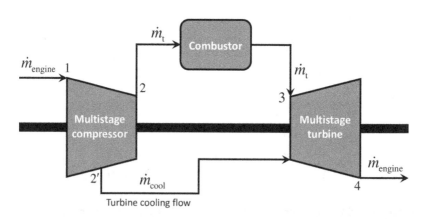

FIGURE 12.4 A simple-cycle air-cooled gas turbine engine for power generation (Problem 12.8).

SOLUTION FOR PROBLEM 12.8

(a) BASELINE DESIGN

For $\gamma_c = 1.4$ and $R = 287 \text{ J}/(\text{kg K})$, we obtain

$$c_{pc} = R\left(\frac{\gamma_c}{\gamma_c - 1}\right) = 287 \times \left(\frac{1.4}{1.4 - 1}\right) = 1004.5 \text{ J}/(\text{kg K})$$

For $\gamma_t = 1.33$ and $R = 287 \text{ J}/(\text{kg K})$, we obtain

$$c_{pt} = R\left(\frac{\gamma_t}{\gamma_t - 1}\right) = 287 \times \left(\frac{1.33}{1.33 - 1}\right) = 1156.7 \text{ J}/(\text{kg K})$$

Polytropic Exponent of Compression in the Compressor

$$\frac{n_c - 1}{n_c} = \frac{\gamma_c - 1}{\gamma_c \eta_{pc}} = \frac{1.4 - 1}{1.4 \times 0.92} = \frac{0.248}{0.92} = 0.311$$

Cooling Air Total Temperature at Extraction Point (2′)

$$\frac{T_{02'}}{T_{01}} = \left(\frac{p_{02'}}{p_{01}}\right)^{\frac{n_c - 1}{n_c}} = (10)^{0.311} = 2.044$$

$$T_{02''} = 2.044 \, T_{01} = 2.044 \times 288 = 588.8 \text{ K}$$

Specific Compressor Work from 1 to 2′

$$W_{c1-2''} = 1004.5 \times (588.8 - 288) = 302{,}131 \text{ W}$$

Air Total Temperature at Compressor Exit or Combustor Inlet (2)

$$\frac{p_{02}}{p_{02'}} = \frac{20}{10} = 2$$

$$\frac{T_{02}}{T_{02'}} = \left(\frac{p_{02}}{p_{02'}}\right)^{\frac{n_c - 1}{n_c}} = (2)^{0.311} = 1.240$$

$$T_{02} = 1.240 \, T_{02'} = 1.240 \times 588.8 = 730.2 \text{ K}$$

Specific Compressor Work from 2' to 2

$$w_{c2'-2} = 1004.5 \times (730.2 - 588.8) = 142,054 \text{ W}$$

Combustor Exit or Turbine Inlet (3)

$$p_{03} = p_{02} - 0.1 p_{t2} = 0.99 \times 20 \times 101.325 = 2006.235 \text{ kPa}$$

Heat Input in Combustor per kg of Air from 2 to 3

$$q_{\text{in}} = c_{pt} T_{03} - c_{pc} T_{02'} = 1156.7 \times 2000 - 1004.5 \times 730.2 = 1,579,913 \text{ W}$$

Turbine Exit (4)

$$\frac{n_t - 1}{n_t} = \eta_{pt}\left(\frac{\gamma_t - 1}{\gamma_t}\right) = 0.92 \times \frac{1.33 - 1}{1.33} = 0.248 \times 0.92 = 0.228$$

$$\frac{T_{03}}{T_{04}} = \left(\frac{p_{03}}{p_{04}}\right)^{\frac{n_t - 1}{n_t}} = \left(\frac{2006.235}{101.325}\right)^{0.228} = 1.977$$

$$T_{04} = \frac{T_{03}}{1.977} = \frac{2000}{1.977} = 1011.7 \text{ K}$$

Specific Turbine Work from 3 to 4

$$w_{t3-4} = 1156.7(2000 - 1011.7) = 1,143,197 \text{ W}$$

Engine Flow Rate $\left(\dot{m}_{\text{engine}}\right)$

With $\dot{m}_{\text{cool}} = 0.20\,\dot{m}_{\text{engine}}$ and $\dot{m}_t = 0.80\,\dot{m}_{\text{engine}}$, we can write

$$\dot{m}_t w_{t3-4} - \left(\dot{m}_{\text{engine}} w_{c1-2'} + 0.80 \dot{m}_{\text{engine}} w_{c2'-2}\right) = 200 \times 10^6$$

$$\dot{m}_{\text{engine}}\left(0.80 \times w_{t3-4} - w_{c1-2'} - 0.80 w_{c2'-2}\right) = 200 \times 10^6$$

giving

$$\dot{m}_{\text{engine}} = \frac{200 \times 10^6}{0.80 w_{t3-4} - w_{c1-2'} - 0.80 w_{c2'-2}} = \frac{200 \times 10^6}{498,783.14} = 401 \text{ kg/s}$$

Cycle Thermodynamic Efficiency (η_{th})

TABLE 12.1

Summary of Results for All Cases (Problem 12.8)

Quantity	(a) Case	(b) Case	(c) Case	(d) Case
η_{pc}, η_{pt}	92%	93%	92%	92%
$p_{02'}/p_{01}$	10	10	9	10
$\dot{m}_{cool}$	$0.2\dot{m}_{engine}$	$0.2\dot{m}_{engine}$	$0.2\dot{m}_{engine}$	$0.18\dot{m}_{engine}$
$\dot{m}_{engine}$ (kg/s)	401	390	398	386
η_{th}	39.46%	40.36%	39.76%	40.05%

$$\eta_{th} = \frac{200 \times 10^6}{0.80 \times 401 \times q_{in}} = \frac{200 \times 10^6}{0.80 \times 401 \times 1{,}579{,}913} = 39.46\%$$

The results for all four cases are summarized in Table 12.1.

PROBLEM 12.9: VELOCITY DIAGRAMS FROM GIVEN FLOW COEFFICIENT, LOADING COEFFICIENT, AND STAGE REACTION

In (1) Case 1: $\varphi = 0.73$, $\psi = 2.0$, and $R = 0$; (2) Case 2: $\varphi = 0.577$, $\psi = 1.0$, and $R = 0.5$; and (3) Case 3: $\varphi = 0.546$, $\psi = 2.0$, and $R = 0.5$, compute absolute and relative flow angles and absolute and relative flow velocities at both inlet and outlet of an axial-flow turbine blade at its meanline. Also, using the quick graphical method presented in Appendix B, draw velocity diagrams in each case.

SOLUTION FOR PROBLEM 12.9

(a) CASE 1: $\varphi = 0.73$, $\psi = 2.0$, AND $R = 0$

Using appropriate equations from Appendix B, the computed absolute and relative flow angles and absolute and relative flow velocities at blade inlet and outlet are summarized in Table 12.2. Figure 12.5 shows the dimensionless velocity diagrams.

(b) CASE 2: $\varphi = 0.577$, $\psi = 1.0$, AND $R = 0.5$

Using appropriate equations from Appendix B, the computed absolute and relative flow angles and absolute and relative flow velocities at blade inlet and outlet are summarized in Table 12.2. Figure 12.6 shows the dimensionless velocity diagrams.

TABLE 12.2

Summary of Computed Values for All Three Cases (Problem 12.9)

Quantity	(a) Case 1	(b) Case 2	(c) Case 3
φ	0.73	0.577	0.546
ψ	2.0	1.0	2.0
R	0	0.5	0.5
α_1	70°	60°	70°
β_1	53.9°	0°	42.5°
V_1/U	2.129	1.155	1.6
W_1/U	1.238	0.577	0.74
α_2	0°	0°	−42.5°
β_2	−53.9°	−60°	−70°
V_2/U	0.73	0.577	0.74
W_2/U	1.238	1.155	1.6

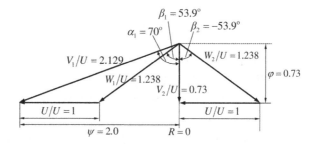

FIGURE 12.5   Dimensionless velocity diagrams for Case 1 (Problem 12.9).

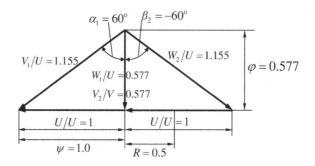

FIGURE 12.6   Dimensionless velocity diagrams for Case 2 (Problem 12.9).

(c) CASE 3: $\varphi = 0.546$, $\psi = 2.0$, AND $R = 0.5$

Using appropriate equations from Appendix B, the computed absolute and relative flow angles and absolute and relative flow velocities at blade inlet and outlet are summarized in Table 12.2. Figure 12.7 shows the dimensionless velocity diagrams.

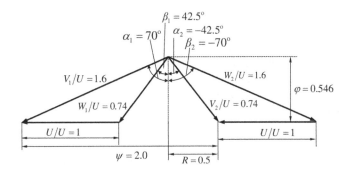

FIGURE 12.7 Dimensionless velocity diagrams for Case 3 (Problem 12.9).

NOMENCLATURE

c_p	Specific heat of gas at constant pressure
D_h	Hub diameter
D_m	Diameter at mean position between hub and tip of blade
D_{tip}	Tip diameter
E	Specific energy transfer $= U\Delta V_\theta = U\Delta W_\theta$
$\dot{E}$	Rate of energy transfer $= P$
h	Blade height $= r_{tip} - r_h$
h_{01}	Total enthalpy of gas at stator inlet
h_{02}	Total enthalpy of gas at rotor inlet
h_{03}	Total enthalpy of gas at rotor exit
$h_{2'}$	Static enthalpy after isentropic expansion to p_2
$h_{3'}$	Static enthalpy after isentropic expansion to p_3
$\dot{m}$	Mass flow rate of gas through annulus
n_{st}	Number of stages in a multistage turbine
N	Rotor speed
p_2	Static pressure at stator exit
p_3	Static pressure at rotor exit
p_{01}	Total pressure at stator inlet
p_{02}	Total pressure at stator exit or at rotor inlet
p_{03}	Total pressure at rotor exit
p_{02R}	Relative total pressure at rotor inlet
p_{03R}	Relative total pressure at rotor exit
P	Power produced by turbine stage
r_h	Hub radius or radius at root of blade
r_m	Mean blade radius $= r_m = D_m/2$
r_{tip}	Radius at blade tip
R	Degree of reaction
T_2	Static temperature at rotor inlet
T_3	Static temperature at rotor exit
$T_{2'}$	Static temperature at state $2'$
$T_{3'}$	Static temperature at state $3'$

T_{01}	Total temperature at stator inlet
T_{02}	Total temperature at stator outlet or rotor inlet
T_{03}	Total temperature at rotor outlet
T_{in}	Turbine inlet temperature for multistage turbine
U	Blade speed at mean radius r_m of blade
U_h	Blade speed at hub radius r_h of blade
U_{tip}	Blade speed at tip radius r_{tip} of blade
V	Absolute velocity of fluid
V_1	Absolute velocity of fluid entering stator
V_2	Absolute velocity of fluid leaving stator or entering rotor
V_3	Absolute velocity of gas leaving rotor
$V_{3'}$	Absolute velocity of gas after expansion to state $3'$
V_a	Component of V in axial direction
V_r	Component of V in radial direction
V_θ	Component of V in tangential direction
$V_{\theta 2}$	Tangential component of V_2
$V_{\theta 3}$	Tangential component of V_3
ΔV_θ	$V_{\theta 2} - V_{\theta 3}$
w_t	Specific turbine work
W	Velocity of fluid relative to blade
W_2	Relative velocity entering rotor
W_3	Relative velocity leaving rotor
W_θ	Tangential component of W
ΔW_θ	$W_{\theta 2} - W_{\theta 3} = \Delta V_\theta$
$W_{\theta 2}$	Tangential component of W_2
$W_{\theta 3}$	Tangential component of W_3
α_1	Angle between V_1 and V_a
α_2	Angle between V_2 and V_a
α_3	Angle between V_3 and V_a
α_m	Mean absolute gas angle
β_2	Angle between W_2 and V_a
β_3	Angle between W_3 and V_a
β_m	Mean relative gas angle
φ	Flow coefficient $= V_a/U$
γ	Ratio of specific heats
η_{st}	Stage efficiency $= \eta_{tt}$ or η_{ts}
η_t	Overall efficiency for a multistage turbine
η_{tt}	Total-to-total efficiency
η_{ts}	Total-to-static efficiency
ρ	Fluid density
ψ	Loading coefficient

REFERENCES

Sultanian, B.K. 2019. *Logan's Turbomachinery: Flowpath Design and Performance Fundamentals*, 3rd edition. Boca Raton, FL: Taylor & Francis.

BIBLIOGRAPHY

Aungier, R.H. 2006. *Turbine Aerodynamics: Axial-Flow and Radial-Flow Turbine Design and Analysis*. New York: ASME Press.

Horlock, J.H. 1973. *Axial Flow Turbines*. Huntington: Krieger.

Kacker, S.C., and U. Okapuu. 1982. A mean line prediction method for axial flow turbine efficiency. *ASME Journal of Engineering for Power*. 104: 111–119.

Saravanamutto, H.I.H., G.F.C. Rogers, H. Cohen, P.V. Straznicky, A.C. Nix. 20017. *Gas Turbine Theory*, 7th edition. Harlow: Pearson.

Shepherd, D. G. 1956. *Principles of Turbomachinery*. New York: Macmillan.

Sultanian, B.K. 2015. *Fluid Mechanics: An Intermediate Approach*. Boca Raton, FL: Taylor & Francis.

Vavra, M.H. 1960. *Aerothermodynamics and Flow in Turbomachines*. New York: John Wiley & Sons.

Vincent, E.T. 1950. *The Theory and Design of Gas Turbines and Jet Engines*. New York: McGraw-Hill.

Wilson, D.G. 1987. New guidelines for the preliminary design and performance prediction of axial-flow turbines. *Proceedings of the Institution of Mechanical Engineers*. 201: 279–290.

13 Diffusers

REVIEW OF KEY CONCEPTS

We concisely present here some key concepts diffuser flow and performance. More details on each topic are given, for example, in Sultanian (2015, 2019). Using results from computational fluid dynamics (CFD), a designer can generate an entropy map, discussed in Sultanian (2015), to delineate diffuser flow zones of excess entropy production for design improvement.

Diffusion is the conversion of dynamic pressure into the stream static pressure. For both incompressible (liquid) flow and subsonic gaseous flow, diffusion is typically achieved through increase in the downstream flow area. This is consistent with the main tenet of the Bernoulli equation, namely "low velocity, high pressure." Sultanian (2015) discusses interesting features of a flow with and without swirl in a sudden pipe expansion. In gas turbines used for power generation, exhaust diffusers play an important role in turbine power output by increasing the pressure ratio across the last stage turbine by making the turbine exit static pressure subambient.

Ideal diffusers are characterized by uniform axial velocities with no swirl at both inlet and exit and with no loss in total pressure across the diffuser. Actual diffusers, however, often have nonuniform profiles of all three velocity components, pressure, and temperature at both inlet and outlet. We discuss here how to handle these nonuniformities in calculating the performance of a real diffuser.

ISENTROPIC EFFICIENCY (η_D)

Figure 13.1 shows both isentropic and nonisentropic variations of flow properties in a diffuser from its inlet to outlet. For an isentropic diffuser operating along 1–2', we have no loss in total pressure. For the nonisentropic process along 1–2, we have $p_{02} < p_{01}$. However, the total enthalpy, or the total temperature for a constant c_p, in an adiabatic diffuser remains constant ($h_{02} = h_{01}$ or $T_{02} = T_{01}$) both along 1–2 and 1–2'. This figure further shows that, compared to a nonisentropic diffuser, the isentropic

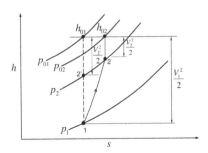

FIGURE 13.1 h-s diagram for a diffuser.

diffuser will have lower flow area and higher dynamic pressure at its exit when the pressure recovery $p_2 - p_1$ is equal in both cases. For the same exit dynamic pressure, the isentropic diffuser will yield a pressure recovery higher than that for a nonisentropic diffuser.

For an equal static pressure recovery, we define the diffuser isentropic efficiency as the ratio of the isentropic (ideal) enthalpy change to the actual enthalpy change. Thus, we write

$$\eta_D = \frac{h_{2'} - h_1}{h_2 - h_1} \tag{13.1}$$

As the total enthalpy remains constant in an adiabatic diffuser—i.e., $h_{01} = h_1 + V_1^2/2 = h_{2'} + V_{2'}^2/2 = h_{02} = h_2 + V_2^2/2$, we can write this equation as

$$\eta_D = \frac{V_1^2 - V_{2'}^2}{V_1^2 - V_2^2} \tag{13.2}$$

For an isentropic process, we have $dh = dp/\rho$, which yields

$$h_{2'} - h_1 = (p_2 - p_1)/\rho \tag{13.3}$$

for an incompressible or low Mach number diffuser flow. Substituting this in Equation 13.1 and using $h_{01} = h_1 + V_1^2/2 = h_{02} = h_2 + V_2^2/2$, we obtain

$$\eta_D = \frac{p_2 - p_1}{\rho V_1^2/2 - \rho V_2^2/2} \tag{13.4}$$

which states that η_D equals the ratio of static pressure change and dynamic pressure change in the diffuser. We can also express η_D given by this equation in terms of only pressures as

$$\eta_D = \frac{p_2 - p_1}{(p_2 - p_1) + (p_{01} - p_{02})} \tag{13.5}$$

The numerator of this equation represents the static pressure recovery, while its denominator represents the sum of the static pressure recovery and the loss in total pressure.

PRESSURE RISE COEFFICIENT (C_p)

We define the pressure rise coefficient of a diffuser as the ratio of static pressure recovery (gain) and inlet dynamic pressure

$$C_p = \frac{p_2 - p_1}{p_{01} - p_1} \tag{13.6}$$

For an incompressible or low Mach number flow through the diffuser with $p_{01} - p_1 = \rho V_1^2/2$, the extended Bernoulli equation, or the mechanical energy equation, yields

$$p_1 + \frac{\rho V_1^2}{2} = p_2 + \frac{\rho V_2^2}{2} + \Delta p_{0\text{ loss}} \tag{13.7}$$

where $\Delta p_{0\text{ loss}}$ is the loss in total pressure from diffuser inlet to outlet. From this equation we obtain

$$p_2 - p_1 = \frac{\rho V_1^2}{2} \frac{\rho V_2^2}{2} - \Delta p_{0\text{loss}} \tag{13.8}$$

which when substituted into Equation 13.6 yields

$$C_p = \frac{p_2 - p_1}{p_{01} - p_1} = \frac{\dfrac{\rho V_1^2}{2} - \dfrac{\rho V_2^2}{2} - \Delta p_{0\text{loss}}}{\dfrac{\rho V_1^2}{2}} = 1 - \left(\frac{V_2}{V_1}\right)^2 - \frac{2\Delta p_{0\text{loss}}}{\rho V_1^2} \tag{13.9}$$

The continuity equation between sections 1 and 2 yields

$$\rho A_1 V_1 = \rho A_2 V_2 \tag{13.10}$$

giving the diffuser area ratio $A_r = A_2/A_1 = V_1/V_2$.

For an ideal diffuser with isentropic flow, which has no loss in total pressure $(\Delta p_{0\text{loss}} = 0)$, the pressure rise coefficient given by Equation 13.9 with the substitution of the velocity ratio in terms of diffuser area ratio becomes

$$C_{pi} = 1 - \left(\frac{V_2}{V_1}\right)^2 = 1 - \frac{1}{A_r^2} \tag{13.11}$$

Thus, we can alternatively express Equation 13.9 as

$$C_p = C_{pi} - \frac{2\Delta p_{0\text{loss}}}{\rho V_1^2} \tag{13.12}$$

To find the relation among η_D, C_p, and C_{pi}, let us divide the numerator and denominator of Equation 13.5 by the inlet dynamic pressure $\rho V_1^2/2$, giving

$$\eta_D = \frac{(p_2 - p_1)/(\rho V_1^2/2)}{(p_2 - p_1)/(\rho V_1^2/2) + (p_{01} - p_{02})/(\rho V_1^2/2)} = \frac{C_p}{C_p + \Delta p_{0\text{loss}}/(\rho V_1^2/2)} \tag{13.13}$$

which with the substitution of Equation 13.12 reduces to

$$\eta_D = \frac{C_p}{C_{pi}} \tag{13.14}$$

PRESSURE RECOVERY IN A DUMP DIFFUSER

Let us consider the pressure recovery in a dump diffuser, shown in Figure 13.2, typically used in space-limited design applications. As shown in this figure, an uniform incompressible flow in the smaller pipe of diameter D_1 and area A_1 with a uniform velocity V_1 enters into the larger pipe of diameter D_2 and area A_2 at section 1, exiting it at section 2 with a uniform velocity V_2.

Continuity Equation

From the continuity equations at sections 1 and 2, we write

$$\rho A_1 V_1 = \rho A_2 V_2$$

giving

$$\frac{A_1}{A_2} = \frac{V_2}{V_1} = \beta^2 \tag{13.15}$$

where

$$\beta = \frac{D_1}{D_2} \tag{13.16}$$

Axial Momentum Equation: Pressure-Rise Coefficient

Assuming a uniform static pressure at section 1 in the incoming flow from the smaller pipe and over the annular area formed by inner diameter D_1 and outer diameter D_2, the axial force and linear momentum balance over the control volume yield

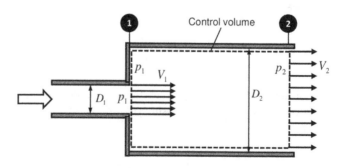

FIGURE 13.2   Sudden-expansion pipe flow: a dump diffuser.

$$p_1 A_1 + p_1 (A_2 - A_1) - p_2 A_2 = \dot{m}(V_2 - V_1)$$

$$A_2 (p_2 - p_1) = \rho V_2 A_2 (V_1 - V_2) \tag{13.17}$$

$$p_2 - p_1 = \rho V_2 (V_1 - V_2)$$

which we can express as

$$C_p = \frac{p_2 - p_1}{\frac{1}{2} \rho V_1^2} = 2\beta^2 \left(1 - \beta^2\right) \tag{13.18}$$

where C_p is the pressure-rise coefficient for the dump diffuser.

Loss in Total Pressure

We compute the total pressure at section 1 as

$$p_{01} = p_1 + \frac{\rho V_1^2}{2} \tag{13.19}$$

and at section 2 as

$$p_{02} = p_2 + \frac{\rho V_2^2}{2} \tag{13.20}$$

From Equations 13.19 and 13.20, we obtain

$$p_{01} - p_{02} = (p_1 - p_2) + \frac{\rho V_1^2}{2} - \frac{\rho V_2^2}{2}$$

Substitution for $p_1 - p_2$ in this equation from Equation 13.17 yields

$$p_{01} - p_{02} = \rho V_2^2 - \rho V_1 V_2 + \frac{\rho V_1^2}{2} - \frac{\rho V_2^2}{2} \tag{13.21}$$

$$p_{01} - p_{02} = \Delta p_{0\,loss} = \frac{\rho (V_1 - V_2)^2}{2}$$

which shows that the theoretical loss in total pressure in a dump diffuser equals the dynamic pressure of the difference in uniform velocities in the two pipes.

NONUNIFORM PROPERTIES AT DIFFUSER INLET AND OUTLET

For a diffuser with uniform axial velocity, with no swirl and radial velocity components, the pressure rise coefficient C_p defined by Equation 13.6 works well to characterize the diffuser performance. For a real diffuser, however, flow properties are generally not uniform over its inlet and outlet sections. This poses a new challenge to compute C_p by Equation 13.6 for both incompressible and compressible flows. The mass-averaging of total pressure to compute C_p in some of the industry practices is not physics-based. In addition, the dynamic pressure (difference between the total pressure and static pressure) used in the definition of C_p includes the contribution of all velocities (both axial and nonaxial) whereas the exhaust diffuser only diffuses the axial velocity. In the absence of detailed measurements needed to compute profiles of velocities, static pressure, total pressure, and total temperature at diffuser inlet and outlet, an accurate evaluation of C_p from engine test data is not possible. For a physics-based computation of C_p for a diffuser with nonuniform flow properties at its inlet and outlet, we propose the following methods for incompressible and compressible flows. These methods are ideally suited for the postprocessing of CFD results to compute C_p.

Incompressible Flow

Without any loss of generality, we assume that the diffuser inlet and outlet sections are normal to the axial direction.

Step 1. Compute inlet (section 1) mass flow rate, which is equal to that at the outlet (section 2)

$$\dot{m} = \iint_{A_1} \rho V_{x1} dA_1 \tag{13.22}$$

Step 2. Compute the section-average value of static pressure by its area-averaging, first at inlet as

$$\bar{p}_1 = \frac{\iint_{A_1} p_1 \, dA_1}{A_1} \tag{13.23}$$

and then at outlet as

$$\bar{p}_2 = \frac{\iint_{A_2} p_2 \, dA_2}{A_2} \tag{13.24}$$

Step 3. Compute kinetic energy flow rate at inlet

$$\dot{E}_{ke} = \iint_{A_1} \left(V_1^2 / 2 \right) \rho V_{x1} dA_1 \tag{13.25}$$

Step 4. Compute the pressure rise coefficient

$$C_p = \frac{\dot{m} \left(\bar{p}_2 - \bar{p}_1 \right)}{\rho \dot{E}_{ke}} \tag{13.26}$$

Compressible Flow

Again, without any loss of generality, we assume that the diffuser inlet and outlet sections are normal to the axial direction. We also before assume that c_p for the gas remains constant.

Step 1. Use Equation 13.22 to compute inlet (section 1) mass flow rate, which is equal to that at the outlet (section 2)

Step 2. Use Equation 13.23 to compute the section-average value of static pressure at inlet and Equation 13.24 at outlet.

Step 3. Compute section-average static temperature at inlet

$$\overline{T}_1 = \frac{\iint_{A_1} T_1 \rho V_{x1} dA_1}{\dot{m}} \tag{13.27}$$

Step 4. First, use Equation 13.25 to compute kinetic energy flow rate at inlet and then compute the section-average specific kinetic energy as

$$\frac{\overline{V_1^2}}{2} = \frac{\dot{E}_{ke}}{\dot{m}} \tag{13.28}$$

Step 5. Compute section-average total temperature at inlet

$$\overline{T}_{01} = \overline{T}_1 + \frac{\overline{V_1^2}}{2} \tag{13.29}$$

Step 6. Compute section-average total pressure at inlet

$$\overline{p}_{01} = \overline{p}_1 \left(\overline{T}_{01} / \overline{T}_1 \right)^{\gamma/\gamma - 1} \tag{13.30}$$

Step 7. Compute the pressure rise coefficient

$$C_p = \frac{\overline{p}_2 - \overline{p}_1}{\overline{p}_{01} - \overline{p}_1} \tag{13.31}$$

The method presented in the foregoing to compute C_p for a compressible flow in a diffuser avoids the controversial design practice of mass-averaging the total pressure at its inlet.

Six Simple Design Rules

We present here six simple rules for an efficient aerodynamic design of a gas turbine exhaust diffuser whose design features are shown in Figure 13.3.

Rule 1. Strive for uniform axial velocity and zero swirl at turbine exit (diffuser inlet).

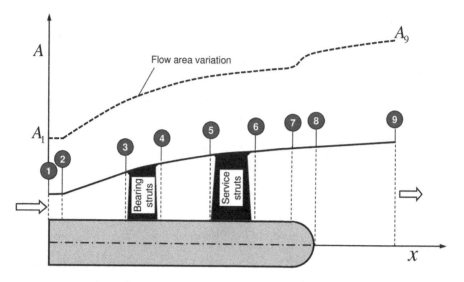

FIGURE 13.3 Design features of an axial-flow gas turbine exhaust diffuser.

The dynamic pressure associated with the swirl and radial velocities is not recovered into static pressure, and it simply adds to the loss in total pressure in the diffuser. Note that the angular momentum associated with the swirl velocity entering the diffuser decays downstream due to the frictional torque of stationary diffuser walls and other intervening components.

> *Rule 2.* As an ideal diffuser operates at constant total pressure, a diffuser design must target a minimum loss in total pressure from its inlet to outlet.
>
>   We can also state this rule in terms of minimizing the production of entropy or maximizing the gain in axial stream thrust in the diffuser flow.
>
> *Rule 3.* Make axial velocity profiles as uniform as possible through a straight (cylindrical) section before starting a conical diffusion.
>
>   When we diffuse a nonuniform axial velocity profile in a constant-area duct, the static pressure increases, and the total pressure decreases. We need the straight diffuser section 1–2, shown in Figure 13.3, to ensure that the axial velocity profile becomes nearly uniform before initiating its further diffusion through area increase. This prevents boundary layer separation at the bounding walls.
>
> *Rule 4.* Design smooth variation in diffuser flow areas without sudden changes in them.
>
>   This is perhaps the most obvious but often ignored rule in exhaust diffuser design. With reference to Figure 13.3, we must increase the outer wall diameters over sections 3–4 and 5–6 to account for the blockage from the bearing and service struts, respectively, so as to ensure a smooth variation of diffuser flow areas. Sudden expansion and contraction of flows are lossy!
>
> *Rule 5.* Move elements (e.g., struts, separations, and steps,) that generate loss in total pressure to diffuser downstream regions of lower dynamic pressure.

We express the total pressure loss by the equation

$$\Delta p_{0\text{loss}} = K \frac{\rho V^2}{2} \qquad (13.32)$$

where K is the empirically determined loss coefficient. According to this equation, to reduce the loss in total pressure, lossy elements of a diffuser should be placed in the region of low dynamic pressure and as much downstream from the inlet as feasible from other (structural) design considerations.

We can also interpret this rule in terms of the drag force of various intervening elements in the diffuser flowpath. We express the drag force by the equation

$$F_D = C_D \frac{\rho V^2}{2} \qquad (13.33)$$

where C_D is the drag coefficient. If the drag producing elements are in the regions of low dynamic pressure, their impact on the axial stream thrust will be lower than when they are in the region of high dynamic pressure.

Rule 6. Minimize drag coefficient of each strut and other obstructions in the diffuser cross-flow.

All the drag-producing elements in the diffuser should be made aerodynamic to produce minimum drag force. They should not be viewed simply as elements blocking the diffuser flow area.

PROBLEM 13.1:   OUTLET-TO-INLET AREA RATIO OF A GAS TURBINE ANNULAR EXHAUST DIFFUSER

Total temperature and total pressure at the inlet to a three-stage axial flow turbine are 1600 K and 10 bar, respectively. The flow at the inlet to the annular exhaust diffuser has a Mach number of 0.6, a static pressure of 0.85 bar, and a total temperature of 673 K. The swirl velocity (tangential velocity) and the radial velocity components at the diffuser inlet are, respectively, 20% and 5% of the total velocity. The flow exits the diffuser fully axially. From inlet to exit, the total temperature in the diffuser decreases by 15 K due to heat transfer, and the total pressure decreases by 5500 Pa due to wall friction and secondary flows. The diffuser design calls for a static pressure of 1.013 bar at the diffuser exit to allow the exhaust gases to discharge into the ambient air. Calculate the annular diffuser exit-to-inlet flow area ratio. Assume $R = 287 \, \text{J}/(\text{kg K})$ and $\gamma = 1.4$ for the exhaust gases.

SOLUTION FOR PROBLEM 13.1

In the solution of this problem, the given total temperature and pressure conditions at the inlet to turbine are not relevant. We need to work only with the data given at the inlet and exit sections of the annular diffuser.

ANNULAR DIFFUSER INLET (SECTION 1)

Flow Velocity Coefficient

$$V_1^2 = V_{a1}^2 + V_{r1}^2 + V_{\theta 1}^2$$

$$C_{V1} = \frac{V_{a1}}{V_1} = \sqrt{1 - (0.05)^2 - (0.2)^2} = 0.9785$$

Static-Pressure Mass Flow Function

$$\hat{F}_{f1} = M_1 \sqrt{\gamma \left(1 + \frac{\gamma - 1}{2} M_1^2\right)} = 0.6 \sqrt{1.4 \left(1 + \frac{1.4 - 1}{2}(0.6)^2\right)} = 0.735$$

Mass Flow Rate per Unit Area

$$\frac{\dot{m}_1}{A_1} = \frac{C_{V1} \hat{F}_{f1} p_1}{\sqrt{RT_{01}}} = \frac{0.9785 \times 0.735 \times 0.85 \times 10^5}{\sqrt{287 \times 673}} = 139.108 \, \text{kg}/(\text{s m}^2)$$

ANNULAR DIFFUSER EXIT (SECTION 2)

Total Pressure

$$p_{01} = p_1 \left(1 + \frac{\gamma - 1}{2} M_1^2\right)^{\gamma/\gamma - 1} = 0.85 \times 10^5 \times \left(1 + \frac{1.4 - 1}{2}(0.6)^2\right)^{1.4/1.4 - 1} = 108,418 \, \text{Pa}$$

$$p_{02} = 108,418 - 5500 = 102,918 \, \text{Pa}$$

Total Temperature

$$T_{02} = 673 - 15 = 658 \, \text{K}$$

Mach Number

$$M_2 = \sqrt{\frac{2}{\gamma - 1} \left\{ \left(\frac{p_{02}}{p_2}\right)^{\gamma - 1/\gamma} - 1 \right\}} = \sqrt{\frac{2}{1.4 - 1} \left\{ \left(\frac{102,918}{1.013 \times 10^5}\right)^{1.4 - 1/1.4} - 1 \right\}} = 0.1506$$

Static-Pressure Mass Flow Function

$$\hat{F}_{f2} = M_2 \sqrt{\gamma \left(1 + \frac{\gamma - 1}{2} M_2^2\right)} = 0.1506 \sqrt{1.4 \left(1 + \frac{1.4 - 1}{2}(0.1506)^2\right)} = 0.1786$$

Mass Flow Rate Per Unit Area

$$\frac{\dot{m}_2}{A_2} = \frac{C_{V2}\hat{F}_{f2}p_2}{\sqrt{RT_{02}}} = \frac{1.0 \times 0.1786 \times 1.013 \times 10^5}{\sqrt{287 \times 658}} = 41.637\,\text{kg}/(\text{s m}^2)$$

As the mass flow rates at annular diffuser inlet and exit are equal, we obtain

$$\frac{A_2}{A_1} = \frac{139.108}{41.637} = 3.341$$

Therefore, the computed exit-to-inlet flow area ratio of the annular exhaust diffuser is 3.341.

PROBLEM 13.2:   EXHAUST DIFFUSER OF A LAND-BASED GAS TURBINE FOR POWER GENERATION

In a land-based gas turbine used for power generation, the turbine exhaust enters an annular diffuser, shown in Figure 13.4, at the total velocity Mach number of 0.60 and the total temperature of 723 K. The swirl velocity (tangential velocity) at the diffuser inlet equals 15% of the total velocity. The flow exits the diffuser fully axially at a Mach number of 0.15 with no change in total temperature. The design calls for the static pressure at the diffuser exit to be 1.018 bar to allow the exhaust gases to discharge into the ambient air via a downstream duct system (not shown in the figure). The exit-to-inlet flow area ratio for the annular diffuser is 3.45. Find the pressure rise coefficient $\left(\tilde{C}_p\right)$ for the annular diffuser. If the last stage turbine is redesigned for zero swirl velocity at its exit, how will the pressure rise coefficient $\left(C_p\right)$ of the annular diffuser change? Assume exhaust gases with $\gamma = 1.4$ and $R = 287\,\text{J}/(\text{kg K})$.

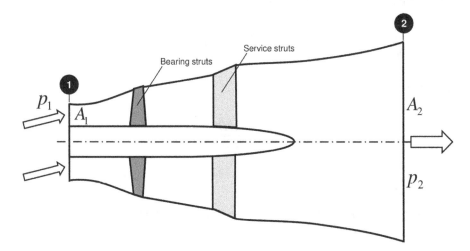

FIGURE 13.4   Exhaust diffuser of a land-based gas turbine for power generation (Problem 13.2).

SOLUTION FOR PROBLEM 13.2

ANNULAR DIFFUSER INLET (SECTION 1)

Flow Velocity Coefficient

$$C_{V1} = \frac{V_{a1}}{V_1} = \sqrt{1-(0.15)^2} = 0.989$$

Static-Pressure Mass Flow Function

$$\hat{F}_{f1} = M_1 \sqrt{\gamma\left(1 + \frac{\gamma-1}{2}M_1^2\right)} = 0.6 \times \sqrt{1.4 \times \left(1 + \frac{1.4-1}{2} \times 0.6 \times 0.6\right)} = 0.7350$$

Mass Flow Rate

$$\dot{m}_1 = \frac{C_{V1}\hat{F}_{f1} A_1 p_1}{\sqrt{RT_{01}}}$$

Total-to-Static Pressure Ratio

$$\frac{p_{01}}{p_1} = \left(1 + \frac{\gamma-1}{2}M_1^2\right)^{\frac{\gamma}{\gamma-1}} = \left(1 + \frac{1.4-1}{2} \times 0.6 \times 0.6\right)^{\frac{1.4}{1.4-1}} = 1.276$$

ANNULAR DIFFUSER EXIT (SECTION 2)

Static-Pressure Mass Flow Function

$$\hat{F}_{f2} = M_2 \sqrt{\gamma\left(1 + \frac{\gamma-1}{2}M_2^2\right)} = 0.15 \times \sqrt{1.4 \times \left(1 + \frac{1.4-1}{2} \times 0.15 \times 0.15\right)} = 0.1779$$

Mass Flow Rate

$$\dot{m}_2 = \frac{\hat{F}_{f2}A_2 p_2}{\sqrt{RT_{02}}}$$

CALCULATION OF $\tilde{C}_p$ WITH INLET SWIRL

Equating mass flow rates at diffuser inlet and exit and noting that $T_{01} = T_{02}$, we can write

$$C_V \hat{F}_{f1}A_1 p_1 = \hat{F}_{f2}A_2 p_2$$

giving

$$\frac{p_2}{p_1} = \frac{C_V \hat{F}_{f1} A_1}{\hat{F}_{f2} A_2} = \frac{0.989 \times 0.7350}{0.1779 \times 3.45} = 1.184$$

Using the values computed in the foregoing, we obtain $\tilde{C}_p$ as

$$\tilde{C}_p = \frac{p_2 - p_1}{p_{01} - p_1}$$

$$\tilde{C}_p = \frac{\frac{p_2}{p_1} - 1}{\frac{p_{01}}{p_1} - 1} = \frac{1.184 - 1}{1.276 - 1} = \frac{0.184}{0.276} = 0.667$$

CALCULATION OF C_p WITH ZERO INLET SWIRL

In this case, $C_{V1} = 1$. Other quantities are calculated as follows:

$$\frac{p_2}{p_1} = \frac{\hat{F}_{f1} A_1}{\hat{F}_{f2} A_2} = \frac{0.7350}{0.1779 \times 3.45} = 1.198$$

$$\frac{p_{01}}{p_1} = \left(1 + \frac{\gamma - 1}{2} M_1^2\right)^{\frac{\gamma}{\gamma - 1}} = \left(1 + \frac{1.4 - 1}{2} \times 0.6 \times 0.6\right)^{\frac{1.4}{1.4-1}} = 1.276$$

$$C_p = \frac{p_2 - p_1}{p_{01} - p_1}$$

$$C_p = \frac{\frac{p_2}{p_1} - 1}{\frac{p_{01}}{p_1} - 1} = \frac{1.198 - 1}{1.276 - 1} = \frac{0.198}{0.276} = 0.717$$

Therefore, the pressure rise coefficient with zero inlet swirl increases to 0.717 from its value 0.667 with the inlet swirl velocity, which is 15% of the total inlet velocity.

PROBLEM 13.3:   SUPERSONIC AIR FLOW DIFFUSER DESIGNS

A designer is considering two alternative designs of a supersonic air flow diffuser shown in Figure 13.5. The first design, shown in Figure 13.5a, uses a convergent-divergent diffuser where the incoming flow passes through a choked throat for transition from the supersonic to subsonic flow at the exit. In the second design, shown in Figure 13.5b, a normal shock ahead of the diffuser converts the supersonic flow into a subsonic flow, being further diffused in a divergent duct. For both designs, the incoming flow Mach number is 1.1 at a static pressure of 0.476 bar,

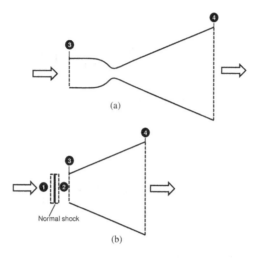

FIGURE 13.5 Supersonic diffuser designs: (a) a convergent-divergent diffuser and (b) a divergent diffuser with a normal shock at its inlet (Problem 13.3).

and the diffuser discharges at the ambient static pressure of 1.0 bar. Perform the necessary calculations for the two designs and evaluate their advantages and disadvantages. Except for the loss in total pressure across the normal shock in the second design, neglect all other losses in both designs. Assume $\gamma = 1.4$ and $R = 287 \text{ J}/(\text{kg K})$ for air.

SOLUTION FOR PROBLEM 13.3

As shown in Figure 13.5, in both designs, section 3 represents the diffuser inlet and section 4 represents the diffuser exit. We assume isentropic air flow is each diffuser.

(a) CONVERGENT-DIVERGENT DIFFUSER

Total Pressure at Diffuser Inlet

$$\frac{p_{03a}}{p_{3a}} = \left(1 + \frac{\gamma-1}{2}M_{3a}^2\right)^{\frac{\gamma}{\gamma-1}} = \left(1 + \frac{1.4-1}{2} \times (1.1)^2\right)^{\frac{1.4}{1.4-1}} = 2.135$$

$$p_{03a} = p_{04a} = 2.135 \times 0.476 = 1.016 \text{ bar}$$

Critical Area Ratio at Diffuser Inlet $\left(A_{3a}/A_a^*\right)$

$$\frac{A_{3a}}{A_a^*} = \frac{1}{M_{3a}}\sqrt{\left(\frac{2+(\gamma-1)M_{3a}^2}{\gamma+1}\right)^{\frac{\gamma+1}{\gamma-1}}} = \frac{1}{1.1}\sqrt{\left(\frac{2+(1.4-1)\times(1.1)^2}{1.4+1}\right)^{\frac{1.4+1}{1.4-1}}} = 1.008$$

Mach Number at Diffuser Exit

We obtain the total-to-static pressure ratio at section 4 as

$$\frac{p_{04a}}{p_{4a}} = \frac{1.016}{1} = 1.016$$

giving

$$M_{4a} = \sqrt{\frac{2}{\gamma-1}\left\{\left(\frac{p_{04a}}{p_{4a}}\right)^{\frac{\gamma-1}{\gamma}} - 1\right\}} = \sqrt{\frac{2}{1.4-1}\left\{(1.016)^{\frac{1.4-1}{1.4}} - 1\right\}} = 0.152$$

Critical Area Ratio at Diffuser Exit $\left(A_{4a}/A_a^*\right)$

$$\frac{A_{4a}}{A_a^*} = \frac{1}{M_{4a}}\sqrt{\left(\frac{2+(\gamma-1)M_{4a}^2}{\gamma+1}\right)^{\frac{\gamma+1}{\gamma-1}}} = \frac{1}{0.152}\sqrt{\left(\frac{2+(1.4-1)\times(0.152)^2}{1.4+1}\right)^{\frac{1.4+1}{1.4-1}}} = 3.854$$

Overall Diffuser Area Ratio

$$\frac{A_{4a}}{A_{3a}} = \frac{A_{4a}}{A_a^*} \times \frac{A_a^*}{A_{3a}} = \frac{3.854}{1.008} = 3.823$$

(b) DIVERGENT DIFFUSER WITH A NORMAL SHOCK AT THE INLET

Conditions Downstream of the Normal Shock

With $M_{1b} = 1.1$ and $p_{1b} = 0.476$ bar, we obtain the Mach number downstream of the normal shock as

$$M_{2b}^2 = \frac{2+(\gamma-1)M_{1b}^2}{2\gamma M_{1b}^2 - (\gamma-1)} = \frac{2+(1.4-1)\times(1.1)^2}{2\times1.4\times(1.1)^2 - (1.4-1)} = 0.831$$

$$M_{2b} = \sqrt{0.831} = 0.912$$

which yields

$$\frac{p_{02b}}{p_{1b}} = \left\{\frac{2\gamma M_{1b}^2 - (\gamma-1)}{(\gamma+1)}\right\}^{\frac{-1}{\gamma-1}}\left\{\frac{(\gamma+1)}{2}M_{1b}^2\right\}^{\frac{\gamma}{\gamma-1}}$$

$$\frac{p_{02b}}{p_{1b}} = \left\{\frac{2\times1.4\times(1.1)^2 - (1.4-1)}{(1.4+1)}\right\}^{\frac{-1}{1.4-1}}\left\{\frac{(1.4+1)}{2}\times(1.1)^2\right\}^{\frac{1.1}{1.4-1}} = 2.133$$

which further yields

$$p_{03b} = p_{02b} = 2.133 \times 0.476 = 1.015 \, \text{bar}$$

Critical Area Ratio at Diffuser Inlet $\left(A_{3b}/A_b^*\right)$

The conditions downstream of the normal shock prevail at the diffuser inlet (section 3). Thus, we obtain

$$\frac{A_{3b}}{A_b^*} = \frac{1}{M_{3b}} \sqrt{\left(\frac{2+(\gamma-1)M_{3b}^2}{\gamma+1}\right)^{\frac{\gamma+1}{\gamma-1}}} = \frac{1}{0.912} \sqrt{\left(\frac{2+(1.4-1)\times(0.912)^2}{1.4+1}\right)^{\frac{1.4+1}{1.4-1}}} = 1.007$$

Mach Number at Diffuser Exit

From the total-to-static pressure ratio computed at section 4

$$\frac{p_{04b}}{p_{4b}} = \frac{1.015}{1} = 1.015$$

we obtain Mach number as

$$M_{4b} = \sqrt{\frac{2}{\gamma-1}\left[\left(\frac{p_{04}}{p_{4b}}\right)^{\frac{\gamma-1}{\gamma}} - 1\right]} = \sqrt{\frac{2}{1.4-1}\left\{(1.015)^{\frac{1.4-1}{1.4}} - 1\right\}} = 0.147$$

Critical Area Ratio at Diffuser Exit A_{4b}/A_b^*

$$\frac{A_{4b}}{A_b^*} = \frac{1}{M_{4b}} \sqrt{\left(\frac{2+(\gamma-1)M_{4b}^2}{\gamma+1}\right)^{\frac{\gamma+1}{\gamma-1}}} = \frac{1}{0.147} \sqrt{\left(\frac{2+(1.4-1)\times(0.147)^2}{1.4+1}\right)^{\frac{1.4+1}{1.4-1}}} = 3.985$$

Overall Diffuser Area Ratio

$$\frac{A_{4b}}{A_{3b}} = \frac{A_{4b}}{A_b^*} \times \frac{A_b^*}{A_{3b}} = \frac{3.985}{1.007} = 3.957$$

The foregoing calculations show that the convergent-divergent diffuser design shown in Figure 13.5a requires an overall area ratio of 3.823, and the air flow exits the diffuser at a Mach number of 0.152. The divergent diffuser design, shown in Figure 13.5b, has an overall area ratio of 3.985 with the exit Mach number of 0.147. Thus, with only a 3.5% increase in the overall area ratio, the divergent diffuser design offers a simpler, shorter, and less expensive diffuser design without the additional feature of a choked throat needed in the convergent-divergent diffuser design. The weak normal shock at the inlet of the diffuser in the second design is nearly isentropic and is an effective means of converting the dynamic pressure into a corresponding rise in the static pressure downstream of the shock over a negligible thickness.

PROBLEM 13.4:   A CONVERGENT-DIVERGENT NOZZLE WITH A NORMAL SHOCK EXHAUSTING INTO A DIFFUSER

A convergent-divergent nozzle, featuring a normal shock, exhausts into a diffuser. Show that the total pressure ratio p_{02}/p_{01} across the shock is equal to the ratio of the first throat area A_1^* to that of the second throat area A_2^*.

SOLUTION FOR PROBLEM 13.4

A normal shock in the C-D nozzle and diffuser system divides the flow into two isentropic flows with equal mass flow rate $(\dot{m})$ and total temperature (T_0). If the critical area with $M = 1$ for the isentropic flow upstream of the normal shock is A_1^* and that for the downstream isentropic flow is A_2^* with the corresponding total pressures p_{01} and p_{02}, the continuity equation yields

$$\dot{m} = \frac{F_{f0}^* A_1^* p_{01}}{\sqrt{T_0}} = \frac{F_{f0}^* A_2^* p_{02}}{\sqrt{T_0}}$$

from which we obtain

$$\frac{p_{02}}{p_{01}} = \frac{A_1^*}{A_2^*}$$

PROBLEM 13.5:   CONVERGENT-DIVERGENT DIFFUSER WITH VARIABLE THROAT AREA TO SWALLOW THE STARTING SHOCK

A convergent-divergent diffuser is to be used at $M_1 = 2.3$. The diffuser uses a variable throat area to swallow the starting shock. What percentage increase in the throat area will be necessary?

SOLUTION FOR PROBLEM 13.5

For $M_1 = 2.3$, we obtain $A_1^*/A_2^* = p_{02}/p_{01} = 0.5833$ from the normal shock table (Sultanian, 2015). Hence, for the isentropic flow downstream of the normal shock, we calculate the percentage increase in the throat area as

$$\frac{A_2^* - A_1^*}{A_1^*} = \frac{A_2^*}{A_1^*} - 1 = \frac{1}{0.5833} - 1 = 71.438\%$$

PROBLEM 13.6:   COMPARING ACTUAL PRESSURE RISE COEFFICIENT WITH THEORETICAL (INCOMPRESSIBLE) VALUE FOR DIFFUSION IN A ROW OF AXIAL-FLOW COMPRESSOR BLADES

Air flow at the inlet to an axial-flow compressor blades, shown in Figure 13.6, has $M_1 = 0.55$, $p_{01} = 300$ kPa, and $T_{01} = 500$ K. The loss in stagnation pressure between inlet (section 1) and outlet (section 2) is 15% of the inlet dynamic head $(p_{01} - p_1)$. The outlet flow area A_2 normal to the outlet absolute velocity V_2 is twice the inlet flow area A_1, which is normal to the inlet absolute velocity V_1. Compute pressure coefficient C_p between sections 1 and 2 from the equation

$$C_p = \frac{p_2 - p_1}{p_{01} - p_1}$$

and compare it to the so-called theoretical (incompressible) pressure rise coefficient C_{pi} computed from the equation

$$C_{pi} = 1 - \frac{V_2^2}{V_1^2}$$

where V_1 and V_2 are mean absolute velocities at inlet and outlet. Assume $\gamma = 1.4$ and $R = 287\,\text{J}/(\text{kg K})$.

SOLUTION FOR PROBLEM 13.6

INLET (SECTION 1)

Static Temperature

$$T_1 = \frac{T_{01}}{1 + 0.5(\gamma - 1)M_1^2} = \frac{500}{1 + 0.5 \times 0.4 \times (0.55)^2} = 471.5\,\text{K}$$

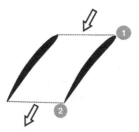

FIGURE 13.6   Diffusing row of axial-flow compressor blades (Problem 13.6).

Static Pressure

$$\frac{p_1}{p_{01}} = \left(\frac{T_1}{T_{01}}\right)^{\frac{\gamma}{\gamma-1}} = \left(\frac{471.5}{500}\right)^{\frac{1.4}{1.4-1}} = 0.814$$

$$p_1 = 0.814 \times 300 = 244.25 \text{ kPa}$$

Absolute Velocity

$$V_1 = M_1\sqrt{\gamma R T_1} = 0.55\sqrt{1.4 \times 287 \times 471.5} = 239.385 \text{ m/s}$$

Total-Pressure Mass Flow Function

$$\hat{F}_{f01} = M_1\sqrt{\frac{\gamma}{\left(1 + \frac{\gamma-1}{2}M_1^2\right)^{\frac{\gamma+1}{\gamma-1}}}} = 0.55\sqrt{\frac{1.4}{\left(1 + \frac{1.4-1}{2}(0.55)^2\right)^{\frac{1.4+1}{1.4-1}}}} = 0.546$$

Mass Flow Rate per Unit Flow Area

$$\frac{\dot{m}_1}{A_1} = \frac{\hat{F}_{f01}p_{01}}{\sqrt{RT_{01}}} = \frac{0.546 \times 300 \times 10^3}{\sqrt{287 \times 500}} = 432.106 \text{ kg}/\left(\text{s} \, \text{m}^2\right)$$

Exit (Section 2)

Total Pressure

$$p_{02} = p_{01} - 0.15(p_{01} - p_1) = 300 - 0.15(300 - 244.25) = 291.637 \text{ kPa}$$

Total-Pressure Mass Flow Function

$$\hat{F}_{f02} = \frac{\dot{m}_2\sqrt{RT_{02}}}{A_2 p_{02}} = \frac{\dot{m}_1\sqrt{RT_{02}}}{A_2 p_{02}} = \frac{432.106 A_1 \times \sqrt{287 \times 500}}{A_2 \times 291.637 \times 10^3} = \frac{331.45\sqrt{287 \times 500}}{2 \times 200.1 \times 10^3}$$

$$= 0.281$$

Mach Number

For $\hat{F}_{f02} = 0.281$, we obtain $M_2 = 0.246$ using an iterative solution method—for example, Goal Seek in MS Excel.

Static Temperature

$$T_2 = \frac{T_{02}}{1 + 0.5(\gamma - 1)M_2^2} = \frac{500}{1 + 0.5 \times 0.4 \times (0.246)^2} = 494.0\,\text{K}$$

Static Pressure

$$\frac{p_2}{p_{02}} = \left(\frac{T_2}{T_{02}}\right)^{\frac{\gamma}{\gamma-1}} = \left(\frac{494}{500}\right)^{\frac{1.4}{1.4-1}} = 0.959$$

$$p_2 = 0.959 \times 291.637 = 279.623 \text{ kPa}$$

Absolute Velocity

$$V_2 = M_2\sqrt{\gamma R T_2} = 0.262 \times \sqrt{1.4 \times 287 \times 494} = 109.552 \text{ m/s}$$

Actual Pressure Rise Coefficient

$$C_p = \frac{p_2 - p_1}{p_{01} - p_1} = \frac{279.623 - 244.25}{300 - 244.25} = 0.634$$

Incompressible Pressure Rise Coefficient

$$C_{pi} = 1 - \left(\frac{V_2}{V_1}\right)^2 = 1 - \left(\frac{109.552}{239.385}\right)^2 = 0.791$$

The actual pressure rise coefficient is less than the theoretical (incompressible) pressure rise coefficient due to the pressure loss in the diffuser.

PROBLEM 13.7:   DEPENDENCE OF TURBINE POWER OUTPUT ON EXHAUST DIFFUSER PRESSURE RISE COEFFICIENT

For a gas turbine with an exhaust-diffuser system, shown in Figure 13.7, compute the increase in turbine power output with the increase in the pressure-rise coefficient (C_p) of the exhaust diffuser from 0.6 to 0.85 in steps of 0.05, and plot the results of turbine power output versus exhaust diffuser C_p.

The following data remain constant during the gas turbine operation:

Flow area at section 3 (exhaust diffuser inlet) $A_3 = 1.0\,\text{m}^2$
Flow area at section 4 (exhaust diffuser exit) $A_4 = 5.0\,\text{m}^2$
Total pressure at section 1 (turbine stage inlet) $p_{01} = 5\,\text{bar}$
Total temperature at section 1 (turbine inlet) $T_{01} = 1000\,\text{K}$

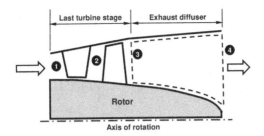

FIGURE 13.7   Last turbine stage with an exhaust diffuser (Problem 13.7).

Mass flow rate $\dot{m} = 150$ kg/s
Flow swirl angle at section 3 (turbine exit or exhaust diffuser inlet) $= 11°$
Static pressure at section 4 (exhaust diffuser exit) $p_4 = 1.013$ bar
Zero flow swirl angle at section 4 (exhaust diffuser exit)
Polytropic efficiency of the turbine stage $\eta_{pt} = 0.9$
Assume $\gamma = 1.4$ and $R = 287\,\mathrm{J}/(\mathrm{kg\,K})$.

SOLUTION FOR PROBLEM 13.7

From the given polytropic efficiency of the last turbine stage, we obtain

$$\frac{n-1}{n} = \frac{\eta_{pt}(\gamma - 1)}{\gamma} = \frac{0.9 \times (1.4 - 1)}{1.4} = 0.257$$

where n is the exponent of polytropic expansion.
     From the definition of the pressure rise coefficient of the exhaust diffuser

$$C_p = \frac{p_4 - p_3}{p_{03} - p_3}$$

we obtain

$$p_{03} = \frac{p_4 - p_3(1 - C_p)}{C_p}$$

For $C_p = 0.6$, we outline here an iterative solution method to compute turbine power output.

Step 1. Assume $p_3 = 84{,}767$ Pa
Step 2. Calculate p_{03} using the equation

$$p_{03} = \frac{p_4 - p_3(1 - C_p)}{C_p} = \frac{1.013 \times 10^5 - 84{,}767 \times (1 - 0.6)}{0.6} = 112{,}322 \text{ Pa}$$

Step 3. Calculate total pressure ratio and total temperature ratio across the turbine stage, and hence calculate T_{03}

$$\frac{T_{01}}{T_{03}} = \left(\frac{p_{01}}{p_{03}}\right)^{\frac{n-1}{n}} = \left(\frac{5\times10^5}{112,322}\right)^{\frac{n-1}{n}} = (4.451)^{0.257} = 1.468$$

$$T_{03} = \frac{1000}{1.468} = 681.15 \text{ K}$$

Step 4. Calculate M_3 from the total-to-static pressure ratio at section 3

$$M_3 = \sqrt{\frac{2}{\gamma-1}\left[\left(\frac{p_{03}}{p_3}\right)^{\frac{\gamma-1}{\gamma}} - 1\right]} = \sqrt{\frac{2}{1.4-1}\left[\left(\frac{112,322}{1.013\times10^5}\right)^{\frac{1.4-1}{1.4}} - 1\right]} = 0.647$$

Step 5. Calculate $\hat{F}_{f03}$ and mass flow rate at section 3

$$\hat{F}_{f03} = M_3 \sqrt{\frac{\gamma}{\left(1+\frac{(\gamma-1)}{2}M_3^2\right)^{(\gamma+1)\big/(\gamma-1)}}} = 0.647\sqrt{\frac{1.4}{\left(1+\frac{(1.4-1)}{2}(0.647)^2\right)^{(1.4+1)\big/(1.4-1)}}}$$

$$= 0.602$$

$$\dot{m} = \frac{\hat{F}_{f03}A_3\cos11°p_{03}}{\sqrt{RT_{03}}} = \frac{0.602\times1.0\times\cos11°\times112,322}{\sqrt{287\times681.15}} = 150\text{ kg/s}$$

Step 6. Repeat Steps 1–5 until the computed mass flow rate equals the specified value of 150 kg/s.

Step 7. Compute turbine power output

$$c_p = \frac{R\gamma}{\gamma-1} = \frac{287\times1.4}{1.4-1} = 1004.5 \text{ J/}(\text{kgK})$$

$$P_t = \dot{m}c_p(T_{01}-T_{03}) = 150\times1004.5\times(1000-681.15) = 48.043 \text{ MW}$$

Figure 13.8 shows the monotonic (almost linear) increase in turbine power output with increasing C_p of the exhaust diffuser—a low-hanging fruit in gas turbine design for power generation!

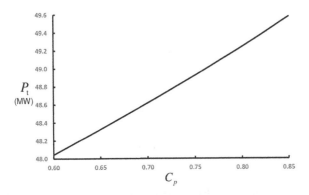

FIGURE 13.8   Variation of turbine power output with the exhaust diffuser C_p (Problem 13.7).

PROBLEM 13.8:   INCOMPRESSIBLE FLOW THROUGH A FRICTIONLESS ADIABATIC DUCT WITH AND WITHOUT A DIFFUSER

Figure 13.9 shows incompressible flows through two ducts: (1) constant-area circular duct and (2) the duct in (1) appended with a short conical diffuser. Both flows with negligible friction and heat transfer have identical inlet total pressure p_0 and exit static pressure p. Will the mass flow rate be equal in each duct? If you expect the mass flow rates to be different, which duct will have higher flow and why?

SOLUTION TO PROBLEM 13.8

The mass flow in the duct shown in Figure 13.9b will be higher than that shown in Figure 13.9a. As both friction and heat transfer are negligible in these ducts, the total pressure will remain constant throughout the duct. Because the exit flow velocity, which only depends on the difference between the total pressure and static, is the same in both ducts, the mass flow rate with the higher exit area for the duct in Figure 13.9b will be higher than that for the duct shown in Figure 13.9a.

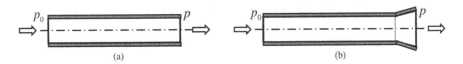

FIGURE 13.9   Incompressible flows through a frictionless adiabatic duct: (a) without a diffuser; (b) with a short diffuser (Problem 13.8).

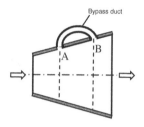

FIGURE 13.10 Incompressible flow in an adiabatic diffuser with a bypass duct (Problem 13.9).

PROBLEM 13.9: INCOMPRESSIBLE FLOW IN AN ADIABATIC DIFFUSER WITH A BYPASS DUCT

Figure 13.10 shows an incompressible flow through an adiabatic diffuser with a bypass duct connecting diffuser sections A and B. Based on how the static and total pressure change in a diffuser, determine the direction of flow in the bypass duct—whether it is from A to B or from B to A.

SOLUTION FOR PROBLEM 13.9

Downstream of an adiabatic diffuser, the static pressure increases (due to conversion of dynamic pressure into static pressure) and the total pressure decreases (due to friction and other losses). If the ends the bypass duct are connected normal to the diffuser wall, the total pressure at its inlet and the static pressure at its exit equal the corresponding diffuser wall static pressures. As the static pressure at B is higher than that at A, the bypass duct flow must take place from B to A.

NOMENCLATURE

A	Flow area
A^*	Critical area with $M = 1$
c_p	Specific heat of gas at constant pressure
c_v	Specific heat of gas at constant volume
C_D	Drag coefficient
C_p	Pressure rise coefficient
C_V	Velocity coefficient
$\tilde{C}_p$	Pressure rise coefficient with inlet swirl
C_{pi}	Static pressure-rise coefficient of an ideal diffuser
CV	Control volume
d	Diameter
E_{ke}	kinetic energy flow rate
F_D	Drag force
$\hat{F}_f$	Static-pressure mass flow function
$\hat{F}_{f0}$	Total-pressure mass flow function
F_{f0}^*	Total-pressure mass flow function at $M = 1$
h	Specific enthalpy of the fluid

h_0	Specific total (stagnation) enthalpy of the fluid
K	Total pressure loss coefficient
$\dot{m}$	Mass flow rate
M	Mach number
p	Static pressure
$\bar{p}$	Section-average static pressure
p_0	Total (stagnation) pressure
$\bar{p}_0$	Section-average total (stagnation) pressure
P_t	Turbine power output
$\Delta p_{0\,loss}$	Loss in total pressure
R	Gas constant
s	Specific entropy
T	Static temperature
$\bar{T}$	Section-average static temperature
T_0	Total (stagnation) temperature
$\bar{T}_0$	Section-average total (stagnation) temperature
V	Absolute flow velocity
$\bar{V}$	Section-average flow velocity
V_r	Absolute flow velocity component in the radial direction
V_x	Component of V in the axial direction
V_θ	Absolute flow velocity component in the tangential direction
x	Cartesian coordinate x
y	Cartesian coordinate y
α	Absolute flow angle to the axial direction
β	Smaller-to-larger diameter ratio in a sudden pipe expansion
η_D	Diffuser isentropic efficiency
η_{pt}	Turbine polytropic efficiency
γ	Ratio of specific heats $= c_p/c_v$
ρ	Fluid density

REFERENCES

Sultanian, B.K. 2015. *Fluid Mechanics: An Intermediate Approach*. Boca Raton, FL: Taylor & Francis.

Sultanian, B.K. 2019. *Logan's Turbomachinery: Flowpath Design and Performance Fundamentals*, 3rd edition. Boca Raton: Taylor & Francis.

BIBLIOGRAPHY

Japikse, D. and N.C. Baines. 2000. *Turbomachinery Diffuser Design Technology*, 2nd edition. White River Junction: Concepts Eti.

Sultanian, B.K., S. Nagao, and T. Sakamoto. 1999. Experimental and three-dimensional CFD investigation in a gas turbine exhaust system. *ASME Journal of Engineering for Gas Turbines and Power*. 121:364–374.

Wilson, D.G. and T. Korakianitis. 2014. *The Design of High-Efficiency Turbomachinery and Gas Turbines*, 2nd edition. Cambridge: MIT Press.

14 Internal Flow around Rotors and Stators

INTRODUCTION

Problems and solutions presented in this chapter deal with flow and heat transfer associated with rotors and stators of turbomachinery, particularly, gas turbine internal flows used for cooling and sealing of various critical components subjected to high temperatures. For better understanding and appreciation of representative problems and their solutions presented here, readers are encouraged to review various topics, which may be new to those not familiar with gas turbine secondary air systems, discussed in detail in Sultanian (2018). They include free disk pumping, disk pumping beneath a forced vortex, rotor disk in an enclosed cavity, rotor-stator cavity with radial outflow, rotor-stator cavity with radial inflow, rotating cavity with radial inflow or outflow, windage and swirl modeling in a general cavity, arbitrary cavity surface orientation: conical and horizontal surfaces, bolts on stator and rotor surfaces, compressor rotor cavity (flow and heat transfer physics, heat transfer modeling with bore flow, and heat transfer modeling of closed cavity), flow and heat transfer modeling in a pre-swirl system, hot gas ingestion (physics of hot gas ingestion, 1-D single-orifice modeling, multiple-orifice spoke modeling), and the computation of axial rotor thrust.

PROBLEM 14.1: PRESSURE AND TEMPERATURE CHANGE IN AN ISENTROPIC COMPRESSIBLE FREE VORTEX FLOW

For a compressible free vortex flow, properties such as T_1, p_1, and $V_{\theta 1}$ are given at point 1 where $r = r_1$. Compute T_2, p_2, and $V_{\theta 2}$ at point 2, where $r = r_2$.

SOLUTION FOR PROBLEM 14.1

We compute the change in entropy in a fluid flow using the equation

$$ds = c_p \frac{dT}{T} - R \frac{dp}{p}$$

which for an isentropic $(ds = 0)$ flow yields

$$\frac{dp}{p} = \frac{c_p}{R} \frac{dT}{T}$$

which along with the radial equilibrium equation $\left(dp/dr = \rho V_\theta^2/r\right)$ and the equation of state $\left(p/\rho = RT\right)$ for a perfect gas yields

$$\frac{dp}{dr} = \frac{pV_\theta^2}{RTr}$$

$$\frac{dp}{p} = \frac{V_\theta^2}{RT}\frac{dr}{r}$$

Substitution for dp/p in this equation from the equation

$$\frac{dp}{p} = \frac{c_p}{R}\frac{dT}{T}$$

we obtain

$$dT = \frac{V_\theta^2}{c_p}\frac{dr}{r}$$

As $V_\theta = C/r = \left(r_1 V_{\theta 1}\right)/r$ for a free vortex, this equation becomes

$$dT = \frac{\left(r_1 V_{\theta 1}\right)^2}{c_p}\frac{dr}{r^3}$$

which, upon integration between points 1 and 2, yields

$$\int_{T_1}^{T_2} dT = \frac{\left(r_1 V_{\theta 1}\right)^2}{c_p}\int_{r_1}^{r_2}\frac{dr}{r^3}$$

$$T_2 = T_1 + \frac{\left(r_1 V_{\theta 1}\right)^2}{2c_p}\left(\frac{1}{r_1^2} - \frac{1}{r_2^2}\right)$$

Using the isentropic compressible flow relation between the pressure ratio and the temperature ratio, we write

$$\frac{p_2}{p_1} = \left(\frac{T_2}{T_1}\right)^{\frac{\gamma}{\gamma-1}} = \left\{1 + \frac{\left(r_1 V_{\theta 1}\right)^2}{2c_p T_1}\left(\frac{1}{r_1^2} - \frac{1}{r_2^2}\right)\right\}^{\frac{\gamma}{\gamma-1}}$$

$$\frac{p_2}{p_1} = \left\{1 + \frac{V_{\theta 1}^2}{2c_p T_1}\left(1 - \frac{r_1^2}{r_2^2}\right)\right\}^{\frac{\gamma}{\gamma-1}}$$

$$p_2 = p_1\left\{1 + \frac{V_{\theta 1}^2}{2c_p T_1}\left(1 - \frac{r_1^2}{r_2^2}\right)\right\}^{\frac{\gamma}{\gamma-1}}$$

Finally, for a free vortex, we obtain $V_{\theta 2}$ as

$$V_{\theta 2} = \frac{r_1 V_{\theta 1}}{r_2}$$

Note that the change in static temperature in an isentropic free vortex is equal and opposite to the change in its dynamic temperature. As a result, its total temperature and total pressure remain constant. Using these facts, we can easily compute changes in the temperature and pressure in a free vortex flow, see Sultanian (2019).

PROBLEM 14.2: PRESSURE AND TEMPERATURE CHANGE IN AN ISENTROPIC FORCED VORTEX

For a compressible forced vortex flow, properties such as T_1, p_1, and $V_{\theta 1}$ are given at point 1 where $r = r_1$. Compute T_2, p_2, and $V_{\theta 2}$ at point 2, where $r = r_2$.

SOLUTION FOR PROBLEM 14.2

For a forced vortex with its constant angular velocity Ω, we write $V_\theta = r\Omega$. From the given $V_{\theta 1}$ at $r = r_1$, we obtain $\Omega = V_{\theta 1}/r_1$, giving $V_\theta = r V_{\theta 1}/r_1$. In the solution for Problem 1.15, we obtained

$$dT = \frac{V_\theta^2}{c_p} \frac{dr}{r}$$

which with the substitution $V_\theta = r V_{\theta 1}/r_1$ becomes

$$dT = \frac{V_{\theta 1}^2}{r_1^2 c_p} r\, dr$$

which, upon integration between points 1 and 2, yields

$$\int_{T_1}^{T_2} dT = \frac{V_{\theta 1}^2}{r_1^2 c_p} \int_{r_1}^{r_2} r\, dr$$

$$T_2 - T_1 = \frac{V_{\theta 1}^2}{2 r_1^2 c_p} \left(r_2^2 - r_1^2 \right)$$

$$T_2 = T_1 + \frac{V_{\theta 1}^2}{2 r_1^2 c_p} \left(r_2^2 - r_1^2 \right)$$

Using the isentropic compressible flow relation between pressure ratio and temperature ratio, we write

$$\frac{p_2}{p_1} = \left(\frac{T_2}{T_1}\right)^{\frac{\gamma}{\gamma-1}} = \left\{1 + \frac{V_{\theta 1}^2}{2 r_1^2 c_p}\left(r_2^2 - r_1^2\right)\right\}^{\frac{\gamma}{\gamma-1}}$$

$$\frac{p_2}{p_1} = \left\{1 + \frac{V_{\theta 1}^2}{2 c_p T_1}\left(\frac{r_2^2}{r_1^2} - 1\right)\right\}^{\frac{\gamma}{\gamma-1}}$$

$$p_2 = p_1 \left\{1 + \frac{V_{\theta 1}^2}{2 c_p T_1}\left(\frac{r_2^2}{r_1^2} - 1\right)\right\}^{\frac{\gamma}{\gamma-1}}$$

Finally, we obtain $V_{\theta 2}$ as

$$V_{\theta 2} = \frac{r_2 V_{\theta 1}}{r_1}$$

Note that the change in static temperature in an isentropic forced vortex equals the change in its dynamic temperature. Using this fact, we can easily compute changes in the temperature and pressure in a forced vortex flow, see Sultanian (2019).

PROBLEM 14.3:   PRESSURE AND TEMPERATURE CHANGE IN A NONISENTROPIC GENERALIZED VORTEX

In a nonisentropic generalized vortex the variations in the swirl factor and total temperature are given by the functions $S_f = f(r)$ and $T_0 = g(r)$, which are piecewise polynomial functions between radii r_1 and r_2. Compute the static temperature and pressure in this vortex flow at $r = r_2$ from their values known at $r = r_1$.

SOLUTION FOR PROBLEM 14.3

The static pressure variation in this vortex is governed by the radial equilibrium equation, which we can express in terms of rotational Mach number as

$$\frac{dp}{p} = \frac{V_\theta^2}{RT}\frac{dr}{r} = \gamma \frac{M_\theta^2}{r}dr$$

where $M_\theta = V_\theta/\sqrt{\gamma RT}, V_\theta = S_f \Omega r = f \Omega r$, and $T = T_0 - V_\theta^2/(2c_p) = g - f^2 \Omega^2 r^2/(2c_p)$. Integration of this equation between radii r_1 and r_2 yields

$$\int_{p_1}^{p_2} \frac{dp}{p} = \gamma \int_{r_1}^{r_2} \frac{M_\theta^2}{r} dr$$

$$\ln\left(\frac{p_2}{p_1}\right) = \gamma G$$

where we obtain G by numerically integrating $\int_{r_1}^{r_2} \left(M_\theta^2 / r \right) dr$, for example, using the Simpson's one-third rule. Thus, we finally obtain

$$p_2 = p_1 e^{\gamma G}$$

PROBLEM 14.4:   RADIALLY OUTWARD FLOW THROUGH A ROTATING CONSTANT-AREA DUCT

As shown in Figure 14.1, an incompressible flow enters a rotating duct at 1 and exits it at 2. Using quantities shown in the figure, determine the increase in static pressure from duct inlet to exit for a fluid of density ρ.

SOLUTION FOR PROBLEM 14.4

As the duct cross-section is uniform, the radial flow velocity remains constant. As a result, on a small control volume shown in the figure, the pressure force must balance the centrifugal force, giving

$$\left(p + \frac{dp}{dr} \Delta r \right) A - pA = A \Delta r \rho r \Omega^2$$

$$\frac{dp}{dr} = \rho r \Omega^2$$

Integrating this equation from $r = 0$ to $r = r_2$ yields

$$\int_{p_1}^{p_2} dp = \int_0^{r_2} \rho r \Omega^2 \, dr$$

$$p_2 - p_1 = \frac{\rho r_2^2 \Omega^2}{2}$$

which we can directly obtain from the radial equilibrium equation for an incompressible forced vortex.

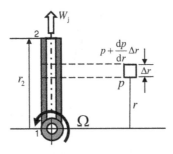

FIGURE 14.1   Incompressible flow through a rotating constant-area duct (Problem 14.4).

PROBLEM 14.5:   RADIALLY OUTWARD FLOW THROUGH
A ROTATING VARIABLE-AREA DUCT

Figure 14.2 shows a radially outward isentropic incompressible flow in a variable-area rotating duct. For the given mass flow rate and other parameters shown in the figure, find the change in static pressure and total pressure (in the rotating reference frame) between section 1 (inlet) and section 2 (outlet). Neglect any shear stress on the duct wall and assume that all properties are uniform over each duct cross-section. The fluid density is ρ.

SOLUTION FOR PROBLEM 14.5

In this case, the incompressible flow through the rotating duct is isentropic (zero heat transfer with no frictional loss). As a result, the total pressure without rotation remains constant over the duct. Under rotation, the centrifugal force in the rotor reference frame generates an opposing pressure force, increasing the static and total pressures radially outward.

CONTINUITY EQUATION

We obtain the mass flow rate at section 1 as $\dot{m} = \rho A_1 V_1$ and that at section 2 as $\dot{m} = \rho A_2 V_2$.

The mass conservation (the continuity equation) yields

$$\frac{A_2}{A_1} = \frac{V_1}{V_2}$$

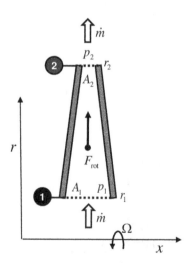

FIGURE 14.2   Radially outward flow through a rotating duct with variable area (Problem 14.5).

MOMENTUM EQUATION

The linear momentum balance on the nonrotating duct control volume yields

$$p_1 + \frac{1}{2}\rho V_1^2 = p_2 + \frac{1}{2}\rho V_2^2$$

giving

$$\left(p_2 - p_1\right)_{\text{non-rotating}} = \frac{1}{2}\rho V_1^2 - \frac{1}{2}\rho V_2^2$$

For the rotating duct, by integrating the radial equilibrium equation, we calculate the change in the static pressure between sections 1 and 2 as

$$\left(p_2 - p_1\right)_{\text{rotation}} = \int_{r_1}^{r_2} \rho \Omega^2 r \, dr = \frac{1}{2}\rho \Omega^2 \left(r_2^2 - r_1^2\right)$$

which states that the corresponding pressure force balances the centrifugal force due to rotation. Hence, we write

$$p_2 - p_1 = \left(p_2 - p_1\right)_{\text{non-rotating}} + \left(p_2 - p_1\right)_{\text{rotation}}$$

which with the substitutions from the foregoing yields

$$p_2 - p_1 = \frac{1}{2}\rho V_1^2 - \frac{1}{2}\rho V_2^2 + \frac{1}{2}\rho \Omega^2 \left(r_2^2 - r_1^2\right)$$

Rearranging terms in this equation, we obtain

$$\left(p_2 + \frac{1}{2}\rho V_2^2\right) - \left(p_1 + \frac{1}{2}\rho V_1^2\right) = \frac{1}{2}\rho \Omega^2 \left(r_2^2 - r_1^2\right)$$

$$p_{02} - p_{01} = \frac{1}{2}\rho \Omega^2 \left(r_2^2 - r_1^2\right)$$

which shows that, between any two sections of a frictionless variable-area rotating duct, the change in total pressure relative to the rotor equals the change in the static pressure due to rotation only.

PROBLEM 14.6:   IMPINGEMENT AIR COOLING OF A CYLINDRICAL SURFACE WITH A ROTARY ARM WITH THREE JETS

As shown in Figure 14.3, a high-pressure rotary arm is used for air impingement cooling of a cylindrical surface. The total pressure and total temperature inside the rotary

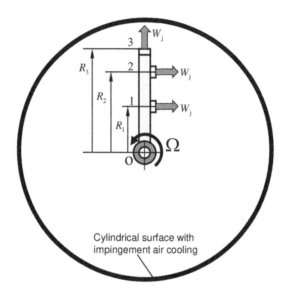

FIGURE 14.3   Impingement air cooling of a cylindrical surface with a rotary arm with three jets (Problem 14.6).

arm are 3.0 bar and 407.5 K, respectively. The static pressure outside the rotary arm is 1.0 bar. At the maximum RPM, the rotary arm needs to overcome a frictional torque of 12.5 Nm. For the geometric data: jet diameter $(d_j) = 7$ mm, $R_1 = 50$ cm, $R_2 = 100$ cm, and $R_3 = 140$ cm, calculate the maximum rpm of the rotary arm. Assume $\gamma = 1.4$ and $R = 287$ J/$(kg\,K)$ for air. (Hint: Each air nozzle operates under choked flow condition with identical jet velocity relative to the rotary arm.)

SOLUTION FOR PROBLEM 14.6

With the nozzle flow area

$$A_j = \frac{\pi d_j^2}{4} = \frac{\pi(0.007)^2}{4} = 3.8485 \times 10^{-5} \text{ m}^2$$

and the total-pressure mass flow function $\hat{F}_{f0}^* = 0.6847$ at $M = 1$, we calculate the mass flow rate through each air nozzle as

$$\dot{m} = \frac{A_j \hat{F}_{f0}^* p_0}{\sqrt{RT_0}} = \frac{3.8485 \times 10^{-5} \times 0.6847 \times 300,000}{\sqrt{287 \times 407.5}} = 0.0231 \text{ kg/s}$$

With the static temperature at the nozzle throat

$$T^* = \frac{2T_0}{\gamma + 1} = \frac{2 \times 407.5}{(1 + 1.4)} = 339.6 \text{ K}$$

we calculate the air jet velocity relative to the rotary arm as

$$W_j = \sqrt{\gamma R T^*} = \sqrt{1.4 \times 287 \times 339.6} = 369.384 \text{ m/s}$$

TORQUE AND ANGULAR MOMENTUM BALANCE ON THE ROTARY ARM CONTROL VOLUME

We obtain the net efflux of angular momentum in the counterclockwise direction as

$$\dot{H} = \dot{m}R_1\left(R_1\Omega - W_j\right) + \dot{m}R_2\left(R_2\Omega - W_j\right) + \dot{m}R_3\left(R_3\Omega - 0\right)$$

We calculate the torque due to the pressure force acting on the fluid control volume in the counterclockwise direction as

$$\Gamma_p = A_j R_1\left(p^* - p_{\text{amb}}\right) + A_j R_2\left(p^* - p_{\text{amb}}\right)$$

where

$$p^* = \frac{p_0}{\left(\dfrac{\gamma+1}{2}\right)^{\frac{\gamma}{\gamma-1}}} = \frac{300,000}{1.893} = 158,485 \text{ Pa}$$

If $\Gamma_{\text{arm-to-fluid}}$ is the torque from the rotary arm acting in counterclockwise direction on the fluid control volume, then the torque-angular-momentum balance yields

$$\Gamma_{\text{arm-to-fluid}} + A_j R_1\left(p^* - p_{\text{amb}}\right) + A_j R_2\left(p^* - p_{\text{amb}}\right)$$
$$= \dot{m}R_1\left(R_1\Omega - W_j\right) + \dot{m}R_2\left(R_2\Omega - W_j\right) + \dot{m}R_3\left(R_3\Omega - 0\right)$$

At the maximum rpm, the net torque acting on the rotary arm must be zero, giving

$$\Gamma_{\text{fluid-to-arm}} - \Gamma_{\text{friction}} = 0$$
$$\Gamma_{\text{fluid-to-arm}} = -\Gamma_{\text{arm-to-fluid}} = \Gamma_{\text{friction}}$$

which we write as

$$-\Gamma_{\text{friction}} + A_j R_1\left(p^* - p_{\text{amb}}\right) + A_j R_2\left(p^* - p_{\text{amb}}\right)$$
$$= \dot{m}R_1\left(R_1\Omega_{\text{max}} - W_j\right) + \dot{m}R_2\left(R_2\Omega_{\text{max}} - W_j\right) + \dot{m}R_3^2\Omega_{\text{max}}$$

giving

$$\Omega_{max} = \frac{\dot{m}W_j(R_1 + R_2) - \Gamma_{friction} + A_j(R_1 + R_2)(p^* - p_{amb})}{\dot{m}(R_1^2 + R_2^2 + R_3^2)}$$

$$= \frac{\left\{ \dot{m}W_j + A_j(p^* - p_{amb}) \right\}(R_1 + R_2) - \Gamma_{friction}}{\dot{m}(R_1^2 + R_2^2 + R_3^2)}$$

$$= \frac{(0.0231 \times 369.384 + 2.251) \times (0.5 + 1.0) - 12.5}{0.0231 \times \left((0.5)^2 + (1.0)^2 + (1.4)^2\right)}$$

$$= 49.597 \text{ rad/s}$$

Therefore, the rotary arm maximum rpm equals 473.6.

PROBLEM 14.7: AXIAL THRUST OF A CENTRIFUGAL AIR COMPRESSOR

For a small single-stage centrifugal air compressor, schematically shown in Figure 14.4, the key impeller dimensions are $r_{sh} = 20$ mm, $r_1 = 50$ mm, and $r_2 = 75$ mm. At $\Omega = 60{,}000$ rpm test speed and inlet air mass flow rate $\dot{m} = 1.0$ kg/s at $p_{01} = 1$ bar and $T_{01} = 290$ K, the compressor discharge occurs at $p_{02} = 4.5$ bar and $T_{02} = 524.5$ K. The slip coefficient at the discharge is 0.94. Assuming that the air in the gap between the stationary casing and impeller, both on forward and aft sides, behaves like a

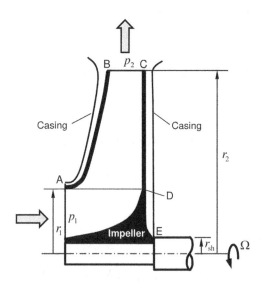

FIGURE 14.4 Schematic of a centrifugal air compressor for axial thrust calculation (Problem 14.7).

forced vortex with the swirl factor $S_f = 0.50$ and constant density corresponding to the compressor discharge conditions, calculate the net axial thrust on the impeller. Neglect the contribution of axial thrust due to the change in stream thrust of the impeller flow.

SOLUTION FOR PROBLEM 14.7

The radial equilibrium equation for a forced vortex with swirl factor S_f becomes

$$\frac{dp}{dr} = \rho r S_f^2 \Omega^2$$

which for constant density yields the following radial variation of static pressure in the gap between the impeller and the casing

$$p = p_2 - \frac{\rho S_f^2 \Omega^2}{2}\left(r_2^2 - r^2\right)$$

Figure 14.4 shows that the axial thrust exerted by the fluid on surface AB is balanced by the equal and opposite thrust on surface CD. Thus, the net contribution to the rotor axial thrust is from the fluid pressure force exerted on surface DE. This rotor thrust is directed to the left and equals

$$F_{rotor} = 2\pi \int_{r_{sh}}^{r_1} p\,r\,dr$$

$$F_{rotor} = 2\pi \int_{r_{sh}}^{r_1} \left\{ p_2 - \frac{\rho S_f^2 \Omega^2}{2}\left(r_2^2 - r^2\right)\right\} r\,dr$$

$$F_{rotor} = \pi\left(r_1^2 - r_{sh}^2\right)\left\{ p_2 - \frac{\rho S_f^2 \Omega^2}{2}\left(r_2^2 - \frac{r_1^2 + r_{sh}^2}{2}\right)\right\}$$

To calculate rotor axial thrust using this equation, we need to first calculate the values of p_2 and ρ as follows:

$$\Omega = \frac{\pi N}{30} = \frac{3.1416 \times 600,000}{30} = 6283.185 \text{ rad/s}$$

$$V_{\theta 2} = 0.94 \times r_2 \times \Omega = 0.94 \times \left(\frac{75}{1000}\right) \times 6283.185 = 442.965 \text{ m/s}$$

Neglecting the contribution of the dynamic temperature associated with the flow radial velocity at compressor discharge, we write

$$T_2 = T_{02} - \frac{V_{\theta 2}^2}{2c_p} = 524.5 - \frac{(442.965)^2}{2 \times 1004.5} = 427 \text{ K}$$

From isentropic relations we obtain

$$\frac{p_{02}}{p_2} = \left(\frac{T_{02}}{T_2}\right)^{\frac{\gamma}{\gamma-1}} = \left(\frac{524.5}{427}\right)^{3.5} = 2.058$$

$$p_2 = \frac{450,000}{2.058} = 218,688 \text{ N}$$

and

$$\rho = \frac{p_2}{RT_2} = \frac{218,688}{287 \times 427} = 1.785 \text{ kg/m}^3$$

Now, we calculate the rotor axial thrust as

$$F_{rotor} = \pi\left((0.05)^2 - (0.02)^2\right) \times$$

$$\left\{218,688 - \frac{1.785 \times (0.5)^2 \times (6283.185)^2}{2}\left((0.75)^2 - \frac{(0.05)^2 + (0.02)^2}{2}\right)\right\}$$

$$F_{rotor} = 1200 \text{ N}$$

PROBLEM 14.8:   HEAT TRANSFER IN A ROTATING DUCT
OF ARBITRARY CROSS SECTION

Figure 14.5 shows a four-sided radial duct representing a part of an internal cooling passage of a steam-cooled gas turbine blade with rotational velocity Ω. Each duct wall has a different surface area A_w, wall temperature T_w, and heat transfer coefficient h_w. For a given coolant (steam) mass flow rate $\dot{m}$ and constant specific heat

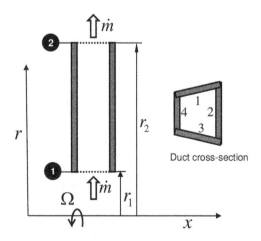

FIGURE 14.5   Heat transfer in a rotating duct of arbitrary cross-section (Problem 14.8).

(at constant pressure) c_p, find expressions to evaluate the rise in coolant total temperature due to heat transfer and blade rotation. As the coolant flow Mach number is less than 0.3, it may be assumed incompressible. Neglect the effect of coolant total temperature change due to rotation on convective heat transfer between the duct wall and the coolant.

SOLUTION FOR PROBLEM 14.8

In this problem, the coolant steam temperature in the gas turbine blade increases by both convective heat transfer and rotational work transfer. The coolant temperature change due to rotation will influence convective heat transfer from the duct walls to the coolant. In the present solution, we calculate separate changes in coolant temperature due to heat transfer and rotation and neglect the coupling between them, see Sultanian (2015) for the coupled heat transfer and rotational work transfer solution.

COOLANT TOTAL TEMPERATURE RISE DUE TO HEAT TRANSFER

Due to convective heat transfer, the difference between the constant wall temperature and the fluid total temperature decays exponentially from duct inlet to exit, that is,

$$\left(T_w - T_{0\_outlet}\right) = \left(T_w - T_{0\_inlet}\right) e^{-\eta}$$

where $\eta = \left(h_w A_w\right)/\left(\dot{m} c_p\right)$. For a four-sided duct, each side with different heat transfer properties, we extend this solution using newly defined average quantities

$$\left(\overline{T}_w - T_{0\_outlet}\right) = \left(\overline{T}_w - T_{0\_inlet}\right) e^{-\overline{\eta}}$$

where

$$\overline{T}_w = \frac{\displaystyle\sum_{i=1}^{i=4} h_{wi} A_{wi} T_{wi}}{\displaystyle\sum_{i=1}^{i=4} h_{wi} A_{wi}}$$

and

$$\overline{\eta} = \frac{\displaystyle\sum_{i=1}^{i=4} h_{wi} A_{wi}}{\dot{m} c_p}$$

Thus, the change in coolant total temperature due to heat transfer becomes

$$\Delta T_{0\_heat\,transfer} = \left(T_{0\_outlet} - T_{0\_inlet}\right)_{heat\,transfer} = \left(\overline{T}_w - T_{0\_inlet}\right)\left(1 - e^{-\overline{\eta}}\right)$$

Note that the coolant total temperatures in the foregoing equations are in the rotor reference frame.

COOLANT TOTAL TEMPERATURE RISE DUE TO ROTATION

As presented in Chapter 8, the fluid rothalpy between any two sections of a rotating duct remains constant under adiabatic conditions. Expressing rothalpy in terms of coolant total temperature in the rotor reference frame, we obtain

$$c_p T_{0\_outlet} - \frac{r_2^2 \Omega^2}{2} = c_p T_{0\_inlet} - \frac{r_1^2 \Omega^2}{2}$$

$$\Delta T_{0\_rotation} = \left(T_{0\_outlet} - T_{0\_inlet}\right)_{rotation} = \frac{\Omega^2}{2c_p}\left(r_2^2 - r_1^2\right)$$

This equation shows that the temperature change due to rotation depends upon the flow direction, but not its magnitude. For a radially outward flow, the work done on the fluid due to rotation is positive, and for a radially inward flow, it is negative.

Thus, the total change in coolant total temperature due to both heat transfer and rotation becomes

$$\Delta T_0 = \Delta T_{0\_heat\ transfer} + \Delta T_{0\_rotation}$$

$$\Delta T_0 = \left(\overline{T}_w - T_{0\_inlet}\right)\left(1 - e^{-\overline{\eta}}\right) + \frac{\Omega^2}{2c_p}\left(r_2^2 - r_1^2\right)$$

If we had considered the inherent coupling between heat transfer and rotation on coolant temperature change, it can be argued that for a radially outward flow, the actual coolant temperature rise will be lower than that obtained by this equation, which uses linear superposition of individual temperature changes. For a radially inward flow, however, the actual coolant temperature will be higher than that obtained by this equation.

PROBLEM 14.9:   WINDAGE TEMPERATURE RISE IN
A ROTOR-STATOR CAVITY

Figure 14.6 shows a typical rotor-stator cavity of a 50-Hertz (3000 rpm) gas turbine engine. The coolant air at 673 K (absolute total) and swirling at 60% of the rotor rpm enters the cavity at the inner radius. It exits the cavity at the outer radius with a swirl of 40% of the rotor rpm. The mass flow rate of the coolant air is 20 kg/s. If the total frictional torque from the stator surface acting on the cavity air is 3008 Nm, find the rotor torque and the exit total temperature of the coolant air. The rotor-stator surfaces are adiabatic (zero heat transfer). All quantities are given in the inertial (stator) reference frame. Assume $c_p = 1004\ \mathrm{J/(kg\,K)}$ for air.

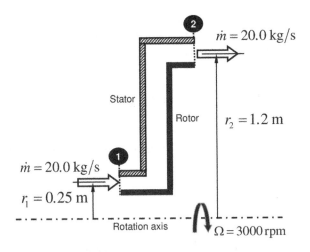

FIGURE 14.6   Windage temperature rise in a rotor-stator cavity (Problem 14.9).

SOLUTION FOR PROBLEM 14.9

In this problem, the coolant air temperature increases not by heat transfer (adiabatic flow) but by work transfer from the rotor. This form of work transfer in a rotor-stator or a rotor-rotor cavity in turbomachinery is often called windage. Note that the air total temperature rise due to windage in a stator cavity is always zero.

The angular momentum equation for the control volume between sections 1 and 2 yields

$$\Gamma_{\text{rotor}} - \Gamma_{\text{stator}} = \dot{m}\left(r_2 V_{\theta 2} - r_1 V_{\theta 1}\right)$$

$$\Gamma_{\text{rotor}} = \Gamma_{\text{stator}} + \dot{m}\left(r_2 V_{\theta 2} - r_1 V_{\theta 1}\right)$$

We can write the work transfer (windage) to the coolant air in the cavity as

$$\dot{W}_{\text{windage}} = \Gamma_{\text{rotor}}\,\Omega$$

Substituting for the rotor torque Γ_{rotor} in this equation in terms of the stator torque Γ_{stator} from the torque and angular momentum balance equation yields the following equation whose first term on the right-hand side is often interpreted as the stator torque doing work to generate windage in the cavity.

$$\dot{W}_{\text{windage}} = \Gamma_{\text{stator}}\,\Omega + \dot{m}\,\Omega\left(r_2 V_{\theta 2} - r_1 V_{\theta 1}\right)$$

As the stator is not capable of doing any work on the cavity air, this interpretation of the term $\Gamma_{\text{stator}}\Omega$ is physically incorrect. In fact, the rotor torque, which is only capable of doing work on the cavity air, consists of two parts: one to balance the stator torque and the other to change the angular momentum of the coolant air.

To bephysically consistent, we should always use the rotor torque to compute the work transfer in a rotor-stator or rotor-rotor cavity. We then compute the rise in air total temperature using the equation

$$\Delta T_{0\_windage} = \frac{\Gamma_{rotor}\ \Omega}{\dot{m}_{air}c_p}$$

Using the given data for this problem, the numerical results obtained using the foregoing equations are as follows:

ROTOR ANGULAR VELOCITY

$$\Omega = \frac{3000 \times 2\pi}{60} = 314.159\ \text{rad/s}$$

AIR TANGENTIAL VELOCITY AT INLET (SECTION 1)

$$V_{\theta 1} = 0.6 \times 314.159 \times 0.25 = 47.124\ \text{m/s}$$

AIR TANGENTIAL VELOCITY AT OUTLET (SECTION 2)

$$V_{\theta 2} = 0.4 \times 314.159 \times 1.2 = 150.796\ \text{m/s}$$

ROTOR TORQUE

$$\Gamma_{rotor} = \Gamma_{stator} + \dot{m}\left(r_2 V_{\theta 2} - r_1 V_{\theta 1}\right)$$

$$= 3008 + 20 \times \left(1.2 \times 150.796 - 0.25 \times 47.124\right)$$

$$= 6391.495\ \text{Nm}$$

WINDAGE TEMPERATURE RISE

$$\Delta T_{0\_windage} = \frac{\Gamma_{rotor}\Omega}{\dot{m}_{air}c_p} = \frac{6391.495 \times 314.159}{20 \times 1004} = 100\ \text{K}$$

EXIT TOTAL TEMPERATURE OF COOLANT AIR

$$T_{02} = T_{01} + \Delta T_{0\,windage} = 673 + 100 = 773\,\text{K}$$

PROBLEM 14.10:   A TWO-TOOTH LABYRINTH SEAL

For a two-tooth labyrinth seal shown in Figure 14.7, the mean radius under each tooth is $r_m = 0.5$ m with a clearance of $s = 3$ mm. The rotor rotates at 3000 rpm, and the swirl factor of the leakage air flow remains constant at $S_f = 0.5$. The seal initially operates under an inlet total pressure of $p_{0in} = 10$ bar and a total temperature of $T_{0in} = 500$ K. The outlet static pressure (back pressure) remains constant at $p_{out} = 5$ bar. Assume that the kinetic energy carry-over factor is zero and the discharge coefficient $C_d = 0.8$. Neglect any change in air total temperature due to heat transfer and rotational work transfer (windage). Compute (1) the leakage air mass flow rate through the seal, and using a noniterative method, (2) the minimum leakage air mass flow rate and inlet total pressure such that the seal just chokes $(M_2 = 1)$ at the second tooth.

SOLUTION FOR PROBLEM 14.10

As the swirl factor remains constant, we may solve this problem in the relative reference frame rotating with the air swirl velocity, using the total pressure and total temperature based only on the axial throughflow velocity. These quantities are computed as follows.

AIR TANGENTIAL VELOCITY

$$V_\theta = S_f r_m \Omega = 0.5 \times 0.5 \times \left(\frac{3000 \times \pi}{30} \right) = 78.540 \text{ m/s}$$

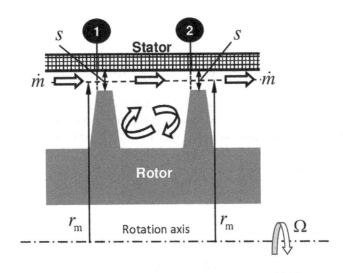

FIGURE 14.7   Schematic of a two-tooth labyrinth seal (Problems 14.10).

Relative Total Temperature at the First Tooth Inlet

$$T_{0x1} = T_{0in} - \frac{V_\theta^2}{2c_p} = 500 - \frac{78.540 \times 78.540}{2 \times 1004.5} = 496.9 \text{ K}$$

Relative Total Pressure at the First Tooth Inlet

$$\frac{p_{0in}}{p_{0x1}} = \left(\frac{T_{0in}}{T_{0x1}}\right)^{\frac{\gamma}{\gamma-1}} = \left(\frac{500}{496.30}\right)^{3.5} = 1.0218$$

$$p_{0x1} = \frac{p_{0in}}{1.0218} = \frac{10 \times 10^5}{1.0218} = 9.787 \times 10^5 \text{ Pa}$$

(a) Leakage Air Mass Flow Rate through the Seal

Leakage Flow Area over Each Tooth

$$A = 2\pi r_m s = 2 \times \pi \times 0.5 \times \left(\frac{3}{1000}\right) = 0.00943 \text{ m}^2$$

Mass Flow Rate through Tooth 1

$$\dot{m}_1 = \frac{AC_d\hat{F}_{f01}p_{0x1}}{\sqrt{RT_{0x1}}}$$

Mass Flow Rate through Tooth 2

$$\dot{m}_2 = \frac{AC_d\hat{F}_{f02}p_{0x2}}{\sqrt{RT_{0x2}}} = \frac{AC_d\hat{F}_{f01}p_{0x1}}{\sqrt{RT_{0x1}}}$$

In this problem, we have $\dot{m}_1 = \dot{m}_2$, $T_{0x1} = T_{0x2}$, and $p_{0x2} = p_1$.

The leakage air mass flow rate can be computed using the following iterative method where the converged values are shown within parentheses:

Step 1. Assume M_1 (0.55)
Step 2. Compute $\hat{F}_{f01}$

$$\hat{F}_{f01} = M_1 \sqrt{\frac{\gamma}{\left(1 + \frac{\gamma-1}{2}M_1^2\right)^{\frac{\gamma+1}{\gamma-1}}}} \quad (0.545)$$

Step 3. Compute $\dot{m}_1$

$$\dot{m}_1 = \frac{AC_d\hat{F}_{f01}p_{0x1}}{\sqrt{RT_{0x1}}} \quad (10.657)$$

Step 4. Compute p_1

$$p_1 = \frac{p_{0x1}}{\left(1 + \frac{\gamma-1}{2}M_1^2\right)^{\frac{\gamma}{\gamma-1}}} \quad \left(7.970 \times 10^5\right)$$

Step 5. Compute Mach number M_2

$$M_2 = \left[\frac{2}{\gamma-1}\left\{\left(\frac{p_1}{p_{out}}\right)^{\frac{\gamma-1}{\gamma}} - 1\right\}\right]^{\frac{1}{2}} \quad (0.844)$$

Step 6. Compute $\hat{F}_{f02}$

$$\hat{F}_{f02} = M_2\sqrt{\frac{\gamma}{\left(1 + \frac{\gamma-1}{2}M_2^2\right)^{\frac{\gamma+1}{\gamma-1}}}} \quad (0.670)$$

Step 7. Compute $\dot{m}_2$

$$\dot{m}_2 = \frac{AC_d\hat{F}_{f02}p_1}{\sqrt{RT_{0x1}}} \quad (10.657)$$

Step 8. Repeat steps from 1 to 7 until $\dot{m}_2 = \dot{m}_1$.

Thus, the seal leakage air mass flow rate computed in this case is $10.657\,\text{kg/s}$.

(b) A Noniterative Method to Compute $\dot{m}$ and p_{0in} for $M_2=1$ and $p_{out} = 5\,\text{bar}$

When the leakage flow through the labyrinth seal is entirely subsonic, the exit static pressure must be equal to the specified static back pressure. When the last tooth chokes, the exit static pressure could be equal to or higher than the back pressure.

In the present case, the inlet total pressure is increased with concurrent increase in the seal leakage air mass flow rate until $M_2 = 1$, and the exit static pressure equals the specified back pressure of $p_{out} = 5$ bar.

To solve for both the mass flow rate and total inlet pressure, we can certainly use an iterative method like the one used in (a). We present here a direct solution method, using backward marching from tooth 2 to tooth 1.

Tooth 2

For $M_2 = 1$ at tooth 2, we compute

$$\frac{p_1}{p_{out}} = \left(\frac{\gamma+1}{2}\right)^{\frac{\gamma}{\gamma-1}} = (1.2)^{3.5} = 1.893$$

which yields

$$p_1 = \left(\frac{p_1}{p_{out}}\right) \times p_{out} = 1.893 \times 5 \times 10^5 = 9.465 \times 10^5$$

and

$$\hat{F}_{f2} = M_2 \sqrt{\gamma\left(1+\frac{\gamma-1}{2}M_2^2\right)} = \sqrt{1.4 \times \frac{1.4+1}{2}} = 1.296$$

which yields the seal leakage mass flow rate as

$$\dot{m} = \frac{AC_d\hat{F}_{f2}p_{out}}{\sqrt{RT_{0x1}}} = \frac{0.00943 \times 0.8 \times 1.296 \times 5 \times 10^5}{\sqrt{287 \times 496.930}} = 12.939 \text{ kg/s}$$

and

$$\hat{F}_{f02} = \frac{\hat{F}_{f2}}{\dfrac{p_1}{p_{out}}} = \frac{1.296}{1.893} = 0.685$$

For directly computing the inlet total pressure at tooth 1, we make use of the following facts:

1. Static pressure at tooth 1 exit equals the total pressure at tooth 2 inlet $(\alpha = 0)$
2. For constant total temperature and seal clearance (equal flow area under both tooth 1 and tooth 2), the static-pressure mass flow function of tooth 1 equals the total-pressure mass function of tooth 2.

Therefore, we obtain $\hat{F}_{f1} = \hat{F}_{f02} = 0.685$, which directly yields M_1 as

$$M_1 = \sqrt{\frac{-\gamma + \sqrt{\gamma^2 + 2\gamma(\gamma - 1)\hat{F}_{f1}^2}}{\gamma(\gamma - 1)}} = \sqrt{\frac{-1.4 + \sqrt{(1.4)^2 + 2 \times 1.4 \times (1.4 - 1) \times (0.685)^2}}{1.4 \times (1.4 - 1)}}$$

$$M_1 = 0.561$$

which from isentropic relations gives the total-to-static pressure ratio

$$\frac{p_{0x1}}{p_1} = \left(1 + \frac{\gamma - 1}{2}M_1^2\right)^{\frac{\gamma}{\gamma - 1}} = \left(1 + 0.2 \times (0.561)^2\right)^{3.5} = 1.239$$

and

$$p_{0x1} = 1.239 \times 9.465 \times 10^5 = 1.172 \times 10^5 \text{ Pa}$$

which with $p_{01}/p_{0x1} = 1.0218$ finally yields

$$p_{01} = p_{0x1} \times 1.0218 = 1.172 \times 10^5 \times 1.0218$$

$$p_{01} = 1.198 \times 10^5 \text{ Pa}$$

Therefore, for the choked second tooth, the direct solution method yields the leakage mass flow rate of 12.939 kg/s at the required inlet total pressure of 1.198×10^5 Pa. It may be verified that, once the seal is choked, the leakage air mass rate varies linearly with the total inlet pressure while the total-to-static pressure ratio across the seal remains constant. The resulting exit static pressure, therefore, becomes higher than the specified back pressure. This behavior of the choked two-tooth labyrinth seal in this problem has significant practical implication. For a given back pressure, we can first compute its leakage mass flow rate and total pressure by using the direct solution method and then scale the values to other inlet and outlet pressure conditions while the seal remains choked.

PROBLEM 14.11:   A TWO-TOOTH LABYRINTH SEAL CHOKED UNDER BOTH TEETH

For the two-tooth seal of Problem 14.10, show that, for the leakage air flow to choke under both teeth, the clearance under the second tooth must be increased to satisfy the following relation:

$$\frac{s_2}{s_1} = \left(\frac{\gamma + 1}{2}\right)^{\frac{\gamma}{\gamma - 1}} = 1.893$$

where s_1 and s_2 are seal clearances under tooth 1 and tooth 2, respectively.

SOLUTION FOR PROBLEM 14.11

From the flow physics of the two-tooth seal of Problem 14.10, we can write $T_{0x1} = T_{0x2}$, $p_{0x2} = p_1$, and $\dot{m}_1 = \dot{m}_2$. We can express the choked mass flow rate under each tooth as follows:

$$\dot{m}_1 = \frac{A_1 C_d \hat{F}_{f1}^* p_1}{\sqrt{RT_{0x1}}}$$

$$\dot{m}_2 = \frac{A_2 C_d \hat{F}_{f02}^* p_{0x2}}{\sqrt{RT_{0x2}}} = \frac{A_2 C_d \hat{F}_{f02}^* p_1}{\sqrt{RT_{0x1}}}$$

where A_1 and A_2 are seal clearance areas under teeth 1 and 2, respectively, and $\hat{F}_{f1}^*$ is the static-pressure mass flow function evaluated at $M_1 = 1$ and $\hat{F}_{f02}^*$ is the total-pressure mass flow function evaluated at $M_2 = 1$.

Equating $\dot{m}_1$ and $\dot{m}_2$ in the foregoing yields

$$A_1 \hat{F}_{f1}^* = A_2 \hat{F}_{f02}^*$$

which for air with $\gamma = 1.4$ results in the following relation:

$$\frac{A_2}{A_1} = \frac{s_2}{s_1} = \frac{\hat{F}_{f1}^*}{\hat{F}_{f02}^*} = \left(\frac{\gamma+1}{2}\right)^{\frac{\gamma}{\gamma-1}} = 1.893$$

PROBLEM 14.12: ROTATING RADIAL PIPES CARRYING COMPRESSOR BLEED AIR FLOW FOR TURBINE COOLING

In some gas turbines, radial pipes are used in the compressor rotor cavity to bleed a part of the high-pressure compressor air flow from rim to bore for downstream cooling of turbine parts. Figure 14.8 schematically shows one such design of a gas turbine operating at 3000 rpm. This design uses 20 radial pipes, each of 1 m length with an inner diameter of 40 mm and an outer diameter of 45 mm. The compressor bleed air at a total mass flow rate of $\dot{m} = 7.5$ kg/s enters the pipe under solid-body rotation of the rotor at a total pressure (p_0) of 7 bar and a total temperature (T_0) of 600 K, both being in the stator (absolute) reference frame. The primary design intent is to have the coolant air at location A with a minimum drop in pressure from its value at the bleed point. In Figure 14.8, $r_1 = 0.15$ m and $r_2 = 0.2$ m. The Darcy friction factor for each adiabatic pipe flow is $f = 0.022$. Assume that the air flow is an isentropic free vortex from the pipe exit to point A. Plot the variations of static pressure along with the total pressure, and the total temperature in stator and rotor reference frames. What are the numerical values of these quantities and the air swirl factor at point A. For air, assume $\gamma = 1.4$ and $R = 287$ J/(kg K).

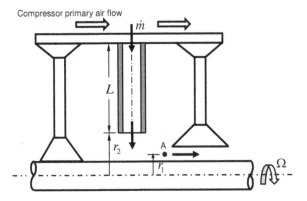

FIGURE 14.8 Schematic of a compressor rotor cavity with radial pipes for supplying compressor bleed air for downstream cooling of turbine parts (Problem 14.12).

SOLUTION FOR PROBLEM 14.11

In this problem, the compressor bleed air flow in each constant-area radial pipe is subjected to both friction and rotation, but no heat transfer. One can easily extend the solution presented here for a variable-area pipe with heat transfer. The flow in the pipe remains in solid-body rotation with the pipe—a forced vortex with a constant angular velocity equal to that of the rotor. Pipe wall friction reduces static pressure and increases entropy in the flow direction. Pipe rotation has two effects on the radially inward flow. First, it decreases the static pressure, which we calculate using the radial equilibrium equation, not by the method of calculating pressure changes in an isentropic forced vortex discussed in Appendix A. Second, the work of centrifugal force due to rotation decreases air total temperature in the rotor reference frame—this decrease equals the decrease in the equivalent dynamic temperature of pipe tangential velocity $\left(=\Omega^{2}\left(r_{2}^{2}-r_{1}^{2}\right)/\left(2c_{p}\right)\right)$.

For computing flow properties in each radial pipe, we divide it into 20 equal segments—one may use more segments until the solutions do not change significantly. Starting with the first pipe segment inlet, which corresponds to the pipe inlet, we compute the remaining properties from the given inlet conditions. Including the effects of pipe friction and rotation, we use an iterative solution method, outlined in the following sections, to compute all flow properties at the outlet of this pipe segment. These properties then become the inlet conditions for the next pipe segment. In this way, we march the numerical solution segment-by-segment for all the remaining pipe segments, where the last segment yields the pipe exit conditions.

Some Preliminary Calculations

$$\frac{\gamma}{\gamma-1}=\frac{1.4}{1.4-1}=3.5$$

$$c_{p}=\frac{\gamma R}{\gamma-1}=3.5\times287=1004.5\ \text{J}/\left(\text{kg K}\right)$$

Pipe Flow Area

$$A = \frac{\pi D^2}{4} = \frac{\pi \times (0.04)^2}{4} = 1.257 \times 10^{-3} \, \text{m}^2$$

Mass Flow Rate Per Pipe

$$\dot{m}_p = \frac{\dot{m}}{n_p} = \frac{7.5}{20} = 0.375 \, \text{kg/s}$$

Pipe Angular Velocity

$$\Omega = \frac{\pi N}{30} = \frac{3.14159 \times 3000}{30} = 314.159 \, \text{rad/s}$$

Inlet Total Temperature in Rotor Reference Frame

$$T_{0Rin} = T_{0in} - \frac{\Omega^2 r_{in}^2}{2 c_p} = 600 - \frac{(314.159)^2 \times (1.2)^2}{2 \times 1004.5} = 529.3 \, \text{K}$$

Inlet Total Pressure in Rotor Reference Frame

$$\frac{T_{0Rin}}{T_{0in}} = \frac{529.3}{600} = 0.882$$

$$p_{0Rin} = p_{0in} \left(\frac{T_{0Rin}}{T_{0in}} \right)^{\gamma/\gamma-1} = 7 \times 10^5 \times (0.882)^{3.5} = 451,235 \, \text{Pa}$$

PIPE SEGMENT 1 INLET (PIPE INLET)

Total-Pressure Mass Flow Function

$$\hat{F}_{f0in} = \frac{\dot{m}_p \sqrt{R T_{0Rin}}}{A p_{0Rin}} = \frac{0.375 \sqrt{287 \times 529.3}}{(1.257 \times 10^{-3}) \times 451,235} = 0.2577$$

Mach Number

For $\hat{F}_{f0in} = 0.2577$, using an iterative method—for example, "Goal Seek" in MS Excel—we obtain $M_{in} = 0.224$ from the equation

$$\hat{F}_{f0in} = M_{in} \sqrt{\frac{\gamma}{\left(1 + \frac{(\gamma - 1)}{2} M_{in}^2 \right)^{(\gamma+1)/(\gamma-1)}}}$$

Static Temperature

$$T_{in} = \frac{T_{0Rin}}{\left(1 + \dfrac{\gamma - 1}{2} M_{in}^2\right)} = \frac{529.3}{1 + 0.5 \times (1.4 - 1) \times (0.224)^2} = 524.0\,\text{K}$$

Static Pressure

$$p_{in} = p_{0Rin} \left(\frac{T_{in}}{T_{0Rin}}\right)^{\frac{\gamma}{\gamma-1}} = 451{,}235 \times \left(\frac{524}{529.3}\right)^{3.5} = 435{,}671\,\text{Pa}$$

Density

$$\rho_{in} = \frac{p_{in}}{RT_{in}} = \frac{435{,}671}{287 \times 524} = 2.897\,\text{kg}/\text{m}^3$$

Flow Velocity

$$V_{in} = M_1 \sqrt{\gamma R T_{in}} = 0.224 \times \sqrt{1.4 \times 287 \times 524} = 103.005\,\text{m/s}$$

PIPE SEGMENT 1 OUTLET (PIPE SECTION 2 INLET)

To compute various flow properties at the outlet of pipe section 1 of length 0.05 m, we use the following iterative solution method—the values within parentheses are the converged values for each quantity:

Step 1. With $r_{in} = 1.2\,\text{m}$ and $r_{out} = 1.15\,\text{m}$, we compute the outlet total temperature as

$$T_{0Rout} = T_{0Rin} + \frac{\Omega^2 \left(r_{out}^2 - r_{in}^2\right)}{2c_p} = 529.3 + \frac{(314.159)^2 \times (1.15 \times 1.15 - 1.2 \times 1.2)}{2 \times 1004.5} = 523.5\,\text{K}$$

Step 2. Assume M_{out} $(M_{out} = 0.232)$

Step 3. Compute $\hat{F}_{f0out}$ from the equation

$$\hat{F}_{f0out} = M_{out} \sqrt{\frac{\gamma}{\left(1 + \dfrac{(\gamma - 1)}{2} M_{out}^2\right)^{(\gamma+1)/(\gamma-1)}}} \qquad \left(\hat{F}_{f0out} = 0.2663\right)$$

Step 4. Compute p_{0Rout} from the equation

$$p_{0Rout} = \frac{\dot{m}_p \sqrt{RT_{0Rout}}}{A\hat{F}_{f0out}} = \frac{0.375 \times \sqrt{287 \times 523.5}}{1.257 \times 10^{-3} \times 0.2663} = 434,293 \, \text{Pa}$$

$$\left(p_{0Rout} = 434,293 \, \text{Pa} \right)$$

Step 5. Compute T_{out}

$$\frac{T_{0Rout}}{T_{out}} = 1 + \frac{\gamma - 1}{2} M_{out}^2 = 1 + \left(\frac{1.4 - 1}{2} \right) \times (0.232)^2 = 1.011$$

$$T_{out} = \frac{T_{0Rout}}{\left(\dfrac{T_{0Rout}}{T_{out}} \right)} = \frac{523.5}{1.011} = 517.9 \, \text{K} \quad \left(T_{out} = 517.9 \, \text{K} \right)$$

Step 6. Compute p_{out}

$$p_{out} = \frac{p_{0Rout}}{\left(\dfrac{T_{0Rout}}{T_{out}} \right)^{\frac{\gamma}{\gamma-1}}} = \frac{434,293}{(1.011)^{3.5}} = 418,255 \, \text{Pa} \quad \left(p_{out} = 418,255 \, \text{Pa} \right)$$

Step 7. Compute V_{out}

$$V_{out} = M_{out} \sqrt{\gamma RT_{out}} = 0.232 \times \sqrt{1.4 \times 287 \times 517.9} = 106.047 \, \text{m/s}$$

$$V_{out} = 106.047 \, \text{m/s}$$

Step 8. Compute ρ_{out}

$$\rho_{out} = \frac{p_{out}}{RT_{out}} = \frac{418255}{287 \times 517.9} = 2.814 \, \text{kg/m}^3 \quad \left(\rho_{out} = 2.814 \, \text{kg/m}^3 \right)$$

Step 9. Compute Δp_f

$$\Delta p_f = -\frac{f}{2} \frac{\Delta L}{D} \frac{\dot{m}_p}{A} \left(\frac{V_{in} + V_{out}}{2} \right)$$

$$= -\frac{0.022}{2} \times \frac{0.05}{0.040} \times \frac{0.375}{1.257 \times 10^{-3}} \left(\frac{103.005 + 106.047}{2} \right) = -0.858 \, \text{Pa}$$

$$\left(\Delta p_f = -858 \, \text{Pa} \right)$$

Step 10. Compute Δp_{rot}

$$\Delta p_{\text{rot}} = \frac{(\rho_{\text{in}} + \rho_{\text{out}})\Omega^2 (r_{\text{out}}^2 - r_{\text{in}}^2)}{4c_p}$$

$$= \frac{(2.897 + 2.814)(314.159)^2 \times (1.15 \times 1.15 - 1.2 \times 1.2)}{4 \times 1004.5} = -16,588\,\text{Pa}$$

$$(\Delta p_{\text{rot}} = -16,588\ \text{Pa})$$

Step 11. Compute p'_{out}

$$p'_{\text{out}} = p_{\text{in}} + \Delta p_f + \Delta p_{\text{rot}} = 43,5671 - 858 - 16,558 = 418,255\,\text{Pa}$$

Step 12. Repeat Steps 2 to 11 until p_{out} computed in Step 6 equals p'_{out} computed in Step 11 within an acceptable tolerance.

We repeat the foregoing solution method for the remaining 19 pipe segments, giving the following values for various flow properties at the pipe exit in the rotor reference frame:

$$p_{\text{exit}} = 247,216\ \text{Pa};\ p_{\text{0Rexit}} = 271,184\ \text{Pa};$$

$$T_{\text{0Rexit}} = 460.5\,\text{K};\ M_{\text{exit}} = 0.366;\ T_{\text{exit}} = 448.5\,\text{K};\ \text{and}\ p_{\text{exit}} = 247,216\,\text{Pa}.$$

Pipe Exit Total Pressure and Total Temperature in Stator Reference Frame

With $r_{\text{exit}} = r_2 = 0.2$, we now calculate the total temperature and total pressure at the pipe exit in the stator reference frame as follows:

$$T_{\text{0exit}} = T_{\text{0Rexit}} + \frac{r_{\text{exit}}^2 \Omega^2}{2c_p} = 460.5 + \frac{(0.2)^2 \times (314.159)^2}{2 \times 1004.5} = 462.4\,\text{K}$$

$$p_{\text{0exit}} = p_{\text{0Rexit}} \left(\frac{T_{\text{0exit}}}{T_{\text{0Rexit}}} \right)^{\gamma/\gamma - 1} = 271,184 \times \left(\frac{460.5}{462.4} \right)^{3.5} = 275,256\,\text{Pa}$$

VARIATION OF FLOW PROPERTIES IN EACH RADIAL PIPE

Figure 14.9 shows how static pressure p, total pressure p_{0R} in rotor reference frame, total pressure p_{0R} in stator reference frame, total temperature T_{0R} in rotor reference frame, and total temperature T_0 in stator reference frame vary along the radial pipe from its inlet to exit.

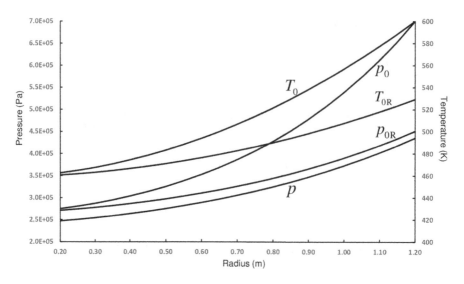

FIGURE 14.9 Variations of properties of the radially inward bleed air flow in the pipe from its inlet to exit (Problem 14.12).

ISENTROPIC FREE-VORTEX FLOW FROM PIPE EXIT TO POINT A

Because the total pressure and total temperature in an isentropic free vortex remain constant, we obtain $T_{0A} = T_{0\text{exit}} = 462.4\,\text{K}$, $T_{0RA} = T_{0R\text{exit}} = 460.5\,\text{K}$, $p_{0A} = p_{0\text{exit}} = 275,256\,\text{Pa}$, and $p_{0RA} = p_{0R\text{exit}} = 271,184\,\text{Pa}$.

Tangential (Swirl) Velocity and Swirl Factor at Point A

$$V_{\theta\text{exit}} = r_{\text{exit}}\Omega = 0.2 \times 314.159 = 62.832 \text{ m/s}$$

$$V_{\theta A} = \frac{V_{\theta\text{exit}} r_{\text{exit}}}{r_A} = \frac{62.832 \times 0.2}{0.15} = 83.776 \text{ m/s}$$

$$S_{fA} = \frac{V_{\theta A}}{r_A \Omega} = \frac{83.776}{0.15 \times 314.159} = 1.778$$

In addition to minimizing the loss of static pressure from the compressor air bleed point to point A, the radial pipes help keep the air swirl factor as close to solid-body-rotation $(S_{fA} = 1)$ as possible.

Increase in Dynamic Temperature from Pipe Exit to Point A

$$(\Delta T)_{\text{dyn}} = \frac{V_{\theta A}^2 - V_{\theta\text{exit}}^2}{2c_p} = \frac{(83.776)^2 - (62.832)^2}{2 \times 1004.5} = 1.53\,\text{K}$$

Static Temperature at Point A

As in an isentropic free vortex, the change in dynamic temperature between two points is equal and opposite to the changes in static temperature between these points, we obtain at A

$$T_A = T_{\text{exit}} - (\Delta T)_{\text{dyn}} = 462.4 - 1.53 = 446.9\,\text{K}$$

Static Pressure at Point A

$$p_A = p_{0A}\left(\frac{T_A}{T_{0A}}\right)^{\gamma/\gamma-1} = 275{,}256 \times \left(\frac{446.9}{462.4}\right)^{3.5} = 244{,}279\,\text{Pa}$$

NOMENCLATURE

A	Flow area
A_w	Wall area
c_p	Specific heat of gas at constant pressure
c_v	Specific heat of gas at constant volume
C_d	Discharge coefficient
d_j	Jet diameter
$\hat{F}_f$	Static-pressure mass flow function
$\hat{F}_{f0}$	Total-pressure mass flow function
F_{f0}^*	Total-pressure mass flow function at $M = 1$
F_{rot}	Centrifugal force due to rotation
F_{rotor}	Rotor axial thrust
h	Specific enthalpy of the fluid
h_w	Heat transfer coefficient at duct wall
$\dot{H}$	Angular momentum flow rate
$\dot{m}$	Mass flow rate
M	Mach number
M_θ	Mach number based on tangential velocity $\left(M_\theta = V_\theta/\sqrt{\gamma RT}\right)$
p	Static pressure
p^*	Critical pressure (static pressure at $M = 1$)
p_0	Total (stagnation) pressure
p_{0R}	Total pressure in rotor reference frame
Δp_f	Change in static pressure due to duct wall friction
Δp_{rot}	Change in static pressure due to duct rotation
r	Radial distance
r_m	Mean radius
r_{sh}	Shaft radius
R	Gas constant; radius
s	Specific entropy
s_1	Clearance under first tooth of a labyrinth seal

s_2	Clearance under second tooth of a labyrinth seal
S_f	Swirl factor (ratio of air swirl velocity to rotor swirl velocity at the same radius)
T	Static temperature
T^*	Critical static temperature (static temperature at $M = 1$)
T_0	Total (stagnation) temperature
T_w	Duct wall temperature
T_{0R}	total temperature in rotor reference frame
V	Absolute flow velocity
V_θ	Absolute flow velocity component in the tangential direction
W_j	Jet velocity
x	Cartesian coordinate x
θ	Coordinate in the tangential direction
α	Kinetic energy carryover factor
γ	Ratio of specific heats $\left(\gamma = c_p/c_v\right)$
ρ	Fluid density
Γ	Torque
Γ_p	Torque due to pressure force
Ω	Angular velocity

REFERENCE

Sultanian, B.K. 2018. *Gas Turbines: Internal Flow Systems Modeling* (Cambridge Aerospace Series #44). Cambridge: Cambridge University Press.

BIBLIOGRAPHY

Abe, T., J. Kikuchi, and H. Takeuchi. 1979. An Investigation of Turbine Disk Cooling: Experimental Investigation and Observation of Hot Gas Flow into a Wheel Space. *13th International Congress on Combustion Engines (CIMAC)*, Vienna, Austria, May 7–10, Paper No. GT30.

Bayley, F.J. and J.M. Owen. 1969. Flow between a rotating and stationary disc. *Aeronautical Quarterly*. 20:333–354.

Bayley, F.J. and J.M. Owen. 1970. Fluid dynamics of a shrouded disk system with a radial outflow of coolant. *ASME Journal of Engineering for Gas Turbines and Power*. 92(3):335–341.

Childs, P.R.N. 2010. *Rotating Flow*. New York: Elsevier.

Dittmann, M., K. Dullenkopf, and S. Wittig. 2004. Discharge coefficients of rotating short orifices with radiused and chamfered inlets. *ASME Journal of Engineering for Gas Turbines and Power*. 126:803–808.

Dittmann, M., K. Dullenkopf, and S. Wittig. 2005. Direct-transfer pre-swirl system: a one-dimensional modular characterization of the flow. *ASME Journal of Engineering for Gas Turbines and Power*. 127:383–388.

Dittmann, M., T. Geis, V. Schramm, S. Kim, and S. Wittig. 2002. Discharge coefficients of a pre-swirl system in secondary air systems. *Journal of Turbomachinery*. 124:119–124.

Dweik, Z., R. Briley, T. Swafford, and B. Hunt. 2009. Computational Study of the Heat Transfer of the Buoyancy-Driven Rotating Cavity with Axial Throughflow of Cooling Air. ASME Paper GT2009-59978.

Hoerner, S.F. 1965. *Fluid-Dynamic Drag: Theoretical, Experimental, and Statistical Information.* Author-published.

Idelchik, I.E. 2005. *Handbook of Hydraulic Resistance*, 3rd edition. New Delhi: Jaico Publishing House.

Lewis, P. 2008. Pre-Swirl Rotor-Stator Systems: Flow and Heat Transfer. PhD thesis. University of Bath.

Lugt, H.J. 1995. *Vortex Flow in Nature and Technology.* Malabar: Krieger Publishing Company.

Miller, D.S. 1990. *Internal Flow Systems*, 2nd edition. Houston: Gulf Publishing Company.

Newman, B.G. 1983. Flow and heat transfer on a disk rotating beneath a forced vortex. *AIAA Journal.* 22(8):1066–1070.

Owen, J.M. 1989. An approximate solution for the flow between a rotating and a stationary disk. *Transactions of the ASME, Journal of Turbomachinery.* 111(3):323–332.

Owen, J. M. and J. Powell. 2006. Buoyancy-induced flow in heated rotating cavities. *ASME Journal of Engineering for Gas Turbines and Power.* 128(1):128–134.

Owen, J.M. and R.H. Rogers. 1989. *Flow and Heat Transfer in Rotating-Disc System. Vol. 1* Rotor-Stator Systems. Taunton: Research Studies Press.

Owen, J.M. and R.H. Rogers. 1995. *Flow and Heat Transfer in Rotating-Disc System. Vol. 2* Rotating Cavities. Taunton: Research Studies Press.

Scanlon, T., J. Wilkes, D. Bohn, and O. Gentilhomme. 2004. A Simple Method of Estimating Ingestion of Annulus Gas into a Turbine Rotor Stator Cavity in the Presence of External Pressure Gradients. ASME Paper No. GT2004-53097.

Sultanian, B.K. 2015. *Fluid Mechanics: An Intermediate Approach.* Boca Raton, FL: Taylor & Francis.

Sultanian, B.K. 2019. *Logan's Turbomachinery: Flowpath Design and Performance Fundamentals*, 3rd edition. Boca Raton, FL: Taylor & Francis.

Sultanian, B.K. and D.A. Nealy. 1987. Numerical modeling of heat transfer in the flow through a rotor cavity. *Heat Transfer in Gas Turbines*, HTD-Vol. 87, ed. D.E. Metzger, 11–24, New York: ASME.

Zografos, A.T., W.A. Martin, and J.E. Sunderland. 1987. Equations of properties as a function of temperature for seven fluids. *Computer Methods in Applied Mechanics and Engineering.* 61:177–187.

Appendix A
Euler's Turbomachinery Equation and Rothalpy

EULER'S TURBOMACHINERY EQUATION

According to the Euler's turbomachinery equation, the aerodynamic power transfer to or from the fluid is simply the product of the torque and the rotor angular velocity in radians per second. In pumps, fans, and compressors, the power transfer occurs into the fluid to increase its rate of angular momentum outflow over its rate of inflow. In turbines, the power transfer occurs from the fluid to the rotor, decreasing the rate of fluid angular momentum outflow over its rate of inflow.

Let us consider a steady adiabatic flow in a rotating passage formed between adjacent blades of the rotor shown in Figure A.1. The velocity vectors V_1 at section 1 (inlet) and V_2 at section 2 (outlet) have components in the axial, radial, and tangential directions. The meridional velocity V_{xr}, which is the resultant of the axial and the radial velocities, is the mass velocity at sections 1 and 2. Also, $V_{\theta 1}$ is the tangential velocity at section 1 and $V_{\theta 2}$ is the tangential velocity at section 2.

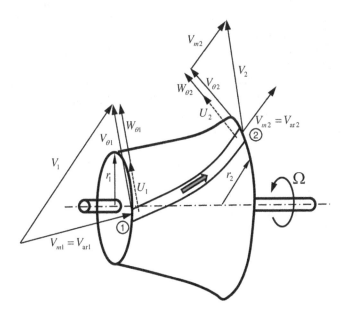

FIGURE A.1 Flow through an axial-radial turbomachinery passage between adjacent blades.

For a constant mass flow rate $\dot{m}$, the angular momentum equation through the blade passage yields the aerodynamic torque as

$$\Gamma = \dot{m}\left(r_2 V_{\theta 2} - r_1 V_{\theta 1}\right) \tag{A.1}$$

from which we determine the aerodynamic power transfer, which is the rate of work transfer due to the aerodynamic torque acting on the fluid control volume, as

$$P = \Gamma\Omega = \dot{m}\left(r_2 V_{\theta 2} - r_1 V_{\theta 1}\right)\Omega = \dot{m}\left(U_2 V_{\theta 2} - U_1 V_{\theta 1}\right) \tag{A.2}$$

where U_1 and U_2 are rotor tangential velocities at section 1 and section 2, respectively. Using the steady flow energy equation in terms of total enthalpy at these sections, we can also write

$$P = \dot{m}\left(h_{02} - h_{01}\right) \tag{A.3}$$

which together with Equation A.2 yields

$$h_{02} - h_{01} = U_2 V_{\theta 2} - U_1 V_{\theta 1} \tag{A.4}$$

which is known as the Euler's turbomachinery equation. This equation reveals that, for turbines, where the work transfer occurs from the fluid to the rotor, we have $h_{02} < h_{01}$ and $U_2 V_{\theta 2} < U_1 V_{\theta 1}$; for compressors, fans, and pumps, where the work transfer occurs from the rotor to the fluid, we have $h_{02} > h_{01}$ and $U_2 V_{\theta 2} > U_1 V_{\theta 1}$. Although the coordinate system attached to a turbomachinery is a noninertial one, featuring the centrifugal and Coriolis forces in this reference frame, Equation A.4 uses quantities in the inertial (absolute) reference frame, see Sultanian (2015, 2019) for additional discussion of this equation.

For axial flow machines, where $r_1 \approx r_2$ and $U_1 \approx U_2$, Equation A.4 reveals that the change in total enthalpy is entirely due to the change in flow tangential velocity—i.e., $\Delta h_0 = \bar{U}\Delta V_\theta$—requiring blades with camber (bow). For radial-flow machines, however, the change in total enthalpy results largely from the change in rotor tangential velocity due the change in radius—i.e., $\Delta h_0 = \Delta U \bar{V}_\theta$.

Substituting $V_\theta = W_\theta + r\Omega$ in Equation A.4 yields

$$h_{02} - h_{01} = \left(U_2 W_{\theta 2} - U_1 W_{\theta 1}\right) + \left(r_2^2 - r_1^2\right)\Omega^2$$

where $\left(U_2 W_{\theta 2} - U_1 W_{\theta 1}\right)$ represents the specific work transfer due to aerodynamic forces and $\left(r_2^2 - r_1^2\right)\Omega^2$ represents the specific work transfer due to Coriolis forces, see Lewis (1996).

ROTHALPY

Rearranging Equation A.4 yields

$$h_{01} - U_1 V_{\theta 1} = h_{02} - U_2 V_{\theta 2} \tag{A.5}$$

which shows that, under adiabatic conditions (no heat transfer), the quantity $(h_0 - UV_\theta)$ at any point in a rotor flow remains constant. We call this quantity rothalpy expressed as

$$I = h_0 - UV_\theta = h + \frac{V^2}{2} - UV_\theta \tag{A.6}$$

where both h_0 and V_θ are in the stator (absolute) reference frame.

For expressing Equation A.6 in the rotor reference frame, we write

$$V^2 = V_x^2 + V_r^2 + V_\theta^2 \tag{A.7}$$

$$W^2 = W_x^2 + W_r^2 + W_\theta^2 \tag{A.8}$$

and

$$V_\theta = W_\theta + U \tag{A.9}$$

where U is the local tangential velocity of the rotor. Substituting for V_θ from Equation A.9 into Equation A.7 and noting that $W_x = V_x$ and $W_r = V_r$, we obtain

$$V^2 = W_x^2 + W_r^2 + W_\theta^2 + 2W_\theta U + U^2 \tag{A.10}$$

Using Equations A.10 and A.9 in Equation A.6 yields

$$I = h + \frac{W_x^2 + W_r^2 + W_\theta^2 + 2W_\theta U + U^2}{2} - U(W_\theta + U)$$

which reduces to

$$I = h + \frac{W^2}{2} - \frac{U^2}{2} = h_{0R} - \frac{U^2}{2} \tag{A.11}$$

where h_{0R} is the specific total enthalpy in the rotor reference frame. For a calorically perfect gas with constant c_p, we write this equation as

$$I = c_p T_{0R} - \frac{U^2}{2} \tag{A.12}$$

where T_{0R} is the fluid total temperature in the rotor reference frame. According to this equation, at any point in a rotor, the rothalpy of the flow measures how much higher the relative total enthalpy of the flow is compared to its dynamic enthalpy under solid-body rotation.

For an isentropic flow, we obtain

$$dh = \frac{dp}{\rho} \tag{A.13}$$

which, when combined with Equation A.11, yields for an incompressible flow

$$\frac{p_1}{\rho} + \frac{W_1^2}{2} - \frac{U_1^2}{2} = \frac{p_2}{\rho} + \frac{W_2^2}{2} - \frac{U_2^2}{2}$$

$$\frac{p_{0R1}}{\rho} - \frac{U_1^2}{2} = \frac{p_{0R2}}{\rho} - \frac{U_2^2}{2} \tag{A.14}$$

$$p_{0R2} - p_{0R1} = \frac{\rho U_2^2}{2} - \frac{\rho U_1^2}{2}$$

where p_{0R1} and p_{0R2} are relative total pressures at points 1 and 2, respectively.

AN ALTERNATE FORM OF EULER'S TURBOMACHINERY EQUATION

For the adiabatic flow in a rotor, the rothalpy remains constant between two points—i.e., $I_1 = I_2$. From Equation A.11, we can write

$$h_1 + \frac{W_1^2}{2} - \frac{U_1^2}{2} = h_2 + \frac{W_2^2}{2} - \frac{U_2^2}{2}$$

$$h_2 - h_1 = \left(\frac{W_1^2}{2} - \frac{W_2^2}{2} \right) + \left(\frac{U_2^2}{2} - \frac{U_1^2}{2} \right) \tag{A.15}$$

which expresses the change in specific static enthalpy in a rotor in terms of the change in the specific kinetic energy of the flow relative velocity and that in the rotor tangential velocity.

From the definition of specific total enthalpy, we can write its change between locations 1 and 2 as

$$h_{02} - h_{01} = (h_2 - h_1) + \left(\frac{V_2^2}{2} - \frac{V_1^2}{2} \right) \tag{A.16}$$

which upon substituting for $(h_2 - h_1)$ from Equation A.15 yields the following alternate form of the Euler's turbomachinery equation

$$h_{02} - h_{01} = \left(\frac{W_1^2}{2} - \frac{W_2^2}{2} \right) + \left(\frac{U_2^2}{2} - \frac{U_1^2}{2} \right) + \left(\frac{V_2^2}{2} - \frac{V_1^2}{2} \right) \tag{A.17}$$

CONVERTING TOTAL PRESSURE AND TEMPERATURE BETWEEN STATOR AND ROTOR REFERENCE FRAMES

In turbomachinery design applications, we often need to convert total temperature and total pressure from the stator reference frame to the rotor reference frame. Let us use the definitions of rothalpy in the two reference frames to perform this task.

To convert total temperature from the rotor reference frame to the stator reference frame, we use Equations A.6 and A.11 to write

$$h_0 - UV_\theta = h_{0R} - \frac{U^2}{2}$$

$$c_p T_0 - UV_\theta = c_p T_{0R} - \frac{U^2}{2} \qquad\qquad (A.18)$$

$$T_0 = T_{0R} + \frac{U}{2c_p}(U + 2W_\theta)$$

which yields $T_0 = T_{0R}$ for $W_\theta = -U/2$.

For an isentropic compressible flow, we compute the total pressure in the stator reference frame as

$$\frac{p_0}{p_{0R}} = \left(\frac{T_0}{T_{0R}}\right)^{\frac{\gamma}{\gamma-1}} = \left\{1 + \frac{U(U + 2W_\theta)}{2c_p T_{0R}}\right\}^{\frac{\gamma}{\gamma-1}} \qquad\qquad (A.19)$$

where we have used the fact that both static pressure and static temperature, being fluid properties, are independent of the reference frame.

For converting total temperature from stator reference frame to rotor reference frame, we rewrite Equation A.18 as

$$T_{0R} = T_0 - \frac{U}{2c_p}(U + 2W_\theta)$$

which upon substitution for W_θ in terms of V_θ and further simplification yields

$$T_{0R} = T_0 + \frac{U}{2c_p}(U - 2V_\theta) \qquad\qquad (A.20)$$

which yields $T_{0R} = T_0$ for $V_\theta = U/2$.

For an isentropic compressible flow, we can compute the total pressure in the rotor reference frame using the equation

$$\frac{p_{0R}}{p_0} = \left(\frac{T_{0R}}{T_0}\right)^{\frac{\gamma}{\gamma-1}} = \left\{1 + \frac{U(U - 2V_\theta)}{2c_p T_0}\right\}^{\frac{\gamma}{\gamma-1}} \qquad\qquad (A.21)$$

REFERENCES

Lewis, R.I. 1996. *Turbomachinery Performance Analysis*. New York: John Wiley and Sons.

Sultanian, B.K. 2015. *Fluid Mechanics: An Intermediate Approach*. Boca Raton, FL: Taylor & Francis.

Sultanian, B.K. 2019. *Logan's Turbomachinery: Flowpath Design and Performance Fundamentals*, 3rd edition. Boca Raton: Taylor & Francis.

Appendix B
Dimensionless Velocity Diagrams for Axial-Flow Compressors and Turbines

INTRODUCTION

In this appendix, we derive equations that relate performance parameters—namely, flow coefficient, blade loading coefficient, and degree of reaction of axial-flow compressors and turbines—for computing all parameters of inlet and exit velocity triangles (diagrams) along a streamline of constant radius and constant blade velocity. Normalizing all absolute and relative flow velocities by the constant blade velocity, we develop here a quick step-by-step method to draw dimensionless velocity diagrams directly from the knowledge of the three performance parameters. The resulting velocity diagram features blade inlet and outlet velocity triangles with a common apex. Once constructed, we can slide each velocity triangle along the tangential direction (horizontal) to obtain the composite inlet-outlet velocity diagram with a common base that corresponds to the dimensionless unit vector for the blade velocity.

SIGN CONVENTION FOR THE VELOCITY DIAGRAMS

For the analysis of turbomachinery designs using velocity diagrams, we need to adopt sign convection and use it consistently. According to the sign convention used here, V_θ and W_θ are positive if they are in the same direction as the blade velocity U; otherwise they are negative. In addition, the absolute flow angle α and the relative flow angle β are positive if they produce tangential velocity components that are positive; otherwise these angles are negative. Figure B.1 depicts this sign convention being used here. For the velocity diagram in Figure B.1a, all quantities are positive. In Figure B.1b, however, both β and W_θ are negative; the rest are positive. For the velocity diagram shown in Figure B.1b, by simply adding U and W_θ without using the present sign convention will result in an incorrect value of V_θ.

Figure B.2 shows the velocity diagrams at the inlet and outlet for an axial-flow compressor blade, and Figure B.3 shows them for an axial-flow turbine blade.

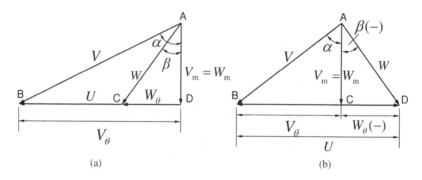

FIGURE B.1   Velocity diagrams showing the present sign convention.

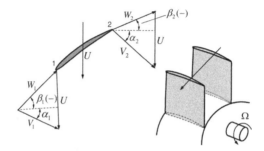

FIGURE B.2   Velocity diagrams at the inlet and outlet of an axial-flow compressor blade.

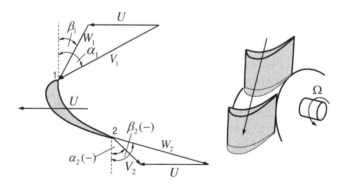

FIGURE B.3   Velocity diagrams at the inlet and outlet of an axial-flow turbine blade.

PERFORMANCE PARAMETERS

Three key performance parameters of a turbomachinery stage are flow coefficient (φ), blade loading coefficient (ψ), and degree of reaction or simply reaction (R). Here we will use the stage definitions for axial-flow compressor and turbine as shown in Figure B.4. Note that in both cases the rotor inlet is designated by 1 and outlet by 2.

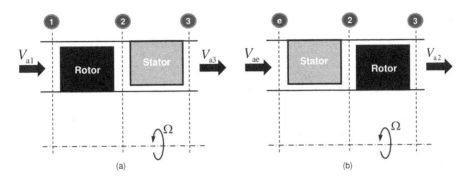

FIGURE B.4   (a) Axial-flow compressor stage and (b) axial-flow turbine stage.

For axial-flow compressors and turbines with $U_1 = U_2 = U$, the three performance parameters are defined as follows.

FLOW COEFFICIENT

We define the flow coefficient as

$$\varphi = \frac{V_a}{U} \tag{B.1}$$

LOADING COEFFICIENT

We define the loading coefficient as

$$\psi = -\frac{\Delta h_0}{U^2} \tag{B.2}$$

which, using Euler's turbomachinery equation (see Appendix A), we express as

$$\psi = -\frac{(U_2 V_{\theta 2} - U_1 V_{\theta 1})}{U^2} \tag{B.3}$$

With $U_1 = U_2 = U$ for an axial-flow turbomachine, Equation B.3 reduces to

$$\psi = -\frac{(V_{\theta 2} - V_{\theta 1})}{U} = -\frac{\Delta V_\theta}{U} \tag{B.4}$$

which renders ψ negative for an axial-flow compressor and positive for an axial-flow turbine.

As $V_\theta = W_\theta + U$, we can rewrite Equation B.4 as

$$\psi = -\frac{\Delta V_\theta}{U} = -\frac{\Delta W_\theta}{U} = -\frac{(W_{\theta 2} - W_{\theta 1})}{U} \tag{B.5}$$

REACTION

We define the degree of reaction of a turbomachine by the equation

$$R = \frac{\Delta h_{\text{rotor}}}{\Delta h_{\text{stage}}} = \frac{\Delta h_{\text{rotor}}}{\Delta h_{\text{stator}} + \Delta h_{\text{rotor}}} \qquad \text{(B.6)}$$

To develop an intuitive understanding of reaction, let us consider changes in static pressure in stator and rotor of a stage. For example, for the Pelton wheel shown in Figure B.5a, the entire change in static pressure occurs in the nozzle, and the static pressure remains constant in the bucket. According to Equation B.6, we have $R = 0$ or zero reaction in this case. For flow situation shown in Figure B.5b, change in static pressure occurs in both the nozzle and blade passage. If these two changes are equal, this turbomachine will have $R = 0.5$ or 50% reaction. Finally, Figure B.5c shows a lawn sprinkler, where the entire change in static pressure occurs in each rotary arm, and hence has $R = 1.0$ or 100% reaction.

For the axial-compressor stage shown in Figure B.1a, we can write

$$R = \frac{h_2 - h_1}{(h_3 - h_2) + (h_2 - h_1)} \qquad \text{(B.7)}$$

As the total (stagnation) enthalpy across the stator (adiabatic with no work transfer) remains constant, we obtain

$$h_{02} = h_{03}$$

$$h_2 + \frac{V_2^2}{2} = h_3 + \frac{V_3^2}{2} \qquad \text{(B.8)}$$

$$h_3 - h_2 = \frac{V_2^2}{2} - \frac{V_3^2}{2}$$

As the rothalpy remains constant in a rotor (see Appendix A), we write

$$I_1 = I_2$$

$$h_1 + \frac{W_1^2}{2} - \frac{U_1^2}{2} = h_2 + \frac{W_2^2}{2} - \frac{U_2^2}{2}$$

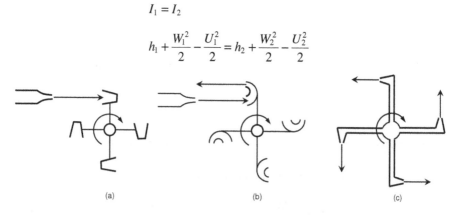

(a) (b) (c)

FIGURE B.5 (a) Zero reaction (impulse turbine), (b) 50% reaction, and (c) 100% reaction.

which for an axial-flow compressor with $U_1 = U_2$ reduces to

$$h_2 - h_1 = \frac{W_1^2}{2} - \frac{W_2^2}{2} \tag{B.9}$$

Substituting Equations B.8 and B.9 in Equation B.7, we obtain

$$R = \frac{W_1^2 - W_2^2}{\left(V_2^2 - V_3^2\right) + \left(W_1^2 - W_2^2\right)}$$

which with equal absolute velocities at the compressor stage inlet and outlet $\left(V_1 = V_3\right)$ for a repeating stage reduces to

$$R = \frac{W_1^2 - W_2^2}{\left(V_2^2 - V_1^2\right) + \left(W_1^2 - W_2^2\right)} \tag{B.10}$$

If we also assume that the axial velocity remains constant at the rotor inlet and outlet $\left(V_{a1} = V_{a2}\right)$, Equation B.10 becomes

$$R = \frac{W_{\theta 1}^2 - W_{\theta 2}^2}{\left(V_{\theta 2}^2 - V_{\theta 1}^2\right) + \left(W_{\theta 1}^2 - W_{\theta 2}^2\right)} \tag{B.11}$$

Substitution of $V_{\theta 1} = W_{\theta 1} + U$ and $V_{\theta 2} = W_{\theta 2} + U$ in Equation B.11 yields

$$R = \frac{W_{\theta 1}^2 - W_{\theta 2}^2}{\left\{\left(W_{\theta 2}^2 + 2W_{\theta 2}U + U^2\right) - \left(W_{\theta 1}^2 + 2W_{\theta 1}U + U^2\right)\right\} + \left(W_{\theta 1}^2 - W_{\theta 2}^2\right)}$$

$$R = \frac{\left(W_{\theta 1} - W_{\theta 2}\right)\left(W_{\theta 1} + W_{\theta 2}\right)}{2U\left(W_{\theta 2} - W_{\theta 1}\right)} \tag{B.12}$$

$$R = -\frac{\left(W_{\theta 1} + W_{\theta 2}\right)}{2U}$$

which, using $W_{\theta 1} = V_{\theta 1} - U$ and $W_{\theta 2} = V_{\theta 2} - U$, we can express in terms of absolute tangential velocities at rotor inlet and outlet as

$$R = 1 - \frac{\left(V_{\theta 1} + V_{\theta 2}\right)}{2U} \tag{B.13}$$

Equations B.12 and B.13, which we derived here to compute the reaction for an axial-flow compressor stage shown in Figure B.4a, are also valid for an axial-flow turbine stage shown in Figure B.4b.

DIMENSIONLESS VELOCITY DIAGRAMS

One can draw a composite inlet-outlet velocity diagram for axial-flow compressors and turbines with equal axial flow velocities at rotor inlet and outlet in two ways: (1)

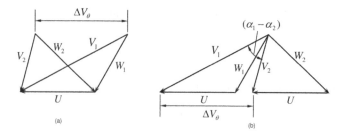

FIGURE B.6 Composite inlet-outlet velocity diagrams: (a) common base and (b) common apex.

using the blade velocity U as the common base for the velocity triangles at rotor inlet and outlet, as shown in Figure B.6a, and (2) using the point where the absolute and relative velocities join together as the common apex for the velocity triangles at rotor inlet and outlet, as shown in Figure B.6b. In the first case, the distance between the peaks of the two triangles measures the magnitude of the loading coefficient times the blade velocity. In the second case, the angle between the absolute velocities gives the flow turning angle $(\alpha_1 - \alpha_2)$ over the rotor blade.

Figures B.7a and b are, respectively, the velocity diagrams for an axial-flow compressor and axial-flow turbine drawn with a common apex where each velocity is made dimensionless by diving it by the blade velocity U. In these velocity diagrams, the dimensional blade velocity becomes unity. Each dimensionless velocity diagram features the three performance parameters—φ, ψ, and R—which we have presented in the foregoing. It is interesting to note that we obtain the velocity diagram of the axial-flow turbine from that of an axial-flow compressor, or vice-versa, by simply exchanging the subscripts 1 and 2 of various quantities involved. This implies that the compressor outlet becomes the turbine inlet, and the compressor inlet becomes the turbine outlet, both having the identical values of φ, $|\psi|$, and R.

DERIVATIONS OF EQUATIONS TO COMPUTE VELOCITIES AND ANGLES OF DIMENSIONLESS VELOCITY DIAGRAMS

Before we present a stepwise method to quickly draw a dimensionless velocity diagram using the performance parameters φ, ψ, and R, let us first derive equations to compute dimensionless absolute and relative velocities at rotor inlet and outlet and their angles measured from the axial direction. For these derivations, we will use the velocity diagrams shown in Figure B.7.

Absolute Velocity Angle at Rotor Inlet (α_1)

From the inlet velocity triangle of Figure B.4a, we can write

$$\tan \alpha_1 = \frac{V_{\theta 1}}{V_a} = \left(\frac{V_{\theta 1}}{U} \right) \frac{1}{\varphi} \tag{B.14}$$

From the definition of loading coefficient ψ given by Equation B.4, we obtain

(a)

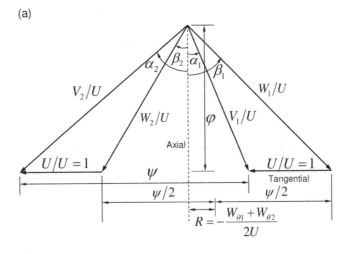

(b)

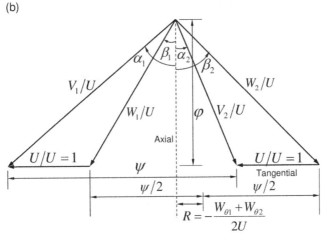

FIGURE B.7 Dimensionless velocity diagrams showing performance parameters: (a) compressor and (b) turbine.

$$\frac{V_{\theta 2}}{U} = -\psi + \frac{V_{\theta 1}}{U} \tag{B.15}$$

Substitution of $V_{\theta 2}/U$ from this into Equation B.13 yields

$$R = 1 + \frac{\psi}{2} - \frac{V_{\theta 1}}{2U} - \frac{V_{\theta 1}}{2U} = 1 + \frac{\psi}{2} - \frac{V_{\theta 1}}{U}$$

$$\frac{V_{\theta 1}}{U} = \frac{\psi}{2} - R + 1 \tag{B.16}$$

Substitution of $V_{\theta 1}/U$ from this equation into Equation B.14 finally yields

$$\tan \alpha_1 = \frac{0.5\psi + (1 - R)}{\varphi}$$

$$\alpha_1 = \tan^{-1}\left\{\frac{0.5\psi + (1 - R)}{\varphi}\right\}$$

(B.17)

ABSOLUTE VELOCITY ANGLE AT ROTOR OUTLET (α_2)

From the outlet velocity triangle of Figure B.7a, we can write

$$\tan \alpha_2 = \frac{V_{\theta 2}}{V_a} = \left(\frac{V_{\theta 2}}{U}\right)\frac{1}{\varphi}$$

(B.18)

From Equation B.4, we obtain

$$\frac{V_{\theta 1}}{U} = \psi + \frac{V_{\theta 2}}{U}$$

(B.19)

Substitution of $V_{\theta 1}/U$ from this equation into Equation B.13 yields

$$R = 1 - \frac{\psi}{2} - \frac{V_{\theta 2}}{2U} - \frac{V_{\theta 2}}{2U} = 1 - \frac{\psi}{2} - \frac{V_{\theta 2}}{U}$$

$$\frac{V_{\theta 2}}{U} = -\frac{\psi}{2} + 1 - R$$

(B.20)

Substitution of $V_{\theta 2}/U$ from this equation into Equation B.18 finally yields

$$\tan \alpha_2 = -\frac{0.5\psi - (1 - R)}{\varphi}$$

$$\alpha_2 = \tan^{-1}\left\{\frac{(1 - R) - 0.5\psi}{\varphi}\right\}$$

(B.21)

RELATIVE VELOCITY ANGLE AT ROTOR INLET (β_1)

From the inlet velocity triangle of Figure B.7a, we can write

$$\tan \beta_1 = \frac{W_{\theta 1}}{V_a} = \left(\frac{W_{\theta 1}}{U}\right)\frac{1}{\varphi}$$

(B.22)

From the definition of the loading coefficient ψ given by Equation B.5, we obtain

$$\frac{W_{\theta 2}}{U} = -\psi + \frac{W_{\theta 1}}{U}$$

(B.23)

Substitution of $W_{\theta 2}/U$ from this equation into Equation B.12 yields

$$R = \frac{\psi}{2} - \frac{W_{\theta 1}}{2U} - \frac{W_{\theta 1}}{2U} = \frac{\psi}{2} - \frac{W_{\theta 1}}{U}$$

$$\frac{W_{\theta 1}}{U} = \frac{\psi}{2} - R$$

(B.24)

Substitution of $W_{\theta 1}/U$ from this equation into Equation B.23 finally yields

$$\tan \beta_1 = \frac{0.5\psi - R}{\varphi}$$

$$\beta_1 = \tan^{-1}\left(\frac{0.5\psi - R}{\varphi}\right)$$

(B.25)

RELATIVE VELOCITY ANGLE AT ROTOR OUTLET (β_2)

From the outlet velocity triangle of Figure B.7a, we can write

$$\tan \beta_2 = \frac{W_{\theta 2}}{V_a} = \left(\frac{W_{\theta 2}}{U}\right)\frac{1}{\varphi}$$

(B.26)

Equation B.5 yields

$$\frac{W_{\theta 1}}{U} = \psi + \frac{W_{\theta 2}}{U}$$

(B.27)

Substitution of $W_{\theta 1}/U$ from this equation into Equation B.12 yields

$$R = -\frac{\psi}{2} - \frac{W_{\theta 2}}{2U} - \frac{W_{\theta 2}}{2U} = -\frac{\psi}{2} - \frac{W_{\theta 2}}{U}$$

$$\frac{W_{\theta 2}}{U} = -\frac{\psi}{2} - R$$

(B.28)

Substitution of $W_{\theta 2}/U$ from this equation into Equation B.26 finally yields

$$\tan\beta_2 = -\frac{0.5\psi + R}{\varphi}$$

$$\beta_2 = \tan^{-1}\left(-\frac{0.5\psi + R}{\varphi}\right)$$

(B.29)

DIMENSIONLESS ABSOLUTE VELOCITY AT ROTOR INLET (V_1/U)

From inlet velocity triangles for axial-flow compressor and turbine shown in Figure B.7, we can write

$$\left(\frac{V_1}{U}\right)^2 = \left(\frac{V_a}{U}\right)^2 + \left(\frac{V_{\theta 1}}{U}\right)^2 = \varphi^2 + \left(\frac{V_{\theta 1}}{U}\right)^2$$

which upon substituting for $V_{\theta 1}/U$ from Equation B.16 yields

$$\frac{V_1}{U} = \left\{\varphi^2 + \left(0.5\psi - R + 1\right)^2\right\}^{1/2} \tag{B.30}$$

DIMENSIONLESS ABSOLUTE VELOCITY AT ROTOR OUTLET $\left(V_2/U\right)$

From outlet velocity triangles for axial-flow compressor and turbine shown in Figure B.7, we can write

$$\left(\frac{V_2}{U}\right)^2 = \left(\frac{V_a}{U}\right)^2 + \left(\frac{V_{\theta 2}}{U}\right)^2 = \varphi^2 + \left(\frac{V_{\theta 2}}{U}\right)^2$$

which upon substituting from $V_{\theta 2}/U$ for Equation B.20 yields

$$\frac{V_2}{U} = \left\{\varphi^2 + \left(0.5\psi + R - 1\right)^2\right\}^{1/2} \tag{B.31}$$

DIMENSIONLESS RELATIVE VELOCITY AT ROTOR INLET $\left(W_1/U\right)$

From inlet velocity triangles for axial-flow compressor and turbine shown in Figure B.7, we can write

$$\left(\frac{W_1}{U}\right)^2 = \left(\frac{V_a}{U}\right)^2 + \left(\frac{W_{\theta 1}}{U}\right)^2 = \varphi^2 + \left(\frac{W_{\theta 1}}{U}\right)^2$$

which upon substituting for $W_{\theta 1}/U$ from Equation B.24 yields

$$\frac{W_1}{U} = \left\{\varphi^2 + \left(0.5\psi - R\right)^2\right\}^{1/2} \tag{B.32}$$

DIMENSIONLESS RELATIVE VELOCITY AT ROTOR OUTLET W_2/U

From outlet velocity triangles for axial-flow compressor and turbine shown in Figure B.7, we can write

$$\left(\frac{W_2}{U}\right)^2 = \left(\frac{V_a}{U}\right)^2 + \left(\frac{W_{\theta 2}}{U}\right)^2 = \varphi^2 + \left(\frac{W_{\theta 2}}{U}\right)^2$$

which upon substituting for $W_{\theta 2}/U$ from Equation B.28 yields

$$\frac{W_2}{U} = \left\{ \varphi^2 + \left(0.5\psi + R \right)^2 \right\}^{\frac{1}{2}} \tag{B.33}$$

USING φ, ψ, AND R TO QUICKLY DRAW DIMENSIONLESS VELOCITY DIAGRAMS

For drawing a dimensionless velocity diagrams for axial-flow compressors and turbines using the performance parameters φ, ψ, and R, we present here a step-by-step procedure for the case of an axial-flow compressor with $\varphi = 0.5$, $\psi = -0.5$, and $R = 0.5$. Note that in the dimensionless velocity diagram the nondimensional blade velocity $\left(U/U \right)$ is always unity, which we represent here by 10 cm for drawing purposes.

STEP 1: FLOW COEFFICIENT $\left(\varphi = 0.5 \right)$

As shown in Figure B.8, we first draw two dotted horizontal parallel lines that are apart by the given value of $\varphi = 0.5$, which scales to 5 cm. We also draw a dotted vertical line for the axial direction, intersecting both horizontal lines. The common apex of the velocity diagram will lie on the top horizontal line, and all the absolute and relative flow velocity vectors will end at the bottom horizontal line, which represents the tangential direction.

STEP 2: STAGE REACTION $\left(R = 0.5 \right)$

We mark on the bottom horizontal line a point that is 5 cm, which corresponds to $R = 0.5$, from the dotted vertical line, as shown in Figure B.9. Connecting this point to the apex gives the mean relative flow velocity through the blade passage.

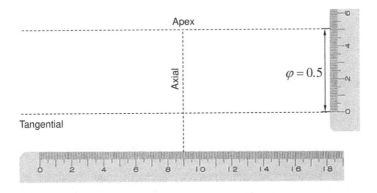

FIGURE B.8 Partially drawn dimensionless velocity diagrams showing flow coefficient.

Step 3: Dimensionless Relative Flow Velocities W_1/U and W_2/U at Blade Inlet and Outlet

On the bottom horizontal line, we mark two points, one on either side of the tip of the mean relative flow velocity, at a distance corresponding to half of the loading coefficient magnitude $(|0.5\psi| = 0.25)$, as shown in Figure B.10. Connecting these points to the apex gives us the dimensionless relative flow velocities W_1/U and W_2/U. Due to diffusion in a compressor rotor, we have $W_2/U < W_1/U$.

Step 4: Dimensionless Absolute Flow Velocities V_1/U and V_2/U at Blade Inlet and Outlet

We now connect the tip of each dimensionless relative flow velocity vector on the bottom horizontal line with the tail of the dimensionless blade velocity $(U/U = 1)$

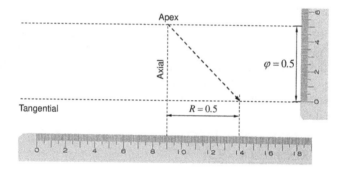

FIGURE B.9 Partially drawn dimensionless velocity diagrams showing flow coefficient, stage reaction, and the mean relative flow velocity through the blade passage.

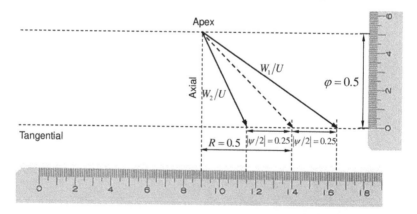

FIGURE B.10 Partially drawn dimensionless velocity diagrams showing flow coefficient, stage reaction, the mean relative flow velocity through blade, and blade inlet and out relative flow velocities.

vector. The line connecting the tip of this vector and the common apex gives the corresponding dimensionless absolute flow velocity, as shown in Figure B.11.

STEP 5: ABSOLUTE AND RELATIVE FLOW ANGLES AT BLADE INLET AND OUTLET

Finally, we remove extra lines and construction aids and mark absolute and relative flow angles at blade inlet and outlet, as shown in Figure B.12. Note that, as per the convention presented here in the foregoing, both the relative flow angles β_1 and β_2 are negative in this case.

For the velocity diagram shown in Figure B.11, we can scale the magnitudes of dimensionless absolute and relative flow velocities at blade inlet and outlet and measure the corresponding absolute and relative flow angles using a protractor. However, the equations developed in the foregoing yield more accurate numerical values of these quantities. Using these equations, we obtain the following values.

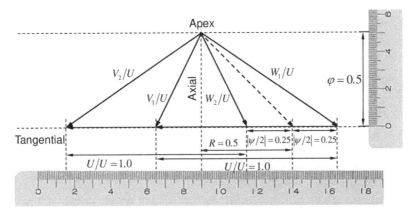

FIGURE B.11 Partially drawn dimensionless velocity diagrams showing flow coefficient, stage reaction, the mean relative flow velocity through blade, blade inlet and out relative flow velocities, and the corresponding absolute flow velocities.

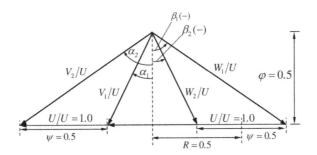

FIGURE B.12 Final dimensionless velocity diagrams for the axial-flow compressor with the given performance parameters φ, ψ, and R.

Rotor Inlet

$$V_1/U = 0.559, \; W_1/U = 0.901, \; \alpha_1 = 26.57°, \text{ and } \beta_1 = -56.31°$$

Rotor Outlet

$$V_2/U = 0.901, \; W_2/U = 0.559, \; \alpha_2 = 56.31°, \text{ and } \beta_2 = -26.57°$$

For $R = 0.5$, these computed values confirm that the velocity diagram is symmetric. Note that the velocity diagram for the axial-flow compressor, shown in Figure B.9, can be easily converted to represent that of an axial-flow turbine with identical performance parameters $(\varphi = 0.5, \psi = 0.5, \text{ and } R = 0.5)$ by simply switching subscripts 1 and 2 for all the quantities involved.

Index